I0055195

Integrated Researches in Agricultural Science

Integrated Researches in Agricultural Science

Editor: Adriana Winkler

www.callistoreference.com

Callisto Reference,
118-35 Queens Blvd., Suite 400,
Forest Hills, NY 11375, USA

Visit us on the World Wide Web at:
www.callistoreference.com

ISBN: 978-1-64116-112-1 (Hardback)

Cataloging-in-Publication Data

Integrated researches in agricultural science / edited by Adriana Winkler.
p. cm.
Includes bibliographical references and index.
ISBN 978-1-64116-112-1
1. Agriculture--Research. 2. Agriculture. 3. Integrated agricultural systems.
I. Winkler, Adriana.
S540.A2 I58 2019
630.7--dc23

Table of Contents

Preface

This book was inspired by the evolution of our times; to answer the curiosity of inquisitive minds. Many developments have occurred across the globe in the recent past which has transformed the progress in the field.

Agricultural science takes a multi-disciplinary approach across disciplines of natural, economic and social sciences, to develop methodologies for understanding and advancing the practices of agriculture. Research in this field includes studies in plant breeding and genetics, soil science, pest-crop dynamics, weed-crop dynamics, entomology, among many others. Improving crop growth, selective breeding of drought-resistant crop and animal varieties, improved pesticides, etc. are also being studied in this discipline. This book highlights the recent studies in the field of agricultural science. It also provides a detailed explanation of the different concepts and applications of the modern agricultural practices developed in recent years. This book will be a resource guide for agronomists, agriculturists, agro-economists and agrobiologists, as well as students and experts working in this domain.

This book was developed from a mere concept to drafts to chapters and finally compiled together as a complete text to benefit the readers across all nations. To ensure the quality of the content we instilled two significant steps in our procedure. The first was to appoint an editorial team that would verify the data and statistics provided in the book and also select the most appropriate and valuable contributions from the plentiful contributions we received from authors worldwide. The next step was to appoint an expert of the topic as the Editor-in-Chief, who would head the project and finally make the necessary amendments and modifications to make the text reader-friendly. I was then commissioned to examine all the material to present the topics in the most comprehensible and productive format.

I would like to take this opportunity to thank all the contributing authors who were supportive enough to contribute their time and knowledge to this project. I also wish to convey my regards to my family who have been extremely supportive during the entire project.

Editor

1

ROLES OF GLUTATHIONE S-TRANSFEREASE IN MAIZE (*Zea mays* L.) UNDER COLD STRESS

Md. Mahfuzur Rahman[1], MA Khaleque Mian[1], Asgar Ahmed and Md. Motiar Rohman*

[1]Bangabandhu Sheikh Mujibur Rahman Agricultural University, Salna, Gazipur-1706, Bangladesh

Breeding Division, Bangladesh Agricultural Research Institute, Joydevpur, Gazipur-1701, Bangladesh

***Corresponding author:** Md. Motiar Rohman, E-mail: motiar_1@yahoo.com

ARTICLE INFO

Key words

Glutathione
glutathione
S-transferase
BARI hybrid maize-7
Chilling stress

ABSTRACT

Glutathione S-transferease (GST) activities involved in antioxidant defense and methylglyoxal detoxification were investigated in the seedlings of a Bangladeshi maize variety, BARI hybrid maize-7, to understand the protecting mechanism under cold stress condition. The activities of glutathione S-transferase (GST) increased, while the activities of catalase (CAT) decreased with the duration of stress. The western blot analysis of the dominant GST revealed that it significantly accumulated during the stress period. The continual increase in H_2O_2 contents along with reduced redox state and activities suggested their roles in maintaining the glutathione homeostasis. The accumulation of GST with the content of H_2O_2 suggested its detoxification roles for organic hydroperoxides during chilling stress. Considering all, glutathione S-transferase (GST) enzymes showed protective role in maize from oxidative damages under Chilling condition.

INTRODUCTION

Cold stress adversely affects plant growth and development. Cold stress is the collective form of Chilling and freezing stresses. Generally chilling stress occurs in between 2 to 10°C temperature. But, some tropical species, rice and sugarcane are highly sensitive to chilling and show injury symptoms up to 15°C (Thomashow, 1999). In freezing stress, ice forms with in plant tissues. Chilling injury is one of the serious problems during germination and early seedling growth in many plant species including maize and rice (Bedi and Basra, 1993). Visible symptoms of chilling damage depend on the species, plant age and the duration of exposure. Seedlings showed wilting, reduced leaf expansion and chlorosis upon exposure to chilling stress in rice (Yoshida et al., 1996). Due to aberrant metabolism, toxic metabolites and reactive oxygen species (ROS) accumulates in the injured cells (Farooq et al., 2009). Chilling stress results accelerated senescence and ultimately the plant death (Sharma et al., 2005).

In recent years, Rangpur, Rajshahi and most part of Dhaka divisions has been experiencing chilling temperatures during winter season for at least 30 to 40 days. During this time, farmers sow maize seeds or the early varieties are in seedling stage. Therefore, chilling stress will be a major headache for farmers to cultivate maize. On the other hand, these areas are major producer of maize. BARI Hybrid Maize-7 (BMH-7) is a good yielder and has stress tolerance capability. Therefore, the farmers are using the variety for more production. In this view, this study was designed to investigate the biomolecular responses of some glutathione dependent enzyme in adaptive response of BMH 7 under chilling in seedling sage in green house condition.

MATERIALS AND METHODS

Plant material

Maize (Zea mays L.) was managed in greenhouse condition as plant material. Bangladesh Agricultural Research Institute (BARI) released hybrid maize variety BARI Hybrid Maize-7 (BMH-7), has stress tolerance capability, and was taken into consideration.

Stress treatment

Maize seedlings were grown in pot in greenhouse condition. After reaching two leaf stages, the seedlings were ready for stress treatment. The plants were kept at 4^0C temperature to induce chilling stress.

Separation and purification of GSTs from crude enzyme solution of maize seedlings

Crude enzyme was extracted by homogenizing 80 g of fresh maize seedlings (except green part) in an equal volume of 25 mMTris-HCl buffer (pH 8.5), which contained 1 mM EDTA and 1% (w/v) ascorbate, with a Waring lender. The homogenate was squeezed through two layers of nylon cloth and centrifuged at 11500×g for 10 minutes, and the supernatant was used as a crude enzyme solution. Proteins were precipitated by ammonium sulfate at 65% saturation of the crude enzyme solution and centrifuged at 11500×g for 10 minutes. The proteins were dialyzed against 10 mM Tris-HCl buffer (pH 8) containing 0.01% (w/v) β-mercaptoethanol and 1 mM EDTA (buffer A) overnight to completely remove low molecular inhibitors. The dialyzate was applied to a column (1.77 cm i.d. × 20 cm) of DEAEcellulose (DE-52; Whatman, U.K.) that had been equilibrated with buffer A and eluted with a linear gradient of 0 to 0.2 M KCl in 600 ml of buffer A. The high active fractions of the highest GST peak were pooled and used to detect their inhibitory substances present in maize seedlings extract.

The pooled GST solution was directly applied on affinity column (0.76 cm i.d. × 4.0 cm) of S-hexylglutathioneagarose (Sigma, St. Louis, MO) that had been equilibrated with 10 mMTris-HCl buffer (pH 8.0) containing 0.01% (v/v) β-mercaptoethanol (buffer B). The column was washed with buffer B containing 0.2 M KCl and eluted with buffer B containing 1.5 mM S-hexylglutathione. The high active protein fractions eluted with S-hexylglutathione were combined and dialyzed against buffer B and the dialyzate was used as the purified GST.

Production of polyclonal antibodies against GST

A rabbit (weighing about 2.5 kg) received subcutaneous injections of a 0.5 mg of purified GST protein in Freund's complete adjuvant at several sites. After two weeks, the rabbit was given a first booster injection of 0.5 mg of the purified GST protein in incomplete adjuvant, and then a second booster injection of 0.5 mg of the purified protein in incomplete adjuvant was given two weeks after the first booster injection. Blood was taken from the ear vein one week after the second booster injection.

SDS-PAGE and Western blotting

SDS-PAGE was done in 12.5% (w/v) gel containing 0.1% (w/v) SDS by the method of Laemmli (1970). The gel was stained with silver. Western blotting was done on the basis of the Amersham ECL detection system.

Protein quantification

Protein was estimated following the method of Bradford (1976).

Preparation of soluble protein extracts

The plant materials were homogenized with proportional volumes of 25 mMTris–HCl buffer (pH 8.0) containing 1 mM EDTA, 1% (w/v) ascorbate and 10% (w/v) glycerol with mortar pestle. The homogenate was centrifuged at 11500 x g for 15 min, and the supernatant was used as the crude enzyme solution. All procedures were performed at 0-4°C. For H2O2 assay, Fresh seedling samples were extracted by homogenizing in 6X volume of 50 mM K-P buffer (pH 6.5). The homogenate was centrifuged at 11500 x g for 15 min, and 2.5 ml supernatant was used for further centrifugation at 11500 x g for 15 min with 833 µl reaction mixture containing 5ml H_2SO_4 and 15 µl TiCl4. The 2nd supernatant was used for the assay of H_2O_2 spectrophotometrically.

Enzyme assay

Glutathione S-transferase (GST, EC: 2.5.1.18) activity was assayed following the methods of Rohman et al. (2010) spectrophotometrically (UV-1800, Shimadzu, Japan). The activity was calculated using the extinction coefficient of 9.6 mM-1Cm-1.

Measurement of H_2O_2

H_2O_2 was assayed according to the method described by Yu et al. (2000). H_2O_2 was extracted by homogenizing 0.5 g of leaf samples with 3 ml of 50 mM K-phosphate buffer pH (6.5) at 40C. The homogenate was centrifuged at 11,500g for 15 min. Three ml of supernatant was mixed with 1 ml of 0.1% $TiCl_4$ in 20% H_2SO_4 (v/v), and the mixture was then centrifuged at 11,500g for 12 min at room temperature. The optical absorption of the supernatant was measured spectrophotometrically at 410 nm to determine the H_2O_2 content (□= 0.28 μM^{-1} Cm^{-1}) and expressed as µmol g^{-1} fresh weight.

SDS-PAGE and Western Blotting to check GST accumulation

To check the accumulation of GSTa SDS-PAGE was done in 12.5% (w/v) gel containing 0.1% (w/v) SDS by the method of Laemmli (1970) followed by western blotting.

RESULTS AND DISCUSSION

Purification of dominant GST from maize seedlings for polyclonal antibody production

To purify the dominant GST for polyclonal antibody production, crude enzyme solution was extracted from maize seedlings. The crude protein was precipitated by 65% $(NH_4)_2SO_4$ and dialyzed overnight in Buffer A (mentioned in Materials and Methods). The dialyzed protein was applied to a DEAE (DE-52, UK) cellulose column chromatography. A 750 ml gradient solution (0-0.4 MKCl) was passed through the column. Total 170 fractions (each contains 5 ml) were collected and activity (A_{340}) and absorbance (A_{280}) were measured. It was found that 3 GST peaks were eluted at 91.67, 162.50 and 234.04 mMKCl of the gradient solution (Fig.1).

Among the GSTs, peak eluted at 91.67 mMKCl contained more than 92.2% of total activity and it was termed as dominant GST, GSTa (Fig. 1) and other two peaks eluted at 162.50 and 234.04 mMKCl termed as minor GSTs, GSTb and GSTc. The dominant GST was further purified by subsequent application of S-hexylglutathione-agarose.

Total 1.5 mg purified dominant GST protein was prepared for production of polyclonal antibody. As basal dose, 0.5 mg protein was injected subcutaneously into a rabbit in complete adjuvant. After two weeks a booster dose was injected in incomplete adjuvant. Again after one week an income dose was injected and after one week blood was collected from rabbit and centrifuged at 3000xg and serum was used as antibody. To test the specificity of the antibody, the cross reactivity the high active GST fractions of each peak (Fig.1) was tested by western blotting. Fraction 87, 88 and 90 (dominant GST) reacted with the antibody and produced thick bands, but 127, 128, 129, 155,156 and 157 number fraction (minor GSTs) did not react with the antibody.

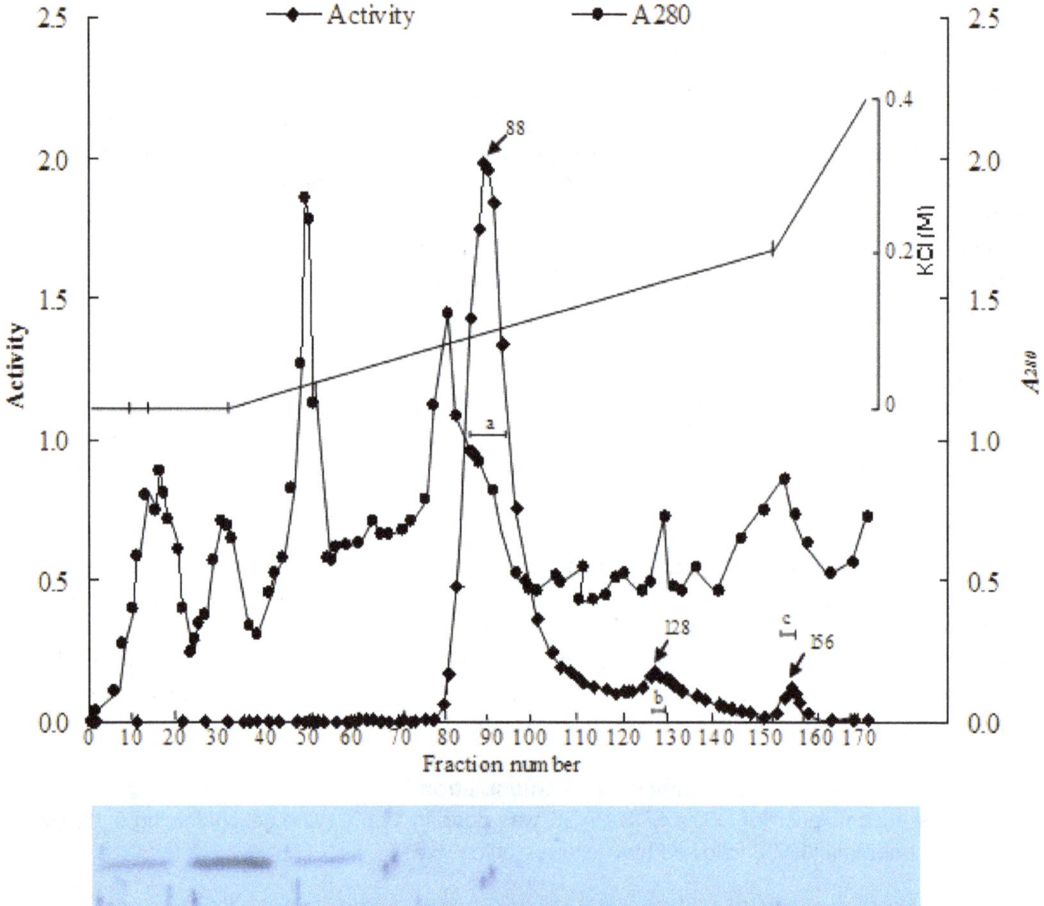

Figure 1. A typical column chromatography of DEAE-cellulose of soluble proteins prepared from 78 g maize seedlings (except green part). For each fraction, absorbance at 280 nm (•) and GST activity toward CDNB (♦) were determined. Activity is expressed as µmol min^{-1} ml^{-1}. Bars indicate the high active peak fractions of three maize GSTs. The fractions under the bar of GSTa peak were pooled for subsequent purification. The curve shows the gradient solution of KCl (0-0.4 mM). The plate shows the cross reactivity of antibody with active fractions of the GST peaks.

This result suggested that the antibody is highly specific to GST against which it was developed.

Glutathione S-transferease (GST) under chilling stress

Expression of maize GST activities was observed to increase under chilling stress at 10 days of stress treatment. Chilling treatment at 10 day showed maximum GST activity (202.19 nmol/min/mg protein) followed by 7 day (138.61 nmol/min/mg protein) and 4 day (106.45 nmol/min/mg protein) of chilling treatment. Minimum GST activity was recorded in controlled plants (99.18 nmol/min/mg protein). The GST activity of untreated and 4 day treated plants was statistically similar (Fig. 2).

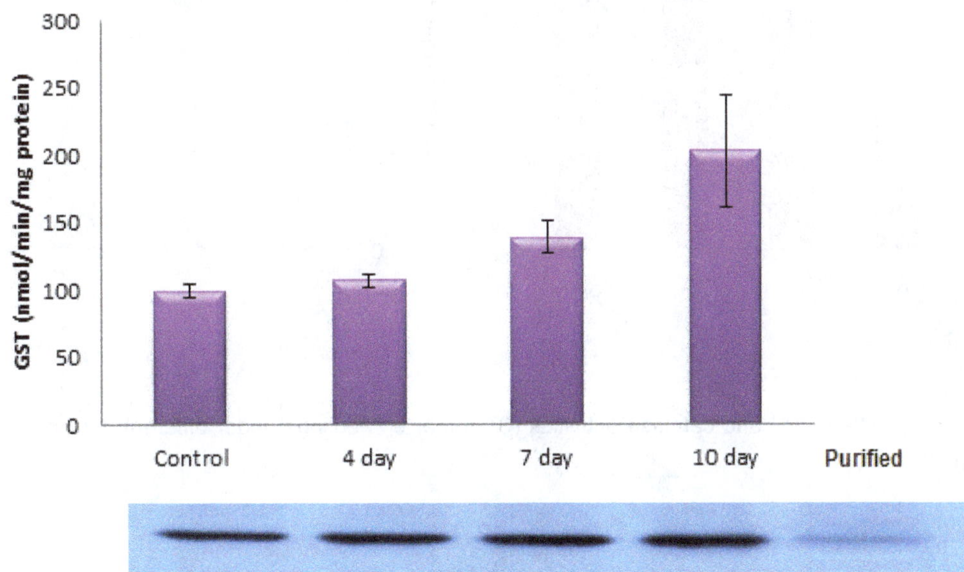

Figure 2. Activities and expressions of maize GSTs at different duration of chilling stress. Bar graph shows the GST activities towards CDNB in soluble protein extracts. Western blotting results show the expressions of GSTa at different duration of chilling stress treatment. Each lane contained 45 µg of the protein.

Western blotting of the GST protein extracts from different duration of chilling treatment was done to recognize the accumulation of GSTa. The expression of GSTa accumulation was detected at all conditions but the levels of GST accumulation varied greatly among different conditions (Fig. 2). Chilling stress caused significant increase in accumulation of GSTa compared to control. The highest expression was found at 10 day followed by 7 day, 4 day of chilling treated seedlings.

H_2O_2 increased continuously with duration of chilling treatments. Plants of 10 day chilling treatment showed highest H_2O_2 activity (11.52 µmol/g FW) followed by 7 day (11.32 µmol/g FW) chilling treatment. Plants of 4 day chilling treatment (7.10 µmol/g FW) and untreated plants showed the lowest (4.49 µmol/g FW) H_2O_2 contents (Fig. 3).

The CAT activity was found to be highest at 4 day (192.31 µmol/min/mg protein) which decreased successively at 7 day (84.35 µmol/min/mg protein) and 10 day (78.07 µmol/min/mg protein) of chilling treatment. Untreated plants showed lowest (57.84 µmol/min/mg protein) CAT activity (Fig. 3).

The chilling tolerance is a complex phenomenon, which entails an array of physiological and biochemical processes at whole plant, organ, cell and subcellular levels. These processes are reduced water loss by stomatal resistance, enhanced water uptake with the development of prolific root systems and synthesis and accumulation of osmolytes (Farooq et al., 2008a & b, 2009). Chilling stress is associated with increased oxidative stress due to enhanced accumulation of ROS, particularly O_2^- and H_2O_2 in chloroplasts , mitochondria , and peroxisomes. Therefore, the induction of antioxidant enzyme activities is a general adaptation strategy which plants use to overcome oxidative stresses (Foyer and Noctor, 2003) and the up-regulation of GST activity over time would be indicating the adaptation of plants to chilling stress. In this study, maximum induction of GST over control was at 10 day stress treatment (103.85%) followed by 7 day (39.75%) and 4 day (7.32%) stress treatment.

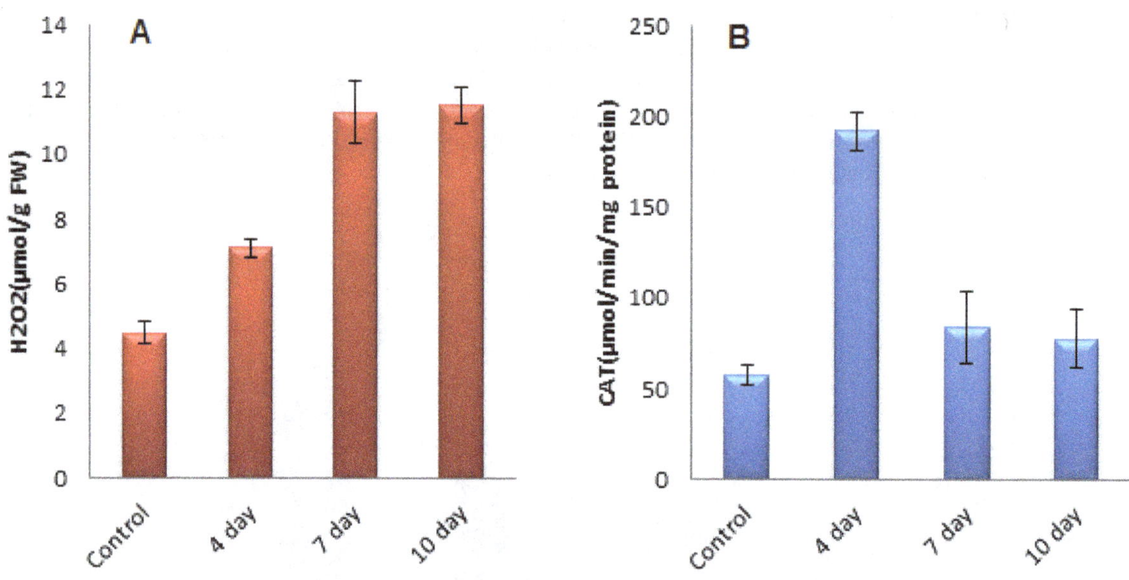

Figure 3. H_2O_2 contents (A) and catalase activities (B) in maize seedlings under different duration of chilling stress condition.

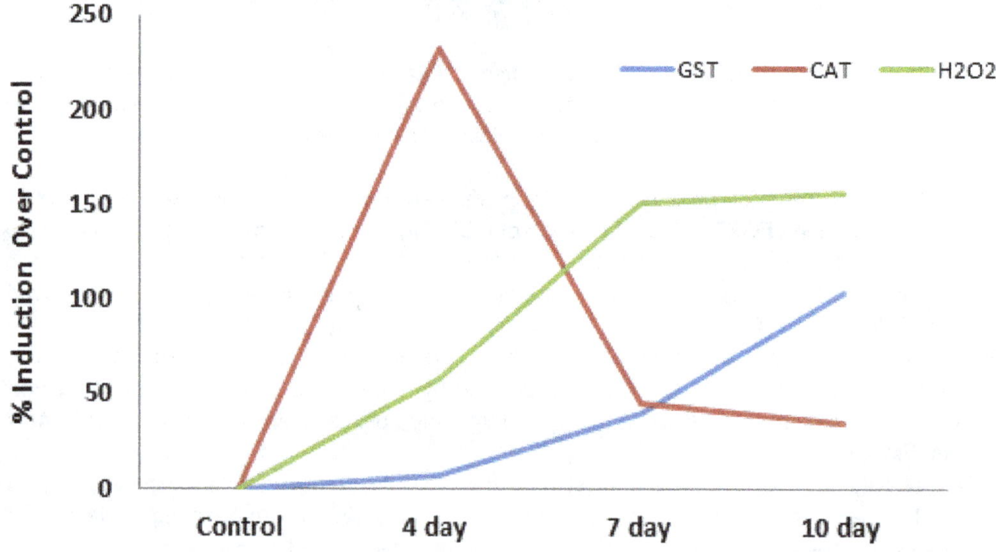

Figure 4. Induction rate of GST, CAT and H_2O_2 over control at different duration of cold stress treatment.

The induction of H_2O_2 over control was maximum at 10 day (156.46%) followed by 7 day (151.69%) and 4 day (58.1%) of stress treatment (Fig. 4). This result supported by the study of Lee and Lee (2000) as they found increased SOD activity in cucumber under chilling stress. Spontaneous dismutation and SOD decomposes ROS to produce H_2O_2. Therefore, the over production of H_2O_2 is supposedly due to the increased activity of spontaneous dismutation and SOD. However, GST usually not found to induce in all plant species.

The activity level and Western blotting of GSTa in maize under chilling stress suggested their protective role of maize under cold. The catalase activity showed the decreasing trend over time (Fig. 4). The induction of CAT over control was maximum at 4 day (232.48%) followed by 7 day (45.82%) and 10 day (34.98%) of stress treatment. Lee and Lee (2000) supported the result as they also found that under chilling stress CAT activity decreased in cucumber over time. The reduction of CAT activity is supposedly due to the inhibition of enzyme synthesis or change in the assembly of enzyme subunits under chilling stress condition.

REFERENCES

1. Bedi S and AS Basra, 1993. Chilling injury in germinating seeds: basic mechanisms and agricultural implications. Seed Science Research, 3: 219–229
2. Bradford MM, 1976. A rapid and sensitive method for the quantitation of microgram quantities of protein utilizing the principle of protein-dye binding. Analytical Biochemistry, 72: 248-254.
3. Farooq M, T Aziz, A Wahid, DJ Lee and KHM Siddique, 2009. Chilling tolerance in maize: agronomic and physiological approaches. Crop Past Science, 60: 501–516.
4. Farooq M, T Aziz, SMA Basra, MA Cheema and H Rehamn, 2008a. Chilling tolerance in hybrid maize induced by seed priming with salicylic acid. Journal of Agronomy and Crop Science, 194: 161–168.
5. Farooq M, T Aziz, SMA Basra, A Wahid, A Khaliq and MA Cheema, 2008b. Exploring the role of calcium to improve the chilling tolerance in hybrid maize. Journal of Agronomy and Crop Science, 194: 350–359.
6. Foyer C and G Noctor, 2003. Redox sensing and signaling associated with reactive oxygen in chloroplasts, peroxisomes and mitochondria. Physiologia Plantarum, 119: 355–364.
7. Laemmli UK, 1970. Cleavage of structural proteins during the assembly of head of bacteriophage T4. Nature, 227: 680-685.
8. Lee DH and CB Lee, 2000. Chilling stress-induced changes of antioxidant enzymes in the leaves of cucumber: in gel enzyme activity assays. Plant science, 159: 75-85.
9. Rohman MM, S Uddin and M Fujita, 2010. Up-regulation of onion bulb glutathione S-transferases (GSTs) by abiotic stresses: A comparative study between two differently sensitive GSTs to their physiological inhibitors. Plant Osmics Journal, 3: 28-34.
10. Sharma P, N Sharma and R Deswal, 2005. The molecular biology of the low-temperature response in plants. Biochemical Essays, 27: 1048–1059.
11. Thomashow MF, 1999. Plant cold acclimation: Freezing tolerance genes and regulatory mechanisms. Annu. Rev. Plant Physiol. Plant Molecular Biology, 50: 571–599.
12. Yoshida R, A Kanno and T Kameya, 1996. Cool temperature induced chlorosis in rice plants. Plant Physiology, 112: 585–590.
13. Yu SJ and GE Abo-Elghar. 2000. Allelochemicals as inhibitors of glutathione *S*-transferases in the fall armyworm. Pesticide Biochemistry and Physiology, 68: 173-183.

2

SHOOT REGENERATION AND ROOT INDUCTION IN BRINJAL BY GROWTH REGULATORS

Md. Abdul Alim[1], Bhabendra Kumar Biswas[2], Md. Hasanuzzaman[2], Pronay Bala[3*] and Santanu Roy[4]

[1]Assistant Production Manager, ACI (Seed) Ltd.; [2]Professor, Department of Genetics and Plant Breeding, Hajee Mohammad Danesh Science and Technology University (HSTU), Dinajpur, Bangladesh; [3]PhD Fellow, Department of Crop Physiology & Ecology, HSTU, Dinajpur; [4]Program officer, Natural Resources Management Project, Caritas Fisheries Program, Mirpur, Dkaka, Bangladesh

***Corresponding author:** Pronay Bala; E-mail: kbdpronay@yahoo.com

ARTICLE INFO

Key words

Regeneration
Induction
Root
Brinjal
Growth regulators

ABSTRACT

The experiment was carried out to study the effect of genotypes and growth regulators on root induction and shoot regeneration of brinjal (*Solanum melongena*) genotypes. Leaf segments of three varieties of *Solanum* were cultured on MS medium with different concentrations and combinations of plant growth regulators. Among the tested genotypes, Protab showed highest percentage of shoot regeneration (65.67%). Early and maximum rate of regeneration was found in MS + 1 mg/L NAA (naphthalene acetic acid) + 0.1 mg/LBAP (6-benzylamino purine) for all the genotypes. The highest number of roots per shoot was counted in Protab (73.33%) on $^1/_2$MS + 2 mg/L IBA (indole butyric acid) + 0.4 mg/L BAP. Based on the overall performance, the variety Protab appeared as the best for shoot and root formation and ultimately successful regeneration of plants.

INTRODUCTION

Brinjal (*Solanum melongena L.),* belongs to the family Solanaceae, is one of the most popular, palatable and nutritious vegetable crop in Bangladesh. It is thought to be originated in Indian sub-continent with the secondary centre of origin in China (Zeven and Zhukovsky, 1975).The area of brinjal cultivation in Bangladesh is 45.57 thousand hectare and production is 3.40 lac metric tons (BBS, 2011). Brinjal is an economically important vegetable comprising an imperative supply of dietary protein, carbohydrate, vitamins and minerals particularly for the vegetarian population of developing countries. It has higher calorie, iron, phosphorus and riboflavin than tomato (Ray et al., 2011). However, the quality of brinjal varies with shape, size and colour of fruits (Bose and Som, 1986). Brinjal can be cultivated as a year round crop but the productivity and quality of this crop suffer due to its susceptibility to a number of diseases and insect pests (Sadilova et al., 2006).

In popular medicine, eggplant is indicated for the treatment of several diseases, including diabetes, arthritis, asthma and bronchitis and reducing blood and liver cholesterol rates in humans (Jorge et al., 1998). Brinjal is cultivated throughout the entire tropics and sub-tropics. Plant regeneration is somehow dependent on the type of explants such as cotyledon (Bardhan et al., 2012, Mir et al., 2011). The type and concentration of growth regulators and varieties can cause significant differences in morphogenetic responses of brinjal (Magioli and Mansur, 2005). The present study was undertaken to develop a suitable regeneration protocol for the tested three brinjal genotypes grown in Bangladesh.

MATERIALS AND METHODS

The experiment was carried out during the period from September, 2011 to January, 2012 in the Biotechnology Laboratory at the Department of Biotechnology of Hajee Mohammad Danesh Science and Technology University, Dinajpur. The varieties of Brinjal used in the present investigation namely-Protab, Green Ball and Ghemma Begun (Local) were collected from A.R. Mallik Seeds Co. Metal Agro Ltd. and locally from Dinajpur.

The surface sterilization of seeds was carried out under a Laminar Air Flow Cabinet. The floated seeds were discarded and others were rinsed in 70% ethyl alcohol for one minute, and then thoroughly washed with sterilized water. The alcohol treated seeds were immersed into 0.1% $HgCl_2$ solution for 8-10 minutes, few drops Tween-20 per 100 ml were also added at that time. The seeds were then washed 5-6 times with sterilized distilled water. Sterilized seeds were placed into seed germination medium in vials. Six seeds were placed in each vial. The culture was then incubated in dark till the germination of seeds. These vials were then transferred to 16 hours light for normal seedling growth. Twenty one days old seedlings were used as source of contamination free explants. The seedlings raised in axenic culture were used as the source of different kinds of explants. The leaf segments were used as explants. Leaf segment from each germinated seedling were cut into small pieces using sterilized scalpel under a Laminar Air Flow cabinet. Four pieces of leaf segments were arranged on each vials and gently pressed into the surface of the sterilized culture medium with various combinations and concentrations of growth regulators viz.,T_1 = MS medium containing with 1 mg/L NAA (Napthaline acetic acid) + 0.1 mg/L BAP (6-benzylamino purine).T_2 = MS medium containing with 1.5 mg/L NAA + 0.5 mg/l BAP,T_3 = MS medium containing with 1.5 mg/L NAA + 1 mg/l BAP,T_4 = MS medium containing with 1 mg/L NAA + 1.5 mg/l BAP and T_5 = MS medium containing with 0.5 mg/L NAA + 2 mg/L BAP. MS media with different concentrations and combinations of BAP and NAA (e.g., T_1 =1/2 MS medium containing 2 mg/LIBA + 0.4 mg/L BAP, T_2 = 1/2 MS medium containing with 1 mg/L IBA + 0.3 mg/l BAP and T_3 = 1/2 MS medium containing with 0.5 mg/L IBA + 0.2 mg/L BAP) were used for root formation. The subculture vials were again incubated at $25\pm2^{\circ}C$ with moderate light intensity. All cultures were examined regularly and the vials showing symptoms of contamination were discarded.

The plantlets with sufficient root system were separated from the vials. Agar was gently washed out with running tap water from root. The plantlets transplanted to small pots containing garden soil, sands and cowdung at the ratio of 1: 2: 1. Immediately after transplantation, the plants along with pots were covered with moist polythene bag to prevent desiccation. To reduce sudden shock, the pots were kept in the controlled environment in a growth room. The interior of the polythene bags were sprayed with water at every 24 hours to maintain high humidity around the plantlets. At the time plantlets were also nourished with Hoagland's solution.

After 2-3 days, the polythene bags were gradually perforated to expose the plants to natural environment. The polythene bags were completely removed after 7-10 days. The plantlets at this stage were placed in natural environment for 3-10 hours daily.

The established plants were calculated based on the number of plantlets placed in the pot and number of plants finally established or survived.

$$\text{Percent plant establishment} = \frac{\text{Number of established plantlets}}{\text{Total number of plantlets}} \times 100$$

Statistical analysis of data

The experiments were arranged in Completely Randomized Design (CRD). The analysis of variance for different characters was analyzed using MSTAT-C and means were compared by the Duncan's Multiple Range Test (DMRT).

RESULTS AND DISCUSSION

The main effect of genotypes, phyto-hormones and combination of these two in brinjal was found to be significant on shoot and root regeneration (Table 1-6).

The present study described development of a rapid and efficient plantlet regeneration protocol using leaf segment obtained from in vitro grown seedling. The type and concentration of a growth regulator was found to have significant impact on morphogenetic responses. The results of shoot regeneration from various brinjal explants at different growth regulators concentration and combinations are given in Table1-3. All three explants initiated callus and formed shoots on all five combinations of growth regulators tested. It was observed that induction and regeneration was quite permissive over a wide range of plant growth regulators.

Table 1. Main effects of genotypes on shoot regeneration

Genotypes	Total no. of calli with shoot	Shoot regeneration (%)	Days required for shoot initiation
Protab	13.13a	65.67 a	40.27 c
Green Ball	11.27 c	56.33 b	47.93 a
Ghemma Begun (Local)	10.47 b	52.33 c	48.33a
LSD	0.6761	3.381	1.089
CV (%)	7.80%	7.80%	3.05%

Table 2. Main effects of treatments on shoot regeneration

Treatment	Total no. of calli with shoot	Shoot regeneration (%)	Days required for shoot initiation
T_1	12.27 a	65.83 a	45.00 c
T_2	12.00 ab	60.00 ab	46.78 c
T_3	12.00 ab	60.00 ab	48.22 b
T_4	11.33 b	56.67 b	49.33 b
T_5	10.22 c	51.11 c	50.89 a
LSD	0.8729	4.365	1.406
CV (%)	7.80%	7.80%	3.05%

T_1=MS medium containing with 1 mg/l NAA + 0.1 mg/l BAP; T_2=MS medium containing with 1.5 mg/l NAA + 0.5 mg/l BAP
T_3=MS medium containing with 1.5 mg/l NAA + 1 mg/l BAP; T_4=MS medium containing with 1 mg/l NAA + 1.5 mg/l BAP
T_5=MS medium containing with 0.5 mg/l NAA + 2 mg/l BAP

Table 3. Combined effect of variety and hormone concentrations on shoot regeneration

Hormone × genotype		Total no. of calli with shoot	% Shoot regeneration	Days required for shoot initiation
T_1	Protab	14.33 a	71.67 a	43.33 g
	Green Ball	12.00 c-f	60.00 c-f	45.67e-g
	Ghemma Begun (Local)	11.33 d-f	56.67 d-f	47.00 c-f
T_2	Protab	14.00 ab	70.00 ab	45.67 e-g
	Green Ball	11.00 e-g	55.00 e-g	46.00 ef
	Ghemma Begun (Local)	11.00 e-g	55.00 e-g	44.27 fg
T_3	Protab	13.00 a-c	65.00 a-c	46.67 d-f
	Green Ball	12.33 c-e	58.67 f	48.67 b-d
	Ghemma Begun (Local)	10.67 f-h	53.33 f-h	49.33 a-c
T_4	Protab	12.67 b-d	63.33 b-d	48.67 b-d
	Green Ball	11.67 c-f	58.33 c-f	48.00 c-e
	Ghemma Begun (Local)	9.667 gh	48.33 gh	51.03 a
T_5	Protab	11.67 c-f	58.33 c-f	50.67 ab
	Green Ball	9.333 h	46.67 h	51.33 a
	Ghemma Begun (Local)	9.667 gh	48.33 gh	50.67 ab
LSD		1.512	7.560	2.435
CV (%)		7.80%	7.80%	3.05%

T_1=MS medium containing with 1 mg/l NAA + 0.1 mg/l BAP; T_2=MS medium containing with 1.5 mg/l NAA + 0.5 mg/l BAP
T_3=MS medium containing with 1.5 mg/l NAA + 1 mg/l BAP; T_4=MS medium containing with 1 mg/l NAA + 1.5 mg/l BAP
T_5=MS medium containing with 0.5 mg/l NAA + 2 mg/l BAP

Table 4. Performance of different genotypes on root induction

Genotypes	No. of shoots with root	% Root induction	Days required to root initiation
Protab	7.333 a	73.33 a	6.00c
Green Ball	6.222 b	62.22 b	7.89 b
Ghemma Begun (Local)	5.000 c	50.00 c	8.222 a
LSD	0.467	4.669	0.4260
CV (%)	7.62%	7.62%	8.07%

Table 5. Performance of growth regulators on different characteristics of root induction

Treatment	No. of shoots with root	Root induction (%)	Days required to root initiation
T_1	6.778 a	67.78 a	6.556 b
T_2	6.111 b	61.11 b	7.778 a
T_3	5.667 b	56.67 b	8.667 a
LSD	0.4666	4.669	0.4260
CV (%)	7.62%	7.62%	8.07%

T_1 =1/2 MS medium containing with 2 mg/l IBA + 0.4 mg/l BAP
T_2=1/2 MS medium containing with 1 mg/l IBA + 0.3 mg/l BAP
T_3=1/2 MS medium containing with 0.5 mg/l IBA + 0.2 mg/l BAP

Table 6. Effect of growth regulators x variety interaction on different characteristics of root induction

Hormone × genotype		No. of shoots with root	Days required to root initiation
T_1	Protab	7.667 ab	6.000 e
	Green Ball	7.000 bc	7.000 cd
	Ghemma Begun (Local)	5.667 d	7.667 de
T_2	Protab	7.332 a	6.000 cd
	Green Ball	5.667 d	7.667 bc
	Ghemma Begun (Local)	4.667 e	8.667 bc
T_3	Protab	6.333 cd	7.667 bc
	Green Ball	6.000 d	7.000 ab
	Ghemma Begun (Local)	4.067 e	9.333 ef
LSD		0.8082	0.738
CV (%)		7.62%	8.07%

Shoot regeneration

Effects of varieties on shoot regeneration

Genotypes showed significant variations for all the characters of shoot regeneration, percent shoot regeneration and days required for shoot initiations (Table1). From the three genotypes, highest number (13.13) of callus with shoot was found in Protab and lowest in Green Ball (11.27). Protab showed best performances (65.67%) on percent shoot regeneration. In contrast, Ghemma Begun (Local) showed lowest performance (52.33%) on percent shoots regeneration. Days required for shoot initiation was early in Protab (40.27 days) compared to Ghemma Begun (Local) (48.33 days). These results were in agreement with those obtained by Jayasree et al. (2001), Hossain et al. (2007).

Effect of growth regulators on shoot regeneration

Different concentrations of BAP, NAA showed significant variations for percent shoot regeneration and days required for shoot initiation (Table 2). Among the treatments, T_1 (MS +1 mg/L NAA + 0.1 mg/L NAA+0.1 mg/L BAP) showed highest percentage of shoot regeneration (65.83%) while T_5 (MS+0.5 mg/L NAA +2 mg/L BAP) showed lowest (51.11%). Days required for shoot initiation was minimum (45.00 days) in the T_1 (MS+1 mg/L NAA + 0.1 mg/L BAP) and maximum (50.89 days) in MS + 0.5 mg/L NAA + 2.0 mg/L BAP (T_5). These results were in agreement with those obtained by Rahman *et al.* (2006). Shivraj and Rao (2011) obtained highest number of shoots on MS medium supplement with 2 mg/L BAP and 0.5mg/l Kinetin using cotyledonary leaf explant.

Growth regulators × genotype interaction on shoot regeneration

Results related to growth regulators x genotype interaction for the characters of shoot regeneration such as total no. of calli with shoot, percent shoot regeneration and days required for shoot initiation in different concentrations of BAP showed significant variations (Table 3). Among the three genotypes, Protab showed highest no. of calli with shoot (14.33), best performance on percent shoot regeneration (71.67%) in MS +1 mg/L NAA + 0.1 mg/L BAP (T_1). In contrast, Ghemma Begun showed lowest number of calli with shoot (9.667) with MS + 0.5 mg/L NAA + 2 mg/L BAP (T_5), Green Ball showed lowest percent shoot regeneration (46.67%) with MS + 0.5 mg/L NAA + 2 mg/L BAP (T_5).Days required for shoot initiation was minimum (43.33 days) on the interactions of MS + 1 mg/L NAA + 0.1 mg/L BAP (T_1) with Protab and maximum (51.33 days) on the interactions MS + 0.5 mg/L NAA + 2 mg/L BAP (T_5) with Green Ball (Table 9). All the genotypes showed satisfactory results with MS+1 mg/l NAA +0.1 mg/l BAP (T_1) treatment. These results were in agreement with those reported by Prakash et al. (2008). Initiation of shoot is shown in Fig. 1.

Root induction from shoot

Root induction was found with wide variations at genotypes and different concentrations of growth regulators (Table 4, 5 and 6).

Effects of genotypes on root induction

Genotypes showed significant variation in producing root. Protab showed the highest number of shoots with root (7.333), highest percentage of root formation (73.33%) and minimum days required (6.00). Ghemma Begun (Local) showed lowest number of shoots with root (5.00), lowest percentage (50.00%) of root formation and maximum days required (8.222) (Table 4).

Effects of growth regulators on root induction

Root initiation varied in a wide range due to difference in growth regulators concentrations and combinations (Table 5). Maximum number of shoots with root (6.778) and highest percentage of root formation (67.78%) were recorded in ½ MS + 2 mg/L IBA + 0.4 mg/L BAP (T_1).Lowest no. of shoots with root (5.667) and lowest percentage of root (56.67%) formation were found in T_3 (Table 5). This result was in agreement with those obtained by Das *et al.* (2002). IBA is widely used for efficient root induction in Brinjal (Zayova et al., 2012; Shivraj and Rao, 2011).

Figure 1: Initiation of shoot from the callus of Protab (left) Variety using MS+1.0 mg/l NAA +0.1 mg/l BAP and Green Ball Variety using MS+1.5 mg/l NAA +0.5 mg/l BAP(right)

Fig. 2: Root initiation from regenerated shoot of Protab Variety in ½ MS +2 mg/l IBA+0.4 mg/l BAP (left) and Green Ball Variety in ½ MS +1 mg/l IBA+0.3 mg/l BAP (right)

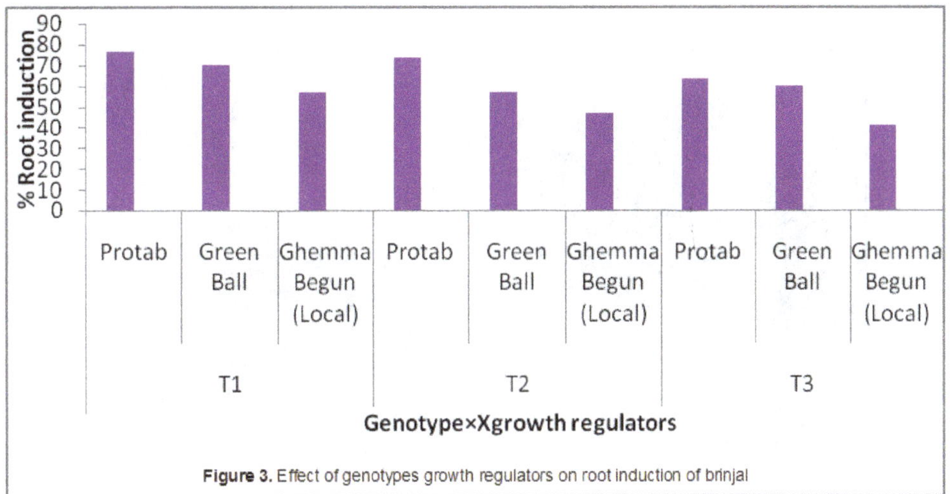

Figure 3. Effect of genotypes growth regulators on root induction of brinjal

Figure 4. Hardening of regenerated plant of Protab (left) and Ghemma Begun (Local) (right) after transplantation into small plastic pot

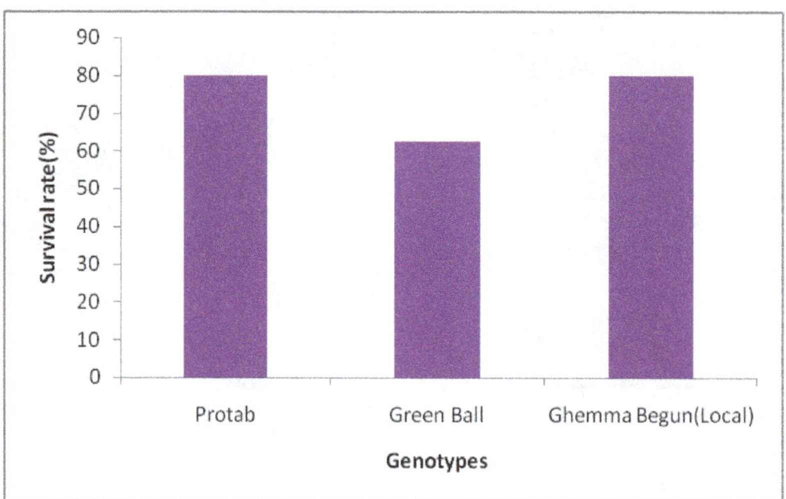

Figure 5. Effect of genotypes on survival rate of brinjal

Figure 6. Survived plant after hardening of Protab (left) and Ghemma Begun (Local) (right) Variety derived from leaf explant

Growth regulator x genotype interaction on root induction

Results related to hormone x genotype interaction for the characters of root regeneration such as number of shoots with root, percent root formation and days required to root initiation in different concentrations of MS, IBA and BAP showed significant variations. The results are presented in Table 6. Among the three genotypes, Protab showed best performance on number of shoots with root (7.667), best performance on percent root regeneration (76.67%) in T_1 (½ MS +2 mg/L IBA+0.4 mg/L BAP). In contrast, Ghemma Begun (Local) showed the lowest performance on number of shoots with root(4.067), lowest performance on percent root regeneration (40.67%) with ½ MS + 0.5 mg/L IBA + 0.2 mg/L BAP (T_3) .Days required for root initiation was minimum (6.000 days) on the interaction of ½ MS + 2 mg/L IBA + 0.4 mg/L BAP (T_1) with Protab and maximum (9.333 days) on the interaction ½ MS+0.5 mg/L IBA+0.2 mg/L BAP (T_3) .

From the above results, it may be concluded that ½ MS +2 mg/l IBA+0.4 mg/l BAP (T_1) with Protab showed the best performance on root regeneration. IBA is widely used for efficient root induction in Brinjal (Zayova et al., 2012). This result was in agreement with those obtained by Hossain et al. (2007), Prabavathi et al. (2007). Root initiation from regenerated shoot is shown in Fig. 2 and 3.

Establishment of plantlet

After sufficient development of root system, the small plantlets were taken out from culture vessels without any damage to roots and shoots .Medium adhered around the roots was removed by washing in running tap water to prevent microbial infection. The plantlets were then transplanted into plastic pots containing sterile soil, sand and cow dung in a 1:2:1 ratio. The pots were then covered with clear polyethylene bag to maintain high humidity conditions and kept in the growth chamber for proper hardening. Gradually the plantlets were adapted to soil and established. The survival rate of transferred regenerated plantlets after hardening was highest in Ghemma Begun (Local) (80%) and lowest in Green ball (62.5 %).Similar result was found Ferdausi et al., 2009. Hardening of regenerated plant and survived plant after hardening from leaf explants are shown in Figure 4 and 5.

CONCLUSION

By considering the overall investigation and comparing the performance of three brinjal genotypes, it was found that Protab was the best cultivar in case of shoot regeneration and root induction. The findings from the present investigation of the effect of genotypes and growth regulators on shoot regeneration and root induction of brinjal (*Solanummelongena.*) could be efficiently utilized for the advanced biotechnological research, as for example gene transfer and crop improvement.

REFERENCE

1. BBS, 2011. Statistical Yearbook of Bangladesh. Bangladesh Bureau of Statistics.Ministry of planning, Government of the people's Repubic of Bangladesh. Dhaka, Bangladesh.
2. Bardhan SK, C Sharma and DK Srivastava, 2012. Invito plant regeneration studies in brinjal. Journal of Cell andTissue Research, 12: 3213-3218.
3. Bose TK and MG Som, 1986. Vegetable crops in India. Nayaprokah, Calcutta.
4. Das GP, S Ramaswamy and MA Bari, 2000. Integrated crop management practices for the control of the brinjal shoot and fruit borer in Bangladesh. DAE-DANIDA Strengthening Plant Protection Services (SPPS) Project. Department of Agriculture Extention, Khamarbari, Dhaka.
5. Ferdausi U, K Nath, BL Das and MS Alam, 2009. *In vitro* regeneration system in brinjal (*Solanum melongena* L.) for stress tolerant somaclone selection Journal of Agricultural University, 7: 253–258.
6. Hossain MJ, M Rahman and MA Bari, 2007. Establishment of cell suspension culture and plantlet regeneration of brinjal (*Solanum melongena* L.). Journal of Plant Science, 2: 407-415.
7. Jayasree T, V Paban, M Ramesh, AV Rao and KJM Reddy, 2001. Somatic embryogenesis from leaf cultures of potato. Plant Cell Tissue Organ Culture, 64: 13-17.

8. Jorge PAR, LC Neyra, RM Osaki, E Almeida and N Bragagnolo, 1998. Efeito da berinjelasobreos lipids plasmáticos, a peroxidaçãolipídica e a reversão da disfunçãoendotelialnahipercolesteromia experimental. ArquivosBrasileiros de Cardiologia, 70: 87-91.
9. Magioli C and E Mansuri, 2005. Eggplant (*Solanum melongena*L.): tissue culture, genetic transformation and use as an alternative model plant. Acta Botanica Brasilica, 19: 139-148.
10. Mir KA, AS Dhatt, JS Sandhu and AS Sidhu, 2011. Effect of genotype, explant and culture medium on organogenesis in brinjal. Indian Journal of Horticulture, 68: 332-335.
11. Prakash DP, BS Deepali, R Asokan, YL Ramachandra, DL Shetti, L Anand, VS Hanur, 2008. Effect of growth regulators on *in vitro* complete plant regeneration in brinjal. Indian Journal of Horticulture, 65: 371-376.
12. Rahman MD and DL Berquam, 2006. Plant cell and Tissue culture: The role of Haberlandt. Botanical Review, 35: 59-88.
13. Ray BP, L Hassan and KM Nasiruddin, 2011. In vitro regeneration of brinjal (*Solanummelongena* L.) Bangladesh Journal of Agricultural Research, 36: 397-406.
14. Sadilova E, FC Stintzing, R Carle, E Van and A Synder, 2006. Anthocyanins, colour and antioxidant properties of eggplant (*Solanum melongena* L.*)* and violet pepper (*Capsicum anmium L.*). Methods of Molecular Biology, 343: 439-447.
15. Shivraj G and S Rao, 2011. Rapid and efficient regeneration of eggplant (*Solanum melongena* L.) Indian Journal of Biotechnology, 10: 125-129.
16. Zayova ER, Vassilevska-Ivanova, B Kraptchev and D Stoeva, 2012. Indirect shoot organogenesis of eggplant (*Solanum melongena* L.) Journal of Central European Agriculture, 13: 446-457.
17. Zeven AC and MP Zhukovsky, 1975. Dictionary of cultivated plants and their centres of Diversity. Wageningen, Netherlands.

3

ASSESSING GENETIC DIVERSITY OF MAIZE (Zea mays L.) GENOTYPES FOR AGRONOMIC TRAITS

Shanjida Rahman*, Md. Mukul Mia, Tamanna Quddus, Lutful Hassan and Md. Ashraful Haque

Department of Genetics and Plant Breeding, Faculty of Agriculture, Bangladesh Agricultural University, Mymensingh- 2202, Bangladesh

***Corresponding author:** Shanjida Rahman; E-mail: shanjidashaki@yahoo.com

ARTICLE INFO

Key words

Genetic diversity
Morphological traits
Genotype
Maize
Assessment

ABSTRACT

Maize is one of the most important cereals globally and a promising cereal supplement in Bangladesh. The current study was undertaken to assess genetic diversity among nine maize genotypes. Data were recorded on seven morphological traits *viz.* plant height (cm), ear height (cm), ear length (cm), ear diameter (cm), number of kernels/ear, 1000-kernel weight (g) and yield/plant (g). Statistical analysis showed significant variation among maize genotypes. Considering plant height, ear length, ear diameter, ear height, number of kernels/ear and yield/plant BHM-7 was observed as the best one. Among all the traits higher phenotypic coefficient of variation and genotypic coefficient of variation were observed for yield/plant. Genetic advance was highest for 1000-kernel weight followed by number of kernels/ear. The correlation study revealed only two positive significant associations: plant height with yield/plant and ear diameter with ear length. Nine genotypes were grouped into three clusters. These all clearly indicated the presence of ample genetic diversity among maize genotypes which can be exploited in future breeding program for better utilization of maize germplasm.

INTRUDUCTION

Maize is an amazing crop throughout the world which occupies a large portion of world economy. Total maize production in the world for the 2013-14 is 959 million tons which is more than that of wheat (709 million tons) and rice (473 million tons) (GMR 2014). In 2009, over 159 million hectares of maize were planted worldwide which gave a yield of 5.12 tons/ha (FAOSTAT, 2009). From nutritional view point it is better than rice and wheat (http://ndb.nal.usda.gov/ndb/search/list). Maize is also important in industrial and medicinal uses. In Bangladesh maize is considered as second most important cereal crop which occupies 165.5 thousand hectares land and produces 1018 thousand metric tons with an average yield of 6.1 tons/ha (BBS, 2011). The main use of maize is as poultry feed (Hossain and Shahjahan, 2007). In Bangladesh current need of maize is 1.60 million metric tons annually (Ahmed, 2013). If the traditional prolonged rice dependent food habit of Bangladesh can be diversified with maize, it would probably be possible to attain foodself sufficiency to a great extent. Because, it is a high yielding and low-cost crop compared to rice and wheat. For satisfying this purpose, diversity assessment of existing genotypes would provide valuable inputs to move forwarding with the current and upcoming breeding needs.

Genetic diversity is the variation of heritable characteristics in genetic makeup present in a population of the same species. Knowledge of diversity in a germplasm is very important for the improvement of crop plants through breeding program (Hallauer et al., 1988). Different methods can be used to assess genetic diversity in plant species, such as pedigree and heterosis data, morphological marker and molecular markers (Melchinger, 1999; Xia et al., 2005; Legesse et al., 2007). Morphological traits are the earliest, convenient and effective genetic markers used for germplasm management (Statonet al., 1994). Morphological markers is of great value in studies of maize landraces (Galarreta and Alvarez, 2001; Lucchin et al., 2003; Ortiz et al., 2008). Both qualitative and quantitative traits have been considered to study phenotypic diversity of maize (Alika et al., 1993; Taba et al., 1998; Lucchin et al., 2003). The most commonly used parameters are related to plant architecture traits, tassel traits, ear and kernel characteristics. The variables contributing to genetic diversity are grain weight and grain yield (Hoque, 2008); kernel weight and days to maturity (Ahmed, 2007); ear height, days to silking, % tryptophan content, cob length and 1000-seed weight (Kadir, 2010); ear length and diameter (Ahmed, 2013). The present investigation was conducted to analyze genetic diversity of maize genotypes using morphological traits.

MATERIALS AND METHODS

Experimental site, duration and materials

The experiment was conducted in the field lab of the Department of Genetics and Plant Breeding, Bangladesh Agricultural University (BAU), Mymensingh from March 2013 to April 2014. Nine maize genotypes were selected as experimental materials, those were: Uttaran, Duranta, BARI Hybrid Maize 5 (BHM-5), BARI Hybrid Maize 7 (BHM-7), BARI Hybrid Maize 9 (BHM-9), V-92, H-981, Pop Corn and Sweet Corn. All of them were collected from Bangladesh Agricultural Research Institute (BARI).

Experimental design and data collection

The experiment was conducted following randomized complete block design (RCBD) with three replications. The plot size was 4.2 m x 2.7 m, row to row and plant to plant distances were 60 cm and 30 cm, respectively. Recommended production packages i.e. application of recommended doses of fertilizers, weeding, thinning, irrigation, pesticide etc. was followed as per BARI recommendation, as and when necessary to ensure the optimum plant growth and development.

At field maturity, five randomly selected plants were used for recording observations on the traits: plant height (cm), ear height (cm), ear length (cm), ear diameter (cm), number of kernels/ear, 1000-kernel weight (g) and yield/plant (g). Analysis of variance (ANOVA) was done on the sample for all the seven character mentioned using MStat-c statistical program. The total variance of each character was partitioned into replication, genotype and error. The differences within the classes of effects were tested by F-test. The mean performance of the nine genotypes for their traits was shown through lettering by DMRT (Duncan's Multiple Range Tests) using the same software.

Genotyping and phenotypic variances were estimated according to the formula given by Johnson et al., (1995).

Genotypic variance, $\sigma^2_g = \frac{GMS-EMS}{r}$

Where,

GMS= Genotypic mean square, EMS= Error mean square

r = Number of replication, Phenotypic variance, $\sigma^2_{ph} = \sigma^2_g + EMS$

Where,

σ^2_g = Genotypic variance

EMS = Error mean square

Heritability in broad sense (h^2_b) was estimated according to the formula suggested by Johnson et al. (1995) and Hanson et al. (1956).

Heritability, $h^2_b = (\sigma^2_g / \sigma^2_{ph}) \times 100$

Where,

$\sigma^2 g$ = Genotypic variance, σ^2_{ph} = Phenotypic variance

Genotypic and phenotypic co-efficient of variations were estimated according to Burton (1952) and Singh and Chaudhary (1985).

Genotypic co-efficient of variations, GCV = $\frac{\sqrt{\sigma^2 g}}{\overline{X}} \times 100$

Where,

$\sigma^2 g$ = Genotypic variance; and $\overline{X}$ = Population mean

Phenotypic co-efficient of variations, PCV = $\frac{\sqrt{\int \sigma^2 ph}}{\overline{X}} \times 100$

Where,

σ^2_{ph} = Phenotypic variance; and

$\overline{X}$ = Population mean

Estimation of genetic advance was done following formula given by Johnson *et al.* (1955) and Allard (1960).

Genetic advance, GA = $h^2_b.K.\sigma_p$

Where, h^2_b = Heritability

K = Selection differential, the value of which is 2.06 at 5% selection intensity

σ_p= Phenotypic standard deviation

Genetic advance in percent of mean was calculated by the formula of Comstock and Robinson (1952) as follows:

Genetic advance in percentage of mean, GA (%) = $\frac{GA}{\overline{X}} \times 100$

Where,

GA = Genetic advance, $\overline{X}$ = Population mean

The correlation coefficient (r) between two variables such as X and Y can be estimated using following formula:

$$r_{xy} = \frac{\sum XY - \frac{\sum X \sum Y}{N}}{\sqrt{\sum Y^2 - \frac{(\sum Y)^2}{N}} \sqrt{\sum X^2 - \frac{(\sum X)^2}{N}}}$$

Where,

$\sum XY - \frac{\sum X \sum Y}{N}$ = Sum of products of X and Y

$\sqrt{\sum Y^2 - \frac{(\sum Y)^2}{N}}$ = Sum of squares of Y

$\sqrt{\sum X^2 - \frac{(\sum X)^2}{N}}$ = Sum of squares of X

The genotypes were arranged in different clusters followed by the method suggested by Ward's Method based on Squared Euclidean distance and hierarchical cluster analysis. The initial cluster distances in Ward's minimum variance method are therefore defined to be the squared Euclidean distance between points:
$d_{ij} = d(\{X_i\}, \{X_j\} = ||X_i - X_j||^2$

RESULT

The analyses of variance of different genotypes of maize for different agronomic traits are shown in Table 1. It indicated that the difference among genotypes for all the traits under study *viz.*, plant height (cm), ear height (cm), ear length (cm), ear diameter (cm), number of kernels/ear, 1000-kernel weight (g) and yield/plant (g) was highly significant.

Table 1. Analysis of Variances (mean squares) for different characters of nine genotypes of maize

Source of variation	Degrees of freedom (df)	Plant height (cm)	Ear height (cm)	Ear length (cm)	Ear dia-meter (cm)	Number of kernels/ ear	1000 kernel weight (g)	Yield/ plant (g)
Replication	2	37.33	5.82	2.22	0.13	81.93	1.79	0.46
Genotype	8	1105.17 **	82.95 **	11.29 **	0.74 **	5459.12**	10016.53 **	1510. 9**
Error	16	3.67	6.57	0.62	0.04	97.01	0.96	1.53

** indicate significant at 1% level of probability

Trait-wise mean performance of the genotypes

Trait-wise mean performance of genotypes gives a clear comparison among genotypes. The mean performance of the nine genotypes for their traits with lettering by DMRT (Duncan's Multiple Range Tests) is shown in Table 2.

Table 2. Mean performance of nine genotypes of maize based on different agronomic traits

Genotype	Plant height (cm)	Ear height (cm)	Ear length (cm)	Ear diameter (cm)	Number of kernels /ear	1000-kernel weight (g)	Yield/ plant (g)
Pop corn	151.6e	53.7ab	7.50b	2.56bc	278a	137h	88.95b
BHM-9	198a	53ab	8.23b	2.80bc	245c	149g	86.86c
BHM-7	182.3b	53.7ab	8.40b	2.84b	264abc	291a	98.35a
BHM-5	182.3b	53.7ab	12.03a	3.81a	282a	214f	78.66e
V-92	145f	45.3c	6.43b	2.26c	153e	227d	75.39f
Uttaran	171.3c	55.3a	11.73a	3.48a	274ab	216e	50.47g
Duranta	164.3d	43.3c	7.97b	2.74bc	251bc	257b	37.80i
H-981	160d	47bc	8.33b	2.73bc	202d	241c	82.39d
Sweet corn	140g	41.7c	7.20b	2.43bc	270abc	122i	39.76h
CV%	**1.15**	**5.16**	**9.12**	**7.28**	**3.99**	**0.48**	**1.74**
Maximum	**200**	**59**	**13.7**	**4.2**	**291**	**292.5**	**99.2**
Minimum	**138**	**40**	**6.2**	**2**	**136**	**121.6**	**36.5**
Mean	**166.1**	**49.6**	**8.7**	**2.8**	**246.9**	**206.7**	**70.9**
F-test	**	**	**	**	**	**	**

Note: CV (%) = Coefficient of variation, ** indicate significant at 1% level of probability
Similar letter indicates there is no significant difference at 5% level of probability as per DMRT
Different letter indicates significant difference at 5% level of probability

Estimation of correlation co-efficient

The correlation value denotes only the nature and extent of association existing among characters. In this experiment, two associations showed positive significant correlation: plant height with yield/plant and ear diameter with ear length (Table 3). In the current study, only one negative association had been observed for number of kernels/ear with 1000-kernel weight which is non-significant.

Table 3. Correlation co-efficient between yield and other yield related characters

Characters	Plant height(cm)	Ear height (cm)	Ear length (cm)	Ear diameter(cm)	Number of kernels/ear	1000-kernel weight(g)
Ear height (cm)	0.606					
Ear length (cm)	0.500	0.515				
Ear diameter (cm)	0.568	0.539	0.969**			
Number of kernels/ear	0.320	0.410	0.530	0.553		
1000-kernel weight (g)	0.246	0.050	0.177	0.198	-0.263	
Yield/plant (g)	0.906**	0.513	0.522	0.574	0.175	0.447

** indicate significant at 1% level of probability

Estimation of genetic parameters of maize genotypes

Genotypic variances, phenotypic variances, heritability, genotypic co-efficient of variation (GCV), phenotypic co-efficient of variation (PCV), genetic advance and genetic advance (GA) as percent of mean (GA %) for all the yield contributing traits are presented in Table 4. Among the all traits yield/plant exhibited high estimates of GCV (31.61%), PCV (31.66%) and highest value of genetic advance in percentage (62.47%). On the other hand highest heritability (99.97%) and genetic advance (113.23) was observer in 1000-kernel weight.

Table 4. Genetic parameters of seven different characters of nine maize genotypes

Traits	Phenotypic variance	Genotypic variance	PCV (%)	GCV (%)	Heritability (%)	GA	GA (%)
Plant Height (cm)	370.84	367.17	11.59	11.54	99.00	37.89	22.81
Ear height (cm)	32.03	25.46	11.40	10.17	79.49	8.92	17.97
Ear length (cm)	6.67	6.05	29.86	28.44	90.70	3.75	43.36
Ear diameter (cm)	0.27	0.23	18.22	16.82	83.94	0.88	30.86
No. of kernels/ear	1884.38	1787.37	17.58	17.14	94.85	81.96	33.19
1000-kernel weight (g)	3339.49	3338.52	27.98	27.97	99.97	113.23	54.82
Yield/plant(g)	504.67	503.14	31.66	31.61	99.97	44.33	62.47

GCV = Genotypic co-efficient of variation, PCV = Phenotypic co-efficient of variation,
GA = Genetic advance, GA% = Genetic advance as percent

Dendrogram

In the present experiment, dendrogram of nine maize lines had been made using mean value of different agronomic traits which results in three clusters (Fig. 1).

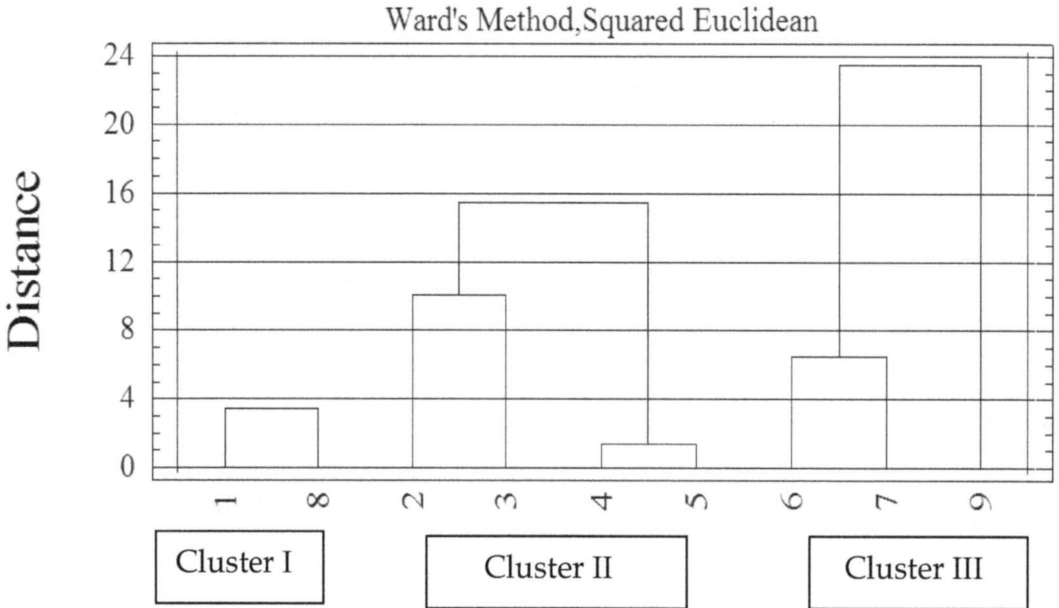

Figure 1. Dendrogram showing distribution of nine genotypes among three clusters

Table 5. Distribution of nine maize genotypes among three clusters

Cluster	No. of Genotypes	Genotypes
I	2	Popcorn, H-981
II	4	BHM-9, BHM-7, BHM-5, V-92
III	3	Uttaran, Duranta, Sweet corn

DISCUSSION

Analysis of genetic diversity is an important step for better understanding and utilization of germplasm. Maize is a diverse and highly cross pollinated crops. There exists a good number of work on genetic diversity of maize but the genotypes experimented in this study were not studied yet. We found wide array of diversity among these genotypes. The ANOVA table suggested presence of ample genetic variability among the genotypes. Shahrokhi and Khorasani (2013) also observed significant variation among genotypes for days to silking, days to anthesis, plant height, ear height, kernel number, rows number, 1000-kernel weight and yield. Ahmed (2013) also observed significant variation among maize genotypes for yield, ear length, ear diameter, number of kernels/ear, 1000-kernel weight, days to maturity, days to silking, plant height and ear height. Prasannaet al. (2001) noted that genetic variability for most of the yield and yield contributing traits in maize were very high and amenable to genetic enhancements.

The presence of a wide range between minimum and maximum values for each trait assures the existence of sustainable variation among the genotypes studied (Table 2). Such variation in the germplasm collection of maize is an opportunity for breeders to improve traits of interest through parent selection, hybridization and recombination of desirable genotypes (Ahmed, 2013). In this experiment, each trait indicated separate genotype as best one, such as BHM-9 considering plant height; BHM-5 considering ear length, ear diameter and number of kernel/ear; and BHM-7 considering 1000-kernel weight and yield/plant.

Correlation is also an important measurement indicating that traits which should be given importance to increase yield. In this experiment, two associations showed positive significant correlation: plant height with yield/plant and ear diameter with ear length (Table 3). Positive significant correlation between plant height and yield/plant had also been recorded by Salami et al. (2007) and Rafiq et al. (2010). This positive and significant association between the traits suggested additive genetic model thereby less affected by the environmental fluctuation. Besides, most of the associations were recorded as positive but non-significant. This type of association referred information of inherent relation among the pairs of combination. Positive and non-significant association between plant height with ear height and ear height with grain yield had also been observed by Olakojo and Olaoye (2011). Munawar et al. (2013) also studied positive, non-significant association between plant height with ear height and ear length. Positive and non-significant association for ear height with ear length and yield/plant was recorded by Rafiq et al. (2010). The negative insignificant association for number of kernels/ear with 1000-kernel weight referred a complex linked of relation among pair of combinations.

Table 4 indicated that variability within the maize genotypes is sufficiently divergent and constitutes potential candidate genotypes on which improvement program can be initiated. Phenotypic variance was higher than the genotypic variances for all the traits thus indicated the influences of environmental factor on these traits. Similar findings were observed by Salami et al. (2007), Bello et al. (2012) and Anshumanet al. (2013).The estimates of genotypic coefficient of variation (GCV) reflect the total amount of genotypic variability. Since most of the economic characters (grain yield) are complex in inheritance and are greatly influenced by several genes interacting with various environmental conditions, the study of phenotypic coefficient of variation (PCV) and genotypic coefficient of variation (GCV) is not only useful for comparing the relative amount of phenotypic and genotypic variations among different traits but also very useful to estimate the scope for improvement by selection. Table 4 indicated that for all traits PCV were higher than GCV. However, the differences between genotypic and phenotypic coefficient of variation indicated the environmental influence. In the present experiment, low value of GCV and PCV for plant height and ear height were recorded which was also observed by Anshuman et al. (2013).

Heritability estimates is of tremendous significance to the breeder, as its magnitude indicates the accuracy with which a genotype can be recognized by its phenotypic expression. Almost all traits studied here showed high heritability (Table 4). High heritability for ear length and 1000-kernel weight was also recorded by Noor et al. (2010). Aminu and Izge (2012) studied high heritability for plant height and yield/plant. Very high heritability (above 90%) was observed for plant height, ear height, number of kernels/ear, 1000-kernel weight and yield/plant by Bello et al. (2012) and Anshuman et al. (2013).

Character exhibiting high heritability may not necessarily give high genetic advance. Johnson et al., (1955) showed high heritability should be accompanied by high genetic advance to arrive at more reliable conclusion. Bello et al. (2012) recorded higher genetic advance for plant height, number of kernels/ear and yield/plant. In the present study high heritability with high genetic advance was found for the trait number of kernels/ear and 1000-kernel weight which indicated the preponderance of additive gene action for the expression of these traits which is fixable in subsequent generations. The author suggested that these parameters were under the control of additive genetic effects. Sumathi et al. (2005) also suggested that these parameters could be manipulated according to requirements, and worthwhile improvement could be achieved through selection. However, high heritability and low genetic advance were observed for ear height, ear length and ear diameter which may be attributed to non-additive gene action governing these traits, and these characters could be improved through the use of hybridization and hybrid vigour.

Maize plant height, yield/plant, number of kernels/ear, 1000-kernel weight can be improved by selection, as these characters exhibited high genotypic and phenotypic coefficient of variations along with high heritability and genetic advance. Ear length, ear height and ear diameter had high heritability but the genetic coefficient of variations was low. This indicated that though, the character was highly heritable, its improvement through early generation selection may not give the desired results. Effective selection for superior genotypes is possible considering yield/plant, number of kernels/ear, 1000-kernel weight, plant and ear heights and can be used as target traits to improve maize grain yield.

Dendrogram grouped nine genotypes into three clusters (Fig 1). Maximum genotypes (four: BHM-9, BHM-7, BHM-5 and V-92) were distributed under cluster II, followed by cluster III and cluster I containing three (Duranta, Uttaran and Sweet corn) and two (Popcorn and H-981) genotypes, respectively (Table 5).

Subramanian and Subbaraman (2010) made dendrogram on 38 genotypes which results in four clusters: 14 genotypes under cluster 1, 13 under cluster 3, 9 under cluster 3 and remainder under cluster 4. 30 maize inbred lines were distributed by Azad et al. (2012) among six clusters. Another study was carried out by Chen et al. (2007) who reported that 186 maize genotypes could be classified into ten clusters.

CONCLUSIONS

Assessing of genetic diversity is the basic need for the utilization of any germplasm. The research findings suggested adequate genetic diversity exists among studied nine genotypes. Though BHM-7 performed well other genotypes were good enough considering different traits. All these genotypes can be utilized for further improvement of maize germplasm for the desired characters.

ACKNOWLEDGEMENT

This research was supported by the special grants of the Ministry of Science and Technology, Bangladesh, grant No.BS-43/2014.

REFERENCES

1. Ahmed S, 2007. Study of genetic diversity in maize inbreds, Annual Research Report, 2006-2007: Maize and Barley improvement, Plant Breeding Division, BARI, Joydebpur, Gazipur, pp: 16-18.
2. Ahmed S, 2013. Study on genetic diversity in maize (*Zea mays* L.) inbred lines for the development of hybrids, PhD Thesis, Department of Genetics and Plant Breeding, Bangladesh Agricultural University, Bangladesh.
3. Alika RW, ME Aken'ova and CA Fatokun, 1993. Variation among maize (*Zea mays* L.) accessions of Bendel State, Nigeria, Multivariate analysis of agronomic data. Euphytica, 66: 65-71.
4. Allard RW, 1960.Principles of Plant Breeding, John Wiley and Sons. Inc., New York.
5. Aminu D and AU Izge, 2012. Heritability and correlation estimates in maize (*Zea mays* L.) under drought conditions in Northern Guinea and Sudan Savannas of Nigeria.World Journal of Agricultural Sciences, 8: 598-602.
6. Anshuman V, NN Dixit, Dipika, SK Sharma and S Marker, 2013.Studies on heritability and genetic advance estimates in Maize genotypes.Bioscience Discovery, 4: 165-168.
7. Azad MAK, BK Biswas, N Alam and SS Alam, 2012. Genetic diversity in maize (*Zea mays* L.) inbred lines. The Agriculturists, 10: 64-70.
8. BBS 2011. Summery crop statistics and crop indices, Agriculture Wing, Bangladesh Bureau of Statistics, Ministry of Planning, Government of People's Republic of Bangladesh, Dhaka, pp. 37.
9. Bello OB, SA Ige, MA Azeez, MS Afolabi, SY Abdulmaliq and J Mahamood, 2012. Heritability and genetic advance for grain yield and its component characters in maize (*Zea Mays* L.).International Journal of Plant Research, 2: 138-145.
10. Burton GW, 1952.Quantitative inheritance in Grasses. Proceeding 6th International Grassland Congress, 1: 277-283.
11. Chen FB, KC Yang, TZ Rong and GT Pan, 2007. Analysis of genetic diversity of maize hybrids in the regional tests of Sichuan and Southwest China, Acta Agronomica Sinica, 33: 991-998.
12. Comstock RE and HF Robinson, 1952. Genetic parameters, their estimate and significance, Proceedings 6th International Grassland Congress, 1: 284-291.
13. Elahi NE, 2005. Effect of whole family training on maize cultivation in Bangladesh, CIMMYT, Bangladesh, House no.18, Road no.4, section 4,Uttara, Dhaka, Bangladesh.
14. Galarreta RJI and A Alvarez, 2001. Morphological classification of maize landraces from northern Spain.Genetic Resources and Crop Evolution, 48: 391–400.
15. GMR (Grain Market Report), 2014: International Grains Council (http://www.lgc.lnt).
16. Hallauer AR and JB Mirinda-Filho, 1988.Quantitative genetic in maize breeding, 2nd edition, Iowa State University Press, Ames, USA.
17. Hanson G, HF Robinson and RE Comstock, 1956. Biometrical studies on yield in segregating population of Korean Lespedza, Agronomy Journal, 48: 268-274.

18. Hoque MM, M Asaduzzaman, MM Rahman, S Zaman and SA Begum, 2008.Genetic divergence in maize (*Zea mays* L.). Bangladesh Journal Agricultural Research, 9: 145-148.
19. Hossain A and M Shahjahan, 2007. Grain quality evaluation of the major varieties or cultivar of maize, Annual Research Report: 2006-07, Post-Harvest Technology Division, Bangladesh Agricultural Research Institute (BARI), Gazipur, pp: 1-6.
20. Johnson HW, HF Robinson and RE Comstock, 1955.Estimates of genetic and environmental variability in soybean. Agronomy Journal, 47: 314-318.
21. Kadir MM, 2010.Development of quality protein maize hybrids and their adaption in Bangladesh, PhD thesis, Department of Genetics and Plant Breeding, Bangladesh Agricultural University.
22. Legesse BW, AA Myburg, KV Pixley and AM Botha, 2007. Genetic diversity of African maize inbred lines revealed by SSR markers. Hereditas, 144: 10-17.
23. Lucchin M, G Barcaccia and P Parrini, 2003. Characterization of a flint maize (*Zea mays* L.convar. *mays*) Italian landrace: I. Morpho-phenological and agronomic traits. Genetic Resources and Crop Evolution, 50: 315–327.
24. Melchinger AE, 1999.Genetic diversity and heterosis. In: JG Coors and S Pandey (Editors), The genetics and Exploitation of Heterosis in Crops. ASA, CSS and SSSA, Madison, Wisconsin. pp: 99-118.
25. Munawar M, M Shahbaz, G Hammad and M Yasir, 2013. Correlation and path analysis of grain yields components in exotic maize (*Zea mays* L.) hybrids. International Journal of Sciences: Basic and Applied Research, 12: 22-27.
26. Noor M, H Rahman, Durrishahwar, M Iqbal, SMA Shah and I Ullah, 2010. Evaluation of maize half sib families for maturity and grain yield attributes. Sarhad Journal Agriculture, 26: 545-549.
27. Olakajo SA and G Olaoye, 2011. Correlation and heritability estimates of maize agronomic traits for yield improvement and *Strigaasiatica* (L) *Kuntze* tolerance. African Journal of Plant Science, 5: 365-369.
28. Ortiz R, R Sevilla, G Alvarado and J Crossa, 2008. Numerical classification of related Peruvian highland maize races using internal ear traits.Genetic Resources and Crop Evolution, 55: 1055–1064.
29. Prasanna BM, SK Vasal, B Kassahun and NN Singh, 2001. Quality protein maize. Current Science, 81: 1308-1319.
30. Rafiq CM, M Rafique,A Hussain and M Altaf, 2010. Studies on heritability, correlation and path analysis in maize (*Zea mays* L.), Journal of Agricultural Research, 48: 35-38.
31. Salami AE, SAO Adegoke and OA Adegbite, 2007.Genetic variability among maize cultivars grown in Ekiti-State, Nigeria,Middle-East Journal of Scientific Research, 2: 9-13.
32. Shahrokhi M and KK Khorasani, 2013. Study of Morphological Traits, Yield and Yield Components on 28 Commercial Corn Hybrids (*Zea mays* L.). International Journal of Agronomy and Plant Production, 4: 2649-2655.
33. Singh RK and BD Chaudhary, 1985. Biometrical method in quantitative genetic analysis, Kalyani Publisher, Ludhiana, New Delhi, pp: 54-57.
34. Subramanian A and N Subbaraman, 2010. Hierarchical cluster analysis of genetic diversity in Maize germplasm, Electronic Journal of Plant Breeding, 1(4): 431-436.
35. Sumathi P, A Nirmalakumariand K Mohanraj, 2005. Genetic variability and traits interrelationship studies in industrially utilized oil rich CIMMYT lines of maize (*Zea mays* L), Madras Agricultural Journal. 92: 612-617.
36. Taba S, J Diaz, J Franco and J Crossa, 1998. Evaluation of Caribbean maize accessions to develop a core subset. Crop Science, 37: 400-405.
37. Xia XC, JC Reif, AE Melchinger, M Frisch, DA Hoisington, D Beck, K Pixley and Warburton, 2005. Genetic diversity among CIMMYT maize inbred lines investigated with SSR markers: 2. Subtropical, tropical mid altitude, and high land maize inbred lines and their relationships with elite U.S. and European maize. Crop Science, 45: 2573-2582.

4

PERFORMANCE OF BITTER GOURD IN ASSOCIATION WITH KARANJA (*Pongamia pinnata* L.) TREE

Md. Nasir Uddin Khan and Mohammad Kamrul Hasan*

Department of Agroforestry, Faculty of Agriculture, Bangladesh Agricultural University, Mymensingh-2202, Bangladesh

***Corresponding author:** M. Kamrul Hasan, E-mail: mkhasanaf@gmail.com

ARTICLE INFO

Key words

Char land
Agroforestry
Growth
Karanja
Bitter gourd

ABSTRACT

The study was conducted at the Char Kalibari which is situated along the bank of Old Brahmaputra River under Sadar Upazila of Mymensingh district during November 2013 to March 2014 to observe the performance of bitter gourd (*Momordica charantia*) as arable crop with karanja (*Pongamia pinnata* L.) trees in an agroforestry system. The experiment was laid out in a Randomized Complete Block Design (RCBD) with three replications having four treatments viz., T_0 (open field condition referred as control), T_1 (< 50 cm distance from the tree base), T_2 (50-100 cm distance from the tree base) and T_3 (>100 cm distance from the tree base). The result showed that all the growth parameters and yield of bitter gourd were significantly influenced by the associated tree component at different distances from the karanja tree base. The highest (1.92 tha^{-1}) fresh yield of bitter gourd was obtained in open field condition compare to any other treatments but no significant different was found from the treatment T_3 (distance >100 cm from the tree) while the lowest (0.8 tha^{-1}) in < 50 cm distance from the tree base. It was found that on an average 58.33%, 29.17% and 14.58% yield of bitter gourd were decreased in <50 cm, 50-100 cm and >100 cm distances from karanja tree base compare to open field condition. On the other hand, the growth performance of karanja trees i.e. both height and girth increment was better in sole tree condition compare to tree with bitter gourd condition. Therefore, it can be concluded that tree-crop combination i.e. >100 cm distance from the tree base would be possible although there was some yield loss (14.58%) which was less significant compare to alone bitter gourd. Through this combination we can get diversified product. So, we can follow this agroforestry system to improve char based farming system of Bangladesh during the early establishment period of trees.

INTRODUCTION

Bangladesh is a densely populated and small country with an area of 147,570 km^2. According to the latest census, the population of the country is over 160 million with an average growing rate of 1.6% and the density of human population is 1033.5/km^2 (Wikipedia, 2015). If the current population growth rate (1.6%) continues, population will increase to 180 million by the year 2025, and the country will face enormous problem for nursing her population. The current forest land of Bangladesh is 2.52 million hectares which is 17.08% of total land area (BFD, 2013).The economy of the country draws its strength and stability mostly from agriculture. The fertility of our land is decreasing day by day due to intensive cropping and use of high input technologies. As a result, the country has been facing acute shortage of food, timber, fruit, vegetable, etc.

Agroforestry plays a vital role in supplying not only the daily necessities of people but also in maintaining ecological balance. In Bangladesh the scope of agroforestry is vast. The major venues of agroforestry are homestead, roadside, railway side, embankment side, char land, coastal area, deforested area, institutional premises, riverside, etc. Among them char land is the most important venue for practicing agroforestry systems. 'Char' a tract of land surrounded by the waters of an ocean, sea, lake, or stream; it usually means any accretion in a river course or estuary (Chowdhury, 1988). Chars in Bangladesh have been distributed into five sub-areas: the Jamuna, the Ganges, the Padma, the Upper Meghna and the Lower Meghna rivers. There are other areas of riverine chars in Bangladesh, along the Old Brahmaputra and the Tista rivers. But compared to the chars in the major rivers, these constitute much less land area. It is estimated in 1993 that the total area covered by chars in Bangladesh was 1.7 thousand sq km. A large number of populations are living in these char areas and maintaining their livelihood through char based farming systems. Therefore, for increasing production, maintaining ecological balance and improving socio-economic condition of the char land people, integrated approach with crops per vegetables and trees is necessary.

Bitter gourd (*Momordica charantia*) is important for its quick growing nature and high yielding potential. It is easily cultivated as a companion crop or inter crop. Bitter gourd is a well-known and a very popular vegetable grown successfully throughout Bangladesh. Even this vegetable can successfully grow in association with agroforestry system to get more diversified output from the same land. This vegetable is very low in calories, providing just 17 calories per 100g. Nevertheless, its fruits are rich in phytonutrients like dietary fiber, minerals and vitamins. Bitter gourd notably contains phyto-nutrient, polypeptide-P; a plant insulin known to lower blood sugar levels. In addition, it composes hypoglycemic agent called charantin. *Charantin* increases glucose uptake and glycogen synthesis in the cells of liver, muscle and adipose tissue. Recently reported that bitter gourd is more effective for treating HIV infection (USDA, 2013). Karanja (*Pongamia pinnata*), a tree species which can survive in the water logged condition and also can stabilize soil. However, the fertility of our land is decreasing rapidly due to intensive cropping and use of high input technologies. During winter season char land is a unique area for vegetables production where land is fertile due to siltation and irrigation requirement is less or easy. For this reason present study was undertaken to observe the performance of bitter gourd in association with karanja tree in the Char Kalibari in the bank of Old Brahmaputra River, Mymensingh.

MATERIALS AND METHODS

Experimental site and geographical position

The experiment was carried out at the experimental farm in the SPGR project field Char Kalibari, Mymensingh Sadar, under the control of Department of Agroforestry, Bangladesh Agricultural University, Mymensingh, during the period from November 2013 to March 2014. The district Mymensingh is located between 24°38′3″ north and 90°16′4″ east latitude. Total area of this district is 4394.57 km^2 and situated on the west bank of Brahmaputra River. The district has total 12 upazila and the study area i.e. Char Kalibari belong to the Mymensingh sadar upazila. The geographical position of Char Kalibari located between 24°45′ to 45′40″ north and north and 90°24′4″ to 90°24′44″ east latitude (Wikipedia, 2014 and Figure 1). Total area of this char land is about 2.57 km^2 where cultivated land is about 175 ha, 10 ha wetland, 40 ha fallow land, 23 ha household and rest forest area. Total population is 2350 of which 1238 male and 1112 female (Source: Six no. Char Ishwardia union parishad office records, 2014).

Characteristics of soil and climate

Char Kalibari is one of the char land area of Mymensingh district located at the Old Brahmaputra river side. The topography of the field is medium high land above flood level belonging to the Old Brahmaputra flood plain agro-ecological zone-9. It is characterized by non-calcareous dark grey flood plain soil having pH values from 6.5 to 6.8 (FAO, 1988). Most of the soil has silty to clay texture. The climate at the locality is sub-tropical in nature. It is characterized by high temperature and heavy rainfall during kharif season (April to September) and a scanty rainfall during rabi season (October to March). The overall relative humidity remains high almost all over the year except the winter.

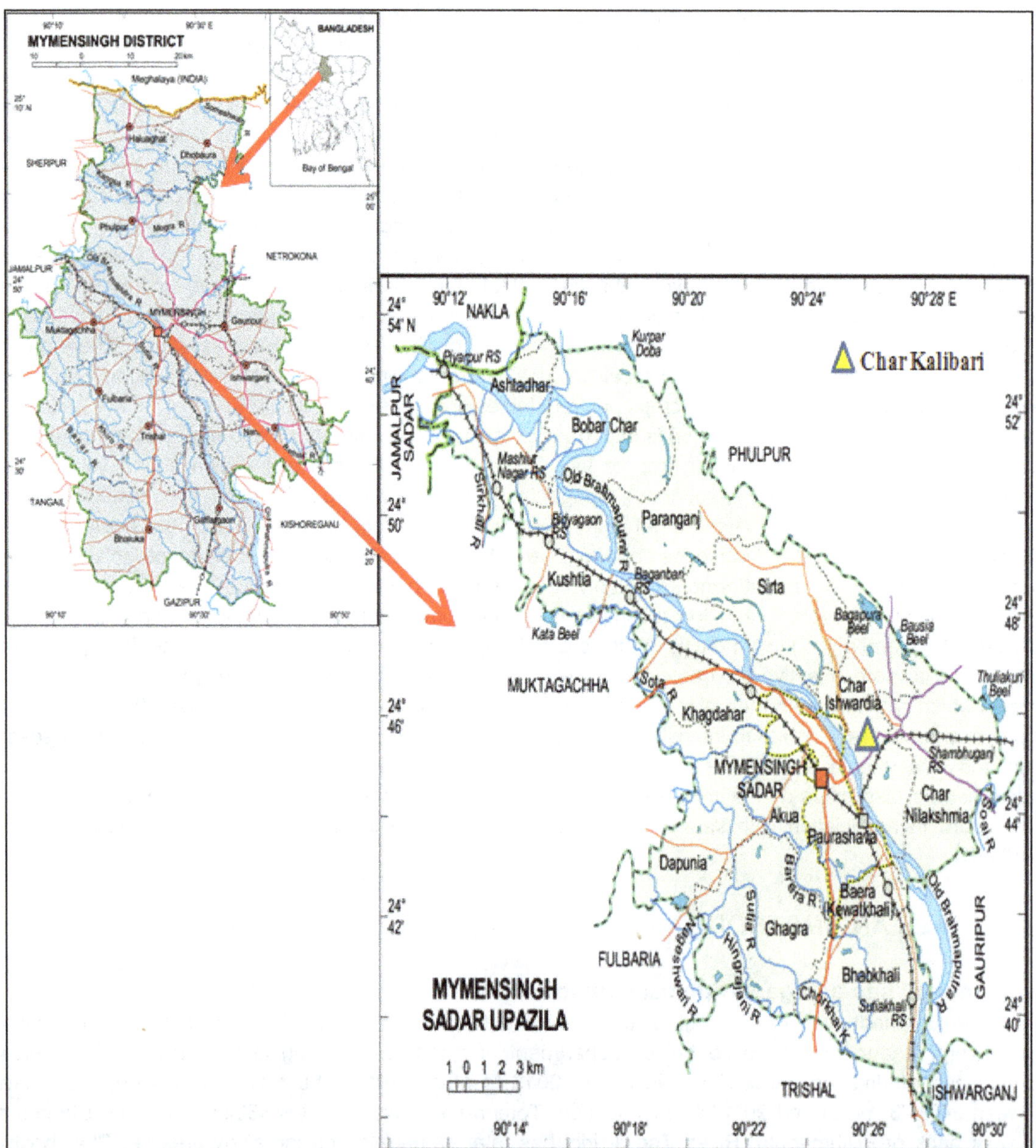

Figure 1. A map of Mymensingh district including Sadar upazila showing the location of the study area (Char Kalibari)

Tree and plant materials

In this study we considered three year's old previously established karanja (*Pongamia pinnata*) tree as test tree component and bitter gourd (*Momordica charantia*) as plant material.

Land preparation

The experimental land was first opened on Ist November 2013 and the operation was done by spade. At first the soils at the base of trees were loosening very well and made friable. Then the land was kept fallow for few days. All the crop residues and weeds were removed from the field and finally the land was properly leveled. Only recommended dose (5 tha^{-1}) of well decomposed cowdung was applied for the crop during final land preparation. No chemical fertilizer was applied.

Experimental design and treatment combination

The experimental design was laid out in a Randomized Complete Block Design (RCBD) with three replications. Four treatments considering the distance from tree to bitter gourd were viz. T_0= Open field referred as control, T_1= <50 cm distance from the tree, T_2= 50-100 cm distance from the tree and T_3= >100 cm distance from the tree.

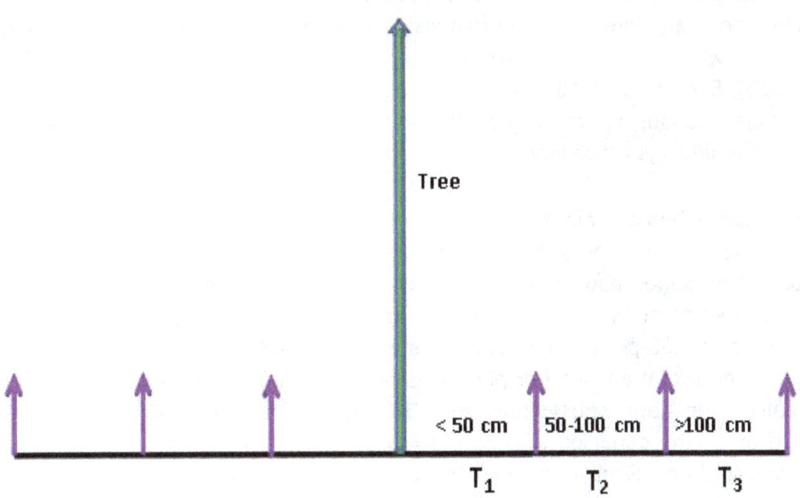

Figure 2. Layout of the experiment

Test crop establishment and management

Bitter gourd seeds were directly sown in the experimental plot on 5th November 2013. After seed sowing, necessary intercultural operations were done properly. To keep the plots free from weeds, weeding was done three times for experimental plots. The plots were irrigated by using water cane to supply sufficient soil moisture for the vegetable. Mulching was used to conserve water. Emergence of bitter gourd seedling was started after two weeks from the date of seed sowing. Bitter gourd seedlings were thinned out at three times. First thinning was done at 15 days after sowing while second and third thinning was done at 5 days interval from every thinning. No pesticide and fungicide were used in the experimental field for tree-vegetable association.

Sampling and data collection

Bitter gourd was harvested at 90 days after seed sowing. It was harvested at several picking. Five bitter gourd plants were randomly selected from each unit plot to record the morphological and yield contributing characters viz. vine length (cm), total leaves per plant, number of primary branches per plant, number of female and male flower per plant, number of fruits per plant, individual fruit weight (g), fresh and dry yield of fruits (tha^{-1}). All the above parameters were taken in three stages of bitter gourd growth like vegetative, flowering and harvesting stage except the fruit yield.

Data analysis

Data were statistically analyzed by using MSTAT-C and wasp2 (Web Agri Stat Package) software package to find out the statistical significance of the experimental results. The analysis of variance for each of the studied parameter was done by F (variance ratio) test. The mean differences were evaluated by Duncan's New Multiple Range Test (DMRT) (Gomez and Gomez, 1984) and also by Least Significant Difference (LSD) test.

RESULTS AND DISCUSSION

Morphological characteristics and yield performance of bitter gourd (*Momordica charantia*)

Vine length (cm)

From the Table 1, it was found that the highest vine length of bitter gourd was 45.33 cm, 70.00 cm and 96.00 cm produced by T_0 (open field without treatment) and the lowest was 36.33 cm, 56.66 cm and 69.33 cm in treatment T_1 at 40, 90 and 120 days after sowing, respectively. Among the distance treatments T_3 (>100 cm distance from the tree base) produced the highest (43.00 cm, 64.00 cm and 88.66 cm) vine length at all growth stages of bitter gourd plant. Masfikha (2013) and Rahman (2013) studies on bitter gourd cultivation in association with fruit trees and three selected tree species during winter season and found the similar findings in case of vine length which were supportive to the present study. Alam et al. (2012), Mallick et al. (2013), Rahman et al. (2013), Bali et al. (2013) and Ahmed et al. (2013) also performed the various experiment on different trees and crops under agroforestry system to evaluate their growth performance and observed that the more or less similar findings which were supported by the present study.

Number of primary branches per plant

The maximum average number of primary branches per plant was 3.33, 5.33 and 6.66 in vegetative, flowering and harvesting stage, respectively which was produced by open field referred as control or without associated tree while treatment T_3 (>100 cm distance from the tree base) produces highest number of primary branches (2.33, 3.66 and 5.33 per plant in all three stages, respectively) among the distance treatment. The minimum average number of branches per plant was obtained from the treatment T_1 (<50 cm distance from the tree base) in all growth stage which was 1.33, 2.33 and 2.33, respectively (Table 1). This result indicated that the open field or control condition noticed the maximum branch number per plant than other distance treatment with associated tree. Similar observation also obtained by Islam et al. (2008) who evaluated that the performance of winter vegetables under guava-coconut based multistrata agroforestry system. Rahman et al. (2013) conducted an experiment to see the performance of sweet gourd grown in association with akashmoni saplings and found that the similar type of results which was supported to the present findings.

Number of leaves per primary branch

Different treatments showed significant effect on number of leaves per primary branch of bitter gourd in all examined stage. Table 1 revealed that the maximum number of leaves per primary branch of bitter gourd in vegetative, flowering and harvesting stage (25.66, 58 and 34.33, respectively) was produced by T_0 treatment (open field or without treatment) while second maximum number of leaves per plant (20, 51 and 32.66) was produced under T_3 treatment (>100 cm distance from the tree base) in all three stage which was also highest among the distance treatments. In contrast, the minimum number of leaves per primary branch was 15.33, 34 and 22.66 in vegetative, flowering and harvesting stage, respectively at T_1 treatment (<50 cm distance from the tree base). Rakib (2013) and Uddin (2013) were conducted studies on radish and carrot in association with akashmoni and fruit trees during winter season and observed the similar findings which were strongly supported by the above findings. Masfikha (2013) and Rahman (2013) found the similar findings in their studies which were supportive to the present study.

Table 1. Morphological characteristics of bitter gourd during vegetative, flowering and harvesting stages in association with karanja tree

Treatments	Vine length (cm)			No. of primary branches/plant			Leaves/primary branch			Female flower/plant	Male flower/plant	No. of fruits/plant	Fresh weight of single fruit
	Vegetative stage	Flowering stage	Harvesting stage	Vegetative stage	Flowering stage	Harvesting stage	Vegetative stage	Flowering stage	Harvesting stage				
T_0	45.33 a	70.00 a	96.00 a	3.33 a	5.33 a	5.66 a	25.66 a	58.00 a	34.33 a	20.66 a	96.00 a	23.00 a	33.50 a
T_1	36.33 b	56.66 d	69.33 d	1.33 c	2.33 c	2.33 b	15.33 c	34.00 d	22.66 c	13.00 c	78.66 d	14.00 d	22.82 c
T_2	41.33 ab	60.66 bc	76.66 c	1.67 bc	2.66 c	4.66 a	18.33 b	44.66 c	30.00 b	17.00 b	83.66 c	18.33 c	29.68 b
T_3	43.00 a	64.00 b	88.66 b	2.33 b	3.66 b	5.33 a	20.00 b	51.00 b	32.66 a	18.66 ab	90.66 b	20.33 b	32.43 a
Level of sig.	*	**	**	**	**	**	*	**	**	**	**	**	**

Note: T_0= Control condition, T_1= <50 cm distance from tree base, T_2= 50-100 cm distance from tree base, T_3= >100 cm distance from tree base; sig.=significance; in a column figures having the same letter (s) do not differ significantly; * & ** Significant at 5% and 1% level of probability.

Number of female flowers per plant

Number of female flowers per plant is the most important yield contributing character which was significantly influenced by different distance of growing bitter gourd under karanja tree. The highest number of female flowers per plant (20.66) was found in T_0 (open field referred as control) while treatment T_3 (>100 cm distance from the tree base) produces the second maximum (18.66) which was statistically similar with control treatment. The treatment T_2 (50-100 cm distance from the tree base) recorded the third maximum number of female flowers per plant (17) which was also statistically similar to the treatment T_3. The lowest number of female flowers per plant (13) was found under close contact of the tree i.e. T_1 treatment and it was probably due to poor photosynthetic capacity and resource pool competition between trees and bitter gourd (Table 1). Masfikha (2013) and Rahman (2013) were conducted studies on bitter gourd cultivation in association with fruit trees and three selected tree species during winter season and obtained the similar findings which were highly supportive to the present study results. Similar observation was obtained by Rahman (2013) who reported that except plant height all others morphological characters of three vegetables (tomato, brinjal and chilli) were performed better in open field condition rather than distance treatments from the akashmoni tree base.

Number of male flowers per plant

Number of male flowers per plant is another important yield contributing character which was significantly influenced by different distance of growing bitter gourd from the test sample karanja tree. The highest number of male flowers per plant (96) was produced in open field referred as control treatment and the lowest number of female flowers per plant (78.66) was in T_1 treatment. On the other hand, among the distance treatments, T_3 (>100 cm distance from the tree base) produces the highest number of male flowers per plant (90.66) (Table 1). Rahman (2013) obtained the similar findings in case of tomato, brinjal and chilli grown in association with akashmoni tree under agroforestry system which was supported to the present study result. Masfikha (2013) and Rahman (2013) were mentioned the similar findings in their studies on bitter gourd cultivation in association with fruit trees and three selected tree species during winter season which were helpful to the present study.

Number of fruits per plant

Number of fruits per plant was significantly influenced by different distance of growing bitter gourd under karanja tree. The highest number of fruits per plant (23.00) was produced in T_0 treatment while treatment T_3 (>100 cm distance from the tree base) produces the second highest number of fruits per plant (20.33). The lowest number of fruits per plant (14.00) was produced under close contact of the tree base i.e. treatment T_1. It was probably due to poor photosynthetic capacity and resource pool competition between tree and bitter gourd (Table 1). Similar results also obtained by Basak et al. (2009) who reported that the yield contributing characters of three vegetables increased gradually with the increase of planting distance from the Lohakat (Xylia *dolabriformis*) tree base. Alam et al. (2012), Mallick et al. (2013), Rahman et al. (2013), Bali et al. (2013) and Ahmed et al. (2013) also conducted the various experiment on different tree and crop grown under agroforestry systems and reported those findings which were supported by the present study.

Fresh weight of single fruit (g)

Fresh weight of single fruit of bitter gourd was also significantly influenced by different distance from the karanja tree base. The highest fresh weight of single fruit (33.50 g) was recorded in T_0 (open field referred as control) while statistically similar fresh weight of single fruit (32.43 g) was produced at >100 cm distance from the sample tree base. The lowest fresh weight of single fruit (22.82 g) was found in <50 cm distance from the tree base (Table 1). Similar findings obtained by Masfikha (2013) and Rahman (2013) in their studies which were highly supportive to the present study. Alam et al. (2012), Mallick et al. (2013) and Rahman (2013) also obtained the helpful findings on different tree and crop association under agroforestry systems which were supported by the present study.

Yield of bitter gourd (*Momordica charantia*)

Fresh yield of bitter gourd (tha^{-1})

The fresh yield of bitter gourd (tha^{-1}) was affected significantly due to effect of different treatments (Figure 3). As evident from the observation of figure 3, the highest fresh (1.92 t/ha) yield of bitter gourd was obtained from the treatment T_0 referred as control or without tree association. While the lowest fresh (0.80 tha^{-1}) yield was obtained from the closest distance treatment T_1 (<50 cm distance from the tree base). Among the distance treatments, >100 cm distance from the tree base produced the highest fresh yield (1.64 tha^{-1}) which was statistically similar to that of the control treatment. It is stated that literally there is some yield loss but statistically there was no significant yield loss in compare to control treatment. Masfikha (2013) and Rahman (2013) found the similar findings in case of fresh yield of bitter gourd in their studies which were supportive to the present study. Basak et al. (2009) mentioned that the yield contributing characters of radish, tomato and soybean were increased gradually with the increase of planting distance from the Lohakat (*Xylia dolabriformis*) tree base which was strongly supported to the present result. Sayed et al. (2009) also found the similar results in their study on interaction effects of vegetables in association with two years old Telsur (*Hopea odorata*) sapling. Alam et al. (2012), Mallick et al. (2013), Rahman et al. (2013), Bali et al. (2013), Rakib (2013), Uddin (2013) and Ahmed et al. (2013) also observed that the more or less similar findings in their studies which were highly supported by the present study results.

Dry yield of bitter gourd (tha^{-1})

From figure 4, the highest dry (0.15 tha^{-1}) yield of bitter gourd were obtained from the treatment T_0 referred as control or without tree association while the lowest dry (0.06 tha^{-1}) yield was obtained from the closest distance treatment T_1 (<50 cm distance from the tree base). Among the distance treatments, >100 cm distance from the tree base produced the highest dry yield (0.13 tha^{-1}) which was statistically similar to that of the control treatment. Masfikha (2013) and Rahman (2013) found the similar findings in case of dry yield of bitter gourd in their studies on bitter gourd cultivation in association with fruit trees and three selected tree species during winter season which were supportive to the present study. Sayed et al. (2009) also obtained the similar results in their study on interaction effects of vegetables in association with two years old Telsur (*Hopea odorata*) sapling. Alam et al. (2012), Mallick et al. (2013), Rahman et al. (2013), Bali et al. (2013), Rakib (2013), Uddin (2013) and Ahmed et al. (2013) also observed the interesting results in their studies which were highly supported by the present study result.

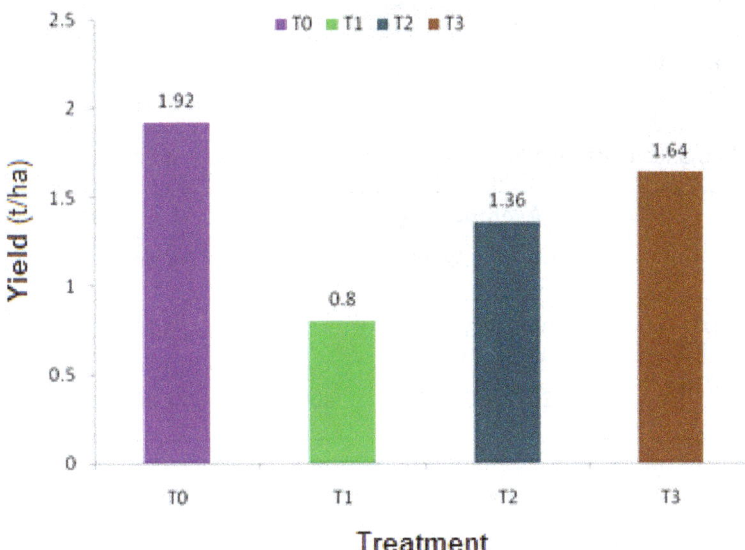

Figure 3. Bar graph showing fresh yield of bitter gourd along with karanja (*Pongamia pinnata*) tree in agroforestry production system.

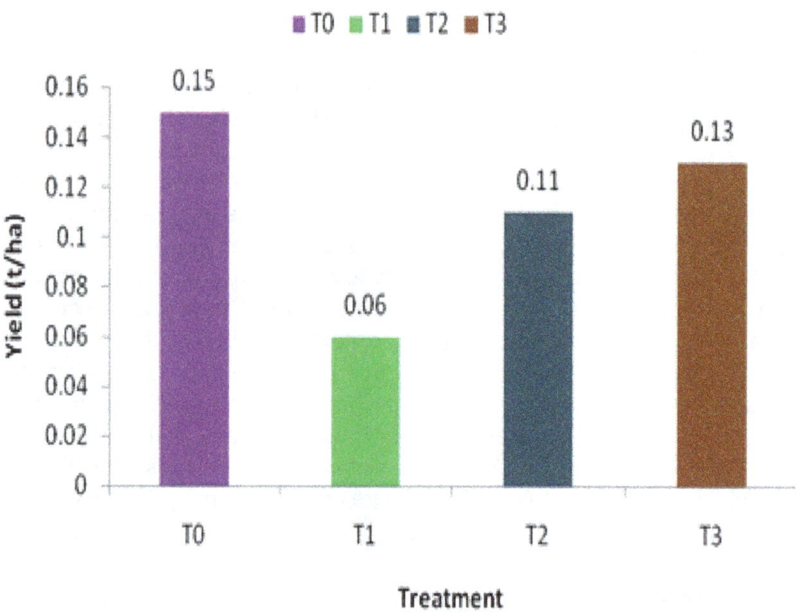

Figure 4. Bar graph showing dry yield (oven dry) of bitter gourd along with karanja (*Pongamia pinnata*) tree in agroforestry production system.

Performance of karanja *(Pongamia pinnata)* tree with and without bitter gourd condition

Tree height (cm)

The height of karanja tree was significantly influenced by the interaction of bitter gourd as arable crops. The highest (45.00 cm) average height increment of sixteen (16) trees was recorded in tree without bitter gourd condition while the lowest (31.00 cm) average tree height increment was found under the tree with bitter gourd condition (Table 2). It might be due to the competition of tree and crop for nutrient, water and light in agroforestry system. The highest tree height was recorded under tree without crop situation in compare to tree with crop association that reported by Sayed et al. (2009), Basak et al. (2009), Islam et al. (2009), Alam et al. (2012), Masfikha (2013), Mallick et al. (2013), Rahman et al. (2013), Bali et al. (2013), Rakib (2013), Uddin (2013) and Ahmed et al. in their studies.

Table 2. Growth performance of karanja tree in association with and without bitter gourd

Condition	Average Height (cm)			Average Girth (cm)		
	Before	After	Increment	Before	After	Increment
Tree with bitter gourd	321.00	352.00	31.00	20.78	23.29	2.51
Tree without bitter gourd	325.20	370.20	45.00	22.10	27.25	5.15

Girth (cm)

The girth of karanja tree was also significantly influenced by the growth of bitter gourd as arable crops. The highest (5.15 cm) average girth increment of sixteen (16) trees was recorded from tree without bitter gourd condition or control condition. On the other hand, the lowest (2.51 cm) average girth increment of sixteen (16) trees was found in tree with bitter gourd condition (Table 2). Similar results also found by Sayed et al. (2009), Islam et al. (2009), Alam et al. (2012), Masfikha (2013), Mallick et al. (2013), Rahman et al. (2013), Bali et al. (2013), Rakib (2013), Uddin (2013) and Ahmed et al. in their studies where they reported that the highest tree girth was recorded under tree without crop condition in compare to tree-crop combination.

CONCLUSION

From the experiment it is observed that growth and yield of bitter gourd gradually decreased with decreasing distance towards the tree base. Growths of any vegetables are directly related with moisture availability in soil but in agroforestry system where a competition occurred between tree and crop for moisture beneath the tree canopy. For this reason, may be growth and yield of bitter gourd remarkably reduced beneath the tree canopy or near the tree base. However, the agroforestry practice is profitable for farmer because from the same land they can produce crop/vegetable and tree/fruit at the same time while other practices like monocropping produces only one product from the same land. Therefore, it can be concluded that agroforestry practices like bitter gourd with karanja tree is better than other practices like agriculture or monocropping system in char based farming system of Bangladesh.

REFERENCES

1. Ahmed MN, MA Mondol, MI Hossain, A Akter and MA Wadud, 2013. Performance of kangkong under two years old akashmoni tree. Journal of Agroforestry and Environment, 7: 89-92.
2. Alam Z, MA Wadud and GMM Rahman, 2012. Performance of summer vegetables in char land based agroforestry system. Journal of Agroforestry and Environment, 6: 1-8.
3. Bali SC, MA Mondol, A Akter, Z Alam and MA Wadud. 2013. Effect of guava and lemon on the yield of okra under agroforestry system. Journal of Agroforestry and Environment, 7: 53-56.
4. Bangladesh Forest Department (BFD), 2013. Ministry of Environment and Forest, Government of the People's Republic of Bangladesh, Dhaka, Bangladesh.
5. Basak S, MK Hasan, MS Islam and MA Wadud, 2009. Performance of radish, tomato and soybean during the first year of Lohakat (*Xylia dolabriformis*) plantation. Journal of Environmental Science and Natural Resources, 2: 185-189.
6. Char Ishwardia union parishad, 2014. Six number Char Ishwardia union parishad office records, Sadar upazila, Mymensingh, Bangladesh.
7. Chowdhury EH, 1988. Human adjustment to river bank erosion hazard in the Jamuna Flood plain, Bangladesh. Human Ecology, 16: 421-437.
8. Food and Agriculture Organization (FAO), 1988. Tropical forestry resources assessment project (GEMD): Tropical Africa, Tropical Asia, Tropical America. Volume 4. Rome, Italy.
9. Gomez KA and AA Gomez, 1984. Statistical Procedures for Agricultural Research. John Wiley and Sons, New York, Chichester, Brisbane, Toronto, Singapore, pp: 139-240.
10. Islam F, KK Islam and MA Rahim, 2008. Performance of winter vegetables in Guava-Coconut based multistrata agroforestry system. Journal of Agroforestry and Environment, 2: 35-38.
11. Islam MS, MA Wadud, MK Hasan, MM Rahman and GMM Rahman, 2009. Performance of three winter vegetables in association with Telsur (*Hopea odorata*). Journal of Agroforestry and Environment, 3: 73-76.
12. Mallick E, MA Wadud and GMM Rahman, 2013. Strawberry cultivation along with Lohakat (*Xylia dolabriformis*) tree as agroforestry system. Journal of Agroforestry and Environment, 7: 1-6.
13. Masfikha M, 2013. Performance of bitter gourd in association with three fruit trees during winter season. M.S. Thesis, Department of Agroforestry, Bangladesh Agricultural University, Mymensingh.

14. Rahaman A, MI Hossain, A Akter, MA Wadud and GMM Rahman, 2013. Performance sweet gourd grown in association with akashmoni saplings. Journal of Agroforestry and Environment, 7: 61-64.
15. Rahman HMS, 2013. Bitter gourd cultivation along with three tree species as char land based agroforestry system. M.S. Thesis, Department of Agroforestry, Bangladesh Agricultural University, Mymensingh, Bangladesh.
16. Rahman M, Z Alam, MA Mondol and GMM Rahman, 2013. Performance of sweet gourd in association with eucalyptus saplings. Journal of Agroforestry and Environment, 7: 19-21.
17. Rahman MM, 2013. Effect of akashmoni tree on three winter vegetables grown in agroforestry system. M.S. Thesis, Department of Agroforestry, Bangladesh Agricultural University, Mymensingh, Bangladesh.
18. Rakib MA, 2013. Radish cultivation in association with fruit trees during winter season. M.S. Thesis, Department of Agroforestry, Bangladesh Agricultural University, Mymensingh.
19. Sayed MKI, MA Wadud, MA Khatun, R Yasmin and GMM Rahman, 2009. Interaction effects of vegetables in association with two years old Telsur (*Hopea odorata*) sapling. Journal of Agroforestry and Environment, 3: 103-106.
20. Uddin MS, 2013. Performance of carrot in association with three years old akashmoni tree. M.S. Thesis, Department of Agroforestry, Bangladesh Agricultural University, Mymensingh.
21. USDA, 2013. National Nutrient data base of USDA (cucumber nutrition facts). Online available at: http:perperwww.nutrition–and–you.comperradish.html.
22. Wikipedia, 2015. The Free Encyclopedia. Demographics of Bangladesh. Online available at: http://en.wikipedia.org/wiki/Demographics_of_Bangladesh
23. Wikipedia, 2014. The Free Encyclopedia. Mymensingh District. Online available at: http://en.wikipedia.org/wiki/Mymensingh_District

5

EFFECTS OF INDIGENOUS MEDICINAL PLANT TULSI (*Ocimum sanctum*) LEAVES EXTRACT AS A GROWTH PROMOTER IN BROILER

Firoj Alom[1], Mahbub Mostofa[2], M. Nurul Alam[2], M.Golam Sorwar[2], Jashim Uddin[2] and M. Mizanur Rahman[2]

[1]Department of Animal Husbandry and Veterinary Science, University of Rajshahi, Rajshahi Sadar-6205; [2]Department of Pharmacology, Bangladesh Agricultural University, Mymensingh-2202, Bangladesh

***Corresponding author:** Mahbub Mostofa; E-mail: mostofa57@yahoo.com

ARTICLE INFO | **ABSTRACT**

The study was conducted to determine the efficacy of Tulsi (*Ocimum sanctum*) leaves extract as a growth promoter in broiler. Thirty (30) day-old broiler chicks were purchased from Kazi hatchery and after seven days of acclimatization in the poultry shed of Pharmacology department randomly divided into two groups I_0 and I_1. No vaccination schedule was practiced and no antibiotics were added in rations. Group was supplemented with Tulsi (*Ocimum sanctum*) leaves extract @ 2ml/litre in drinking water. Weekly observations were recorded for live body weight gain upto 5th weeks and blood test was performed at 17th and 35th day's age of broiler to observe the hematological changes between control (Group) and treatment (Group) group. The treatment group (Group A) recorded statistically non- significant for live body weight at 1st and 2nd weeks than that of control group (Group B) but found statistically significant at 3rd ($p<0.01$), 4th ($P<0.05$) and 5th ($P<0.01$) weeks of age and the Hematological parameters (TEC, PCV, Hb and ESR) showed statistically significant ($p<0.01$) difference as compared to control group.

Key words

Tulsi
Growth promoter
Broiler

INTRUDUCTION

The poultry industry has become an important economic activity in many countries. In large-scale rearing facilities, where poultry are exposed to stressful conditions, problems related to diseases and deterioration of environmental conditions often occur and result in serious economic losses. Prevention and control of diseases have led during recent decades to a substantial increase in the use of veterinary medicines. However, the utility of antimicrobial agents as a preventive measure has been questioned, given extensive documentation of the evolution of antimicrobial resistance among pathogenic bacteria (Ladefoed et *al*, 1996). So, the possibility of antibiotics ceasing to be used as growth stimulants for poultry and the concern about the side-effects of their use as therapeutic agents has produced a climate in which both consumer and manufacturer are looking for alternatives (Trafalska and Grzybowska, 2004; Griggs and Jacob, 2005, Nava *et al.,* 2005, Shivakumar *et al.*,2005, Maity *et al*,2004)). Essentially, there are two main ways in which we can reduce our dependence on antibiotic use in poultry feed. An obvious choice is the development of alternatives to antibiotics that work via similar mechanisms, promoting growth whilst enhancing the efficiency of feed conversion.

The genus Ocimum, typically contain fragrant herbs and small herbs. It has nearly 30 species which are mainly found in the tropics and subtropics (Paton, 1992). Several medicinal properties have been attributed to *Ocimum sanctum L.* Different parts of Tulsi plant e.g. leaves, flowers, stem, root, seeds etc. are known to possess therapeutic potentials and have been used, by traditional medical practitioners, as expectorant, analgesic, anticancer, antiasthmatic, antiemetic, diaphoretic, antidiabetic, antifertility, hepatoprotective, hypotensive, hypolipidemic and antistress agents. Its leaves contain a bright yellow volatile oil. The oil contain eugenol, eugenal, methyl chavicol, limatrol and Caryophylline and a number of sesquiterpenes and monoterpenes viz., barnyl acetate, B-elemense, methylengenol, neral, B-pinene, comphene, A-pinene etc (Jansen, 1981). Tulsi has also been used in treatment of fever, bronchitis, arthritis and convulsions (Afolabi *et al*,2007, Dermi *et al,* 2003, Gupta *et al,* 2007, Saksena *et al,* 1987,Singh *et al,* 2002, Surender singh *et al*, 2003) In modern animal feed formulation many Antimicrobial Growth Promoters (AGP) is being used. But due to the prohibition of most of AGP, plant extracts have gained interest in animal feed strategies (Charis, 2000). The risks of the presence of antibiotic residues in milk and meat and their harmful effects on human health have led to their prohibition for use in animal feed in the European Union. This research work was therefore, designed to study the efficacy of tulsi (*Ocimum sanctum*) leave extract as a growth promoter and its safety evaluation in broilers.

MATERIALS AND METHODS

The experiment was conducted at the Department of Pharmacology, Bangladesh Agricultural University, Mymensingh. Collection and processing of plant material Tulsi (*Ocimum sanctum*) leaves were selected to determine its efficacy as growth promoter on broilers. Mature and disease free Tulsi (*Ocimum sanctum*) leaves were collected from Bangladesh Agricultural University campus.

Preparation of tulsi fresh juice

After washing, the fresh leaves were cut into small pieces by scissors and water was added at 1:10 ratio. Then juice were prepared by blending the leaves with pestle and motor and stored in a refrigerator at 4°C to maintain the active ingredients of juice.

Collection and management of broilers

At first the experimental poultry shed of Pharmacology Department for rearing broiler chicks was properly prepared i.e., the floor and compartment of cages and other surroundings of the shed were cleaned with disinfectant. Day old broiler chicks, 30 (thirty) in number were brought in the experimental shed. Immediately after unloading from the chick boxes the chicks were supplied with Vitamin-C and glucose to prevent the stress occurring during transport. The broiler chicks were kept in the same compartment for 7 days and brooding temperature were maintained accordingly. The litter management was also done very carefully. The starter and finisher broiler rations were supplied to the broiler chicks appropriately.

Experimental design

After 7 days all the 30 broiler chicks were divided into 2 groups (Group A and Group B) for assessing the efficacy of Tulsi (*Ocimum sanctum*) leaves extract as growth promoter on broiler. Chicks of Group B, were kept as control and was not treated. Chicks of Group A, were treated with Tulsi leaves extract (1%) through drinking water @ 2ml/litre for four weeks. All the chicks of treated and control groups were closely observed for 35 days after treatment and following parameters were studied.

Clinical examination

The effect of the Tulsi (*Ocimum sanctum*) leaves extract on body weight of broilers were recorded before and after treatment. Broiler chicks of control and treatment groups were weighed with spring weighing machine. The weight of broiler chicks were taken weekly. The average of these weight was calculated and recorded. Mean live weight gain of each group of chicks on 7th, 14th, 21th, 28th, and 35 th days were recorded.

Hematological parameters

Blood samples were collected from wing vein of Broiler of both control and treated groups at 17th and 35th days to study the effect of the Tulsi (*Ocimum sanctum*) leaves extract and the following parameters were observed: (a) Total erythrocyte count (TEC), (b) Hemoglobin estimation (Hb), (c) Packed cell volume (PCV), and (d) Erythrocyte sedimentation rate (ESR) by using well known methods as described by Lamberg and Rothstein (1977).

Post-mortem examination

Five broilers from each group were slaughtered to observe if there were any pathological changes present on 35th day after treatment. There was no significant pathological changes found in any internal organs of the broilers of treatment group.

Statistical analysis

The data were analyzed statistically between control and treated groups of broiler by the well know student's t test.

RESULTS AND DISCUSSION

The observations for live body weight (gm) means of control group(Group B) for 1st ,2nd ,3rd ,4th and 5th weeks of the experimental period were 217gm, 470gm, 850gm, 1320gm and 1780gm ,respectively and treatment group (Group A) were 220gm, 540gm, 980gm, 1480gm and 2220gm, respectively. It is observed from the results of Table 1, that supplementation of tulsi leaves extract in Group A of broilers at 3rd, 4th and 5th weeks effected significant ($P<0.012$, $P<0.049$ and $P<0.014$ respectively) increase in mean live body weights as compared to Control group (Group B). The treatment group of broilers showed statistically higher body weight gain as compared to control group.

Observation of hematological parameters (TEC, Hb, PCV, ESR) on 17th day and 35th day showed significant difference ($P<0.01$) between the control and Tulsi leaves extract treated groups (Table 2). Observation of birds also revealed low mortality rate among the birds without any vaccination program and also without any antibiotic as growth promoter. The effect of administration of Tulsi leaves extract on TEC was determined on the 17th and 35th day after treatment. The values are shown in Table 2. The administration of Tulsi leaves extract with drinking water increased significantly the number of erythrocytes of chickens in Group A. The highest number of cells was recorded on 35th day after application of extract.

Table 1. Effect of Tulsi (*Ocimum sanctum*) leaves extract on body weight in broiler

	Body weight (gm)		Standard error	P-value	Significance value
	Control	Tulsi			
1	217.00	220.00	1.38	0.330	NS
2	470.00	540.00	20.53	0.078	NS
3	850.00	980.00	31.16	0.012	S
4	1320.00	1480.00	44.05	0.049	S
5	1780.00	2020.00	59.63	0.014	S

S=Significant, NS=Non-significant

Table 2. Effect of Tulsi (*Ocimum sanctum*) leaves extract on hematological parameters of broilers

Age of birds	Parameters	Treatment	Mean	Standard error	P-value	Significance value
17th day	TEC	Tulsi	211.67	5.859	0.048	S
		Control	192.33	3.67		
	Hb	Tulsi	6.47	0.040	0.007	S
		Control	6.00	0.086		
	PCV	Tulsi	18.33	0.191	0.008	S
		Control	17.33	0.085		
	ESR	Tulsi	7.33	0.315	0.0019	S
		Control	10.67	0.336		
35th day	TEC	Tulsi	275.33	9.03	0.013	S
		Control	247.67	6.553		
	Hb	Tulsi	7.47	0.0889	0.034	S
		Control	6.93	0.146		
	PCV	Tulsi	19.00	0.153	0.007	S
		Control	18.00	0.126		
	ESR	Tulsi	6.00	0.289	0.037	S
		Control	7.00	0.153		

Hematological parameters (TEC, Hb, PCV& ESR) on 17th day and 35th day showed significant difference ($P<0.01$) between the control and Tulsi leaves extract

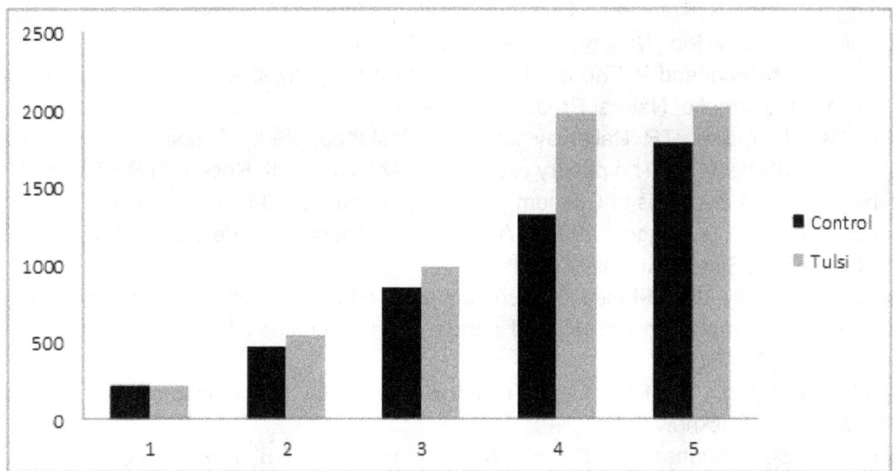

Figure 1. Effect of Tulsi (*Ocimum sanctum*) leaves extract on body weight in broilers

From the findings of the present study, it can be concluded that supplementation of Tulsi (*Ocimum sanctum*) leaves extract @ 2ml/liter drinking water of broiler cause significant increase in live body weight and significant change in hematological parameters. Thus Tulsi leaves extract supplementation in the broiler ration may be useful for the safe, economical and efficient production of broiler and this formulation can be used as an alternative to antibiotic growth promoter.

REFERENCES

1. Afolabi C, EO Akinmoladun , I Ibukun , A Emmanuel , EM Obuotor and EO Farombi, 2007. Phytochemical constituent and antioxidant activity of extract from the leaves of *Ocimum gratissimum.* Scientific Research and Essay. 2: 163-166.
2. Booth NH and LE McDonald, 1986. Veterinary Pharmacology and Therapeutics, 6th Edition, Iowa State University Press, Ames, Iowa.
3. Catala-Gregori P, S Mallet , A Travel , J Orengo and M Lessire, 2008. Efficiency of a prebiotic and a plant extract alone or in combination on broiler performance and intestinal physiology. Canadian Journal of Animal Science, 88: 623-629.
4. Charis K, 2000. A novel look at a classical approach of plant extracts. Feed mix (Special issue on nutraceuticals), 19-21.
5. Demir E, S Sarica, MA Ozcan and M Suicmez, 2003. The use of natural feed additives as alternatives for an antibiotic growth promoter in broiler diets. British Poultry Science. 44: 44- 45.
6. Griggs JP and JP Jacob, 2005. Alternatives to antibiotics for organic poultry production. Journal of Applied Poultry Research, 14: 750–756.
7. Gupta G and S Charan, 2007. Exploring the potentials of *Ocimum sanctum* (Shyama tulsi) as a feed supplement for its growth promoter activity in broiler chickens. Indian Journal of Poultry Science, 42: 140-143.
8. Hernandez F, J Madrid, V Garcia, J Orengo and MD Megias, 2004. Influence of two plant extracts on broiler performance, digestibility, and digestive organ size. Poultry Science, 83: 169-174.
9. Jansen PCM, 1981. Species, condiments and medicinal Plants in Ethiopia, their Taxonomy and Agricultural significance: 85-86.
10. Khan AJ, 1975. Misuse of antibiotics. In: Cento Seminar on Use & Misuse of Antimicrobial Drugs.
11. Ladefoged O, 1996. Drug residues in food of animal origin and related human hazards. In: Proc. Int. Workshop on Rational Applications of Vet. Pharmaceuticals and Biologicals. Balochistan Livestock Devlopment Project, L & DD, Government of Balochistan, Quetta. March 1–3, 1996. pp. 246–253.

12. Lamberg S L and R Rothstein, 1977. Laboratory Manual of Hematology and Urinanalysis. AVi. Publishing Company, Inc, West port Connecticut, U.S.A.
13. Maity TK, SC Mandal and M Pal, 2004. Assessment of antitussive activity of *Ocimum sanctum* root extract. Indian Journal of Natural Products, 20: 9-13.
14. Nava GM, LR Bielke, TR Callaway and MP Castañed, 2005. Probiotic alternatives to reduce gastrointestinal infections: The poultry experience. Animal Health Research Review, 6: 105–118.
15. Paton A (1992). A Synopsis of Ocimum (labiate) in Africa Kew Bull. 47:403-436.
16. Prescott J F and JD Baggot, 1993. Antimicrobial Therapy in Veterinary Medicine, 2nd edition, pp 564-565: Iowa State University Press.
17. Shivakumar MC, BK Javed-Mulla Pugashetti and S Nidgundi, 2005. Influence of supplementation of herbal growth promoter on growth and performance of broilers. Karnataka Journal of Agricultural Sciences, 18: 481-484.
18. Singh N, Y Hotter and R Miller, 2002. Tulsi, the Mother Medicine of Nature. International Institute of Herbal Medicine. Lucknow, India.
19. Singh VK, SS Chauhan, K Ravikanth, S Maini and DS Rekhe, 2009. Effect of dietary supplementation of polyherbal liver stimulant on growth performance and nutrient utilization in broiler chicken. Veterinary World, 2: 350-352.
20. Gupta S, 2005. Efficacy of 'Tulsi' (*Ocimum sanctum*) leaf powder on growth rate and development of *Trogoderma granarium*. Flora and Fauna Jhansi, 11: 237-243.
21. Singh S and DK Majumdar, 2003. *Ocimum sanctum* phytochemical and pharmacological evaluation. Phytochemistry and Pharmacology, 1-81.
22. Trafalska E and K Grzybowska, 2004. Probiotics An alternative for antibiotics? Wiad Lek, 57: 491–498.
23. Vidyarthi VK, RC Gupta and VB Sharma, 2010. Effect of herbal additives on the performance of broiler chicken. Indian Veterinary Journal, 87: 258-260.
24. Vidyarthi VK, K Nring and VB Sharma, 2008. Effect of herbal growth promoters on the performance and economics of rearing broiler chicken. Indian Journal of Poultry Science, 43: 297-300.
25. WHO, 1997. Antibiotic Use in Food-Producing Animals Must Be curtailed to Prevent Increased Resistance in Humans. Press Release WHO, October 20th, 1997.
26. WHO, 1998. Antimicrobial Resistance. Fact Sheet No. 194, May 1998.

6

PRECISION AGRICULTURE IN THE WORLD AND ITS PROSPECT IN BANGLADESH

Mahfuza Afroj[1], Mohammad Mizanul Haque Kazal[2*] and Md. Mahfuzar Rahman[3]

[1]Department of Agribusiness and Marketing and [2]Department of Development and Poverty Studies, Faculty of Agribusiness Management, Sher-e-Bangla Agricultural University, Sher-e-Bangla Nagar, Dhaka-1207, Bangladesh; [3]Department of Agronomy, Faculty of Agriculture, Sher-e-Bangla Agricultural University, Sher-e-Bangla Nagar, Dhaka-1207, Bangladesh

***Corresponding author:** Mohammad Mizanul Haque Kazal; E-mail: mhkazal@gmail.com

ARTICLE INFO

Key words

Agriculture, Geographical Information System (GIS), Global Positioning System (GPS)

ABSTRACT

Precision Farming merges the new technologies borne of the information age with a mature agricultural industry. It is an integrated crop management system that attempts to match the kind & amount of inputs with the actual crop needs for small areas within a farm fields. This study is basically based on the secondary data and it is a review paper. As it is a review paper so, there was less opportunity to follow any specific method in preparing this paper. Valuable information has been collected through internet browsing, journals etc. In this paper, we examine that, GPS, GIS, VRT, profitability, yield mapping etc. are most common precision agriculture techniques where, GIS can help in site-specific applications of fertilizers and soil amendments and help effectively detect and map black fly infestations, making it possible to achieve precision in pest control. Remote sensing combined with GIS and GPS can help in site-specific weed management. From the review we observed that, the global adoption of yield monitors has been predominated in North-America and Europe. The precision technologies have been used on a wider variety of crops in Denmark and UK than in the US. In Bangladesh, precision agriculture has great prospect as our country is highly natural calamity sensitive and through it we can easily take measure to prevent our agricultural products from damage caused by natural calamities. Though, precision agriculture is very costly but the benefit from it is more than its cost for most of the developing countries. So, the precision agriculture has great prospect in Bangladesh as well as in the world and it is the utmost time to adopt this technology in our traditional agriculture.

INTRODUCTION

Agricultural production system is an outcome of a complex interaction of seed, soil, water & agro-chemicals (including fertilizer). The time has now arrived to exploit all the modern tools available by bringing information technology & agricultural science together for improved economic & environmentally sustainable crop production. Precision agriculture merges the new technologies borne of the information age with a mature agricultural industry. As we have known, precision agriculture (PA) was first expounded in 1980's in American based on the demands to solve the agricultural environment such as fertilizer and pesticide pollutions. According to Srinivasan (2001) said that, the scope for funding new hardware, software and consulting industries related to precision agriculture is gradually widening. In Japan, the market in the next 5 years is estimated at about US $ 100 billion for GIS and about US $ 50 billion for GPS and RS. It targets inputs and management practices to variable field conditions such as soil/landscape characteristics, pest presence and microclimate. A more holistic agricultural approach, PA uses information technology to bring data from multiple sources to bear on decisions associated with agricultural production, logistics, marketing, finance, and personnel. Precision agriculture is an information-based approach to farming that is enabled by a collection of rapidly changing technologies. Although adoption of these technologies has been reasonably rapid, it is not clearly understood how to exploit the full power of precision agriculture. Additional research related to precision agriculture should be needs based. Once these key needs are understood, we can develop appropriate research projects to meet these needs. Considering the above facts, the present study was undertaken to know about the precision farming, to be acquainted with the opportunities & need of precision farming and to know the present scenario of precision farming around the world and its prospect in Bangladesh.

MATERIALS AND METHODS

Scientific approach requires a close understanding of the subject matter. This paper mainly depends on the secondary data. Different published reports of different journals mainly supported in providing data in this paper. This paper is completely a review paper. Therefore no specific method has been followed in preparing this paper. It has been prepared by Internet search, comprehensive studies of various articles published in different journals, books and proceedings available in the libraries of SAU, BARI, BRRI and BARC. Valuable information has been collected through personal contact with respective resource personnel to enrich the paper. It compiled the all related information to prepare this paper.

RESULT AND DISCUSSION

Basic concept of precision agriculture

The precision agriculture is a agricultural technique where the crop, soil and climate related data are monitored then mapping the attribute and then taking decision and finally take action. It is fully information based approach.

The Basic Tools of Precision Farming

Precision farming basically depends on measurement and understanding of variability, the main components of precision farming system must address the variability. Precision farming technology enabled, information based and decision focused, the components include, Remote Sensing (RS), Geographical Information System (GIS), Global Positioning System (GPS), Soil Testing, Yield Monitors and Variable Rate Technology.

The Global Positioning System (GPS) is the heart of precision agriculture. A GPS receiver is a location device that calculates its position on earth from radio signals broadcast by satellites orbiting the earth. The U.S. government has 24 satellites that are constantly orbiting the earth. These satellites contain precise atomic clocks, and the exact time is encoded into the signals broadcast from each satellite. A GPS receiver uses this time information to measure the distance to each satellite from which a signal is being received. With at least four satellite signals, the receiver can use triangulation to calculate its position on the ground.

The base station receives the satellite signals and compares its calculated position with its exact position. GPS receivers can be used in a wide range of situations to provide the latitude and longitude of a machine operating in a field, or of a field scout who is making observations and taking samples. Field images, or maps, can be made by recording parameters such as yield or fertilizer application along with position in the field. Mapping software is used to handle, display and analyze data stored as a value and a position. Mapping software is available with a wide range of capabilities. Low-end packages are used primarily for creating maps or graphical images, and have little capability to process or analyze data. High-end products are known as Geographic Information Systems or GIS, and have many data processing capabilities. Because precision farming requires a relatively high level of data processing, software used for this purpose has become known generically as GIS software.

Variable Rate Technologies (VRT) describes machines that can automatically change their application rates in response to their position. VRT systems are available for applying a variety of substances including granular and liquid fertilizers, pesticides, seed and irrigation water.

Yield mapping is another important technique in precision farming. Yield maps show the variability in yield within a field. A yield mapping system measures and records the amount of grain being harvested at any point in the field, along with the position of the harvester. To produce a yield map, the harvester must be equipped with a GPS receiver and a yield monitor. A yield monitor can be a flow meter or a scale.

The profitability of precision farming is as variable as field conditions. Producers who use site-specific management must recognize that information becomes another input to the system, and that it has a cost. With soil, weed, fertility and yield maps for a particular field, the producer can know more about the field's yield potential and determine which areas of a field are creating the largest profit, as well as which areas are not capable of producing as well as others.

Remote sensing is the acquisition of information about an object from a distance, with precision, without coming into contact with the same. Although the use of RS is a decade old, its relevance to agriculture in spatial variability management is relatively new. RS measures visible and invisible properties of a field or a group of fields and converts point measurements into spatial information, to monitor temporally dynamic plant and soil conditions.

Basic Steps in Precision Farming

- The basic steps in precision farming are,
- Assessing variability
- Managing variability and
- Evaluation

I) Assessing variability

Assessing variability is the critical first step in precision farming. Factors and the processes that regulate or control the crop performance in terms of yield vary in space and time. Quantifying the variability of these factors and processes and determining when and where different combinations are responsible for the spatial and temporal variation in crop yield is the challenge for precision agriculture.

ii) Managing variability

Once variation is adequately assessed, farmers must match agronomic inputs to known conditions employing management recommendations. Those are site specific and use accurate applications control equipment. In site-specific variability management, we can use GPS instrument, so that the site specificity is pronounced and management will be easy and economical. For successful implementation, the concept of precision soil fertility management requires that within-field variability exists and is accurately identified and reliably interpreted, that variability influences crop yield, crop quality and for the environment.

iii) Evaluation

There are three important issues regarding precision agriculture evaluation.

- Economics
- Environment and
- Technology transfer

Potential improvements in environmental quality are often cited as a reason for using precision agriculture. Reduced agrochemical use, higher nutrient use efficiencies, increased efficiency of managed inputs and increased production of soils from degradation are frequently cited as potential benefits to the environment. Enabling technologies can make precision agriculture feasible, agronomic principles and decision rules can make it applicable and enhanced production efficiency or other forms of value can make it profitable.The term technology transfer could imply that precision agriculture occurs when individuals or firms simply acquire and use the enabling technologies.

Technology Transition

In precision farming, "Variability of production and quality equals opportunity". For example the magnitude of the variability may be too small to be economically feasible to manage. Alternatively the variability may be highly randomized across the production system making it impossible to manage with current technology. Finally the variability may due to a constraint that is not manageable. Thus the implementation of precision farming is limited by the ability of current variable rate technology (VRT machinery/ technology that allows for differential management of a production system) to cope with the highly variable sites and the economic inability to produce returns from sites with low variability using precision farming (VRT).

Opportunities

Business opportunities for precision farming technologies including GIS, GPS, RS and yield monitor systems are immense in many developing countries. The scope for funding new hardware, software and consulting industries related to precision agriculture is gradually widening. In Japan, the market in the next 5 years is estimated at about US $ 100 billion for GIS and about US $ 50 billion for GPS and RS (Srinivasan, 2001). Punjab and Haryana states in India, where farm mechanization is more common than in others, may be the first to adopt precision farming on a large scale.

Recently, the governments of certain Asian countries initiated special efforts to promote precision farming. In Japan, the Ministry of Agriculture has allocated special funds for research on remote sensing applications of precision farming. A quasi-governmental institute "Bio-oriented Technology Research Advancement Institute (BRAIN)" is also funding research on precision farming. In Malaysia, the Malaysian Agricultural Research and Development Institute (MARDI) is promoting research on precision farming of upland rice.

Precision farming is useful in many situations in developing countries. Rice, wheat, sugar beet, onion, potato and cotton among the field crops and apple, grape, tea, coffee and oil palm among horticultural crops are perhaps the most relevant.

In Sri Lanka, researchers at the Tea Research Institute are examining precision management of soil organic carbon.Nutrient stress management is another area where precision farming can help sub- continent farmers. Most cultivated soils here vary in pH. Detecting nutrient stresses using remote sensing and combining data in a GIS can help in site-specific applications of fertilizers and soil amendments such as lime, manure, compost, gypsum and sulphur. Pests and diseases cause huge losses to many crops. If remote sensing can help in detecting small problem areas caused by pathogens, timing of applications of fungicides can be optimized. Recent studies in Japan show that pre-visual crop stress or incipient crop damage can be detected using radio-controlled aircraft and near-infrared narrow-band sensors. Likewise, GIS have been shown to effectively detect and map black fly infestations in citrus orchards, making it possible to achieve precision in pest control.

The Need for Precision Farming

Green revolution of course contributed a lot in world agriculture. However, even with the spectacular growth in the agriculture, the productivity levels of many major crops are far below than expectation. In many countries of the world the production of their high yielding varieties have not reach in their potential level.Precision farming techniques can improve the economic and environmental sustainability of crop production. Some of the primary impacts are cost reduction and more efficient use of production inputs, use of information technology to increase the size and scope of farming operations without increasing labor requirements, improved site selection and control of production processes that help in the production of higher value or specialty products, improved recordkeeping and production tracking for food safety, and environmental benefits (Lowenberg-DeBoer and Boehlje 1996). When used to precisely control where equipment travels in a field, precision agriculture can also reduce soil compaction and erosion.

At a basic level, precision agriculture can include simple practices such as field scouting and the spot application of pesticides. However, precision agriculture usually brings to mind complex, intensely managed production systems using global positioning system (GPS) technology to spatially reference soil, water, yield, and other data for the variable rate application of agricultural inputs within a field. Precision agriculture methods help farmers recognize areas that have productivity and environmental problems and to select the best solution for each one. At the extreme, precision agriculture may help a producer identify land that should be taken out of the current production system because of economic and environmental considerations.

Present scenario of precision farming around the world

Yield monitors connected to a GPS-receiver was by most farmers the first real attempt to conduct site-specific management on their fields. The global adoption of yield monitors has been predominated in North-America, Europe and Australia but countries like Argentina, Brasilia and some East Asian countries have also adopted some practices. To date, we are in a stationary state between the early adopters and the early majority, mainly since yield increases aren't well enough documented to cover the cost of equipment. According to S. Blackmore (2001), the adoption of precision farming technologies is likely to follow a normal distribution with the innovators and early adopters as the first to adopt the technology and then later on will the majority of farmers follow up. The adoption of precision farming is currently in a stationary phase between the early adopters and the majority. Figure 10 shows a likely adoption pattern for precision farming.

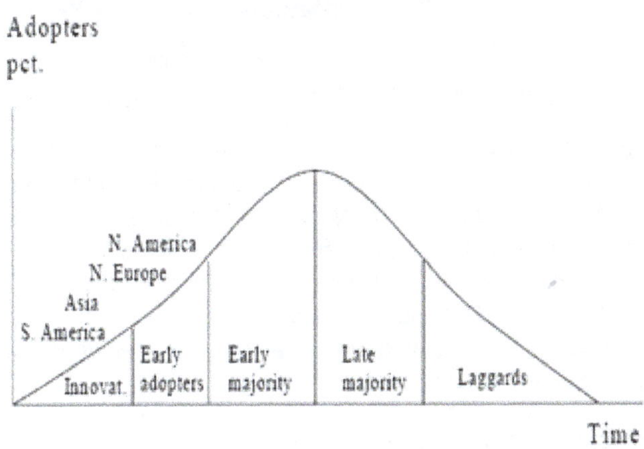

Source: Blackmore, 2008

Figure 1. Global adoption of precision farming practices

The age distribution of farmers of Denmark, USA & UK adopting precision agricultural practices tends to be similar among the three countries (Figure 2). In general, younger farmers may be interested in precision agriculture but have less economic flexibility to invest in equipment, while older farmers may be reluctant to invest in the time necessary to learn new technologies.

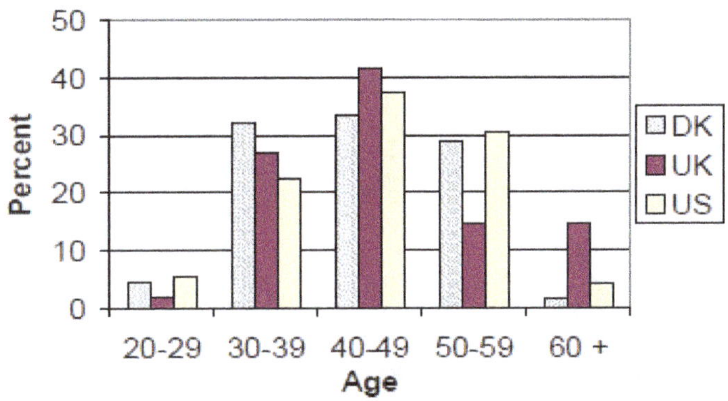

Source: Pedersen *et al.*, 2009

Figure 2. Age distribution of producers using precision agriculture

Farms in the US tend to be larger, followed by the UK, then Denmark (Figure 3). The land area farmed by producers using precision technologies in all three countries tends to be larger than the average in each country.

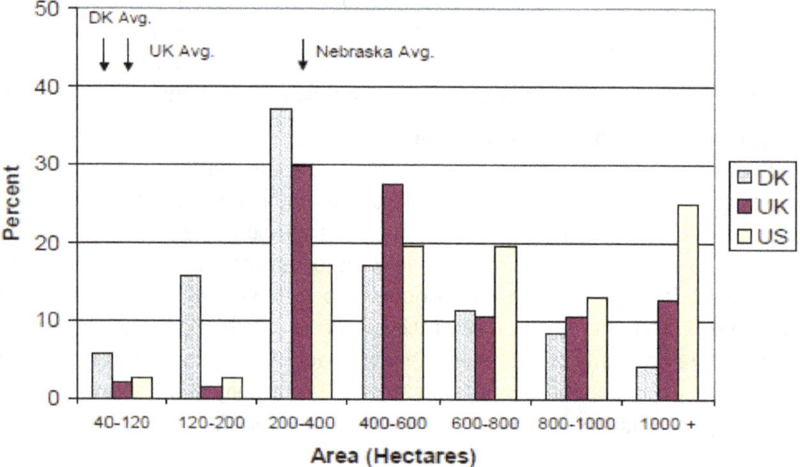

Source: Pedersen *et al.*, 2009

Figure 3. Area farmed by producers using precision agriculture

Precision farming practices have been applied on a large variety of crops, although the most common application is to grain crops that can be harvested by a combine harvester (Table 1). This finding was anticipated given that the UK and Denmark survey participants were customers of companies producing yield-mapping combines. Precision technologies have been used on a wider variety of crops in Denmark and UK. In Denmark and the UK, 91% and 95% respectively use precision practices on wheat, as well as on barley, oilseed rape, grass seed, peas and tubers (beets and potatoes). It should be stressed that table 1 only includes crops where some type of precision practice has been used. It does not address the extent of current farm practices within each country on each crop.

Table 1. Percentage of respondents who have used a precision practice for the given crop

Crop	DK	UK	US
		%	
Wheat	91	95	13
Barley	82	72	0
Rye	16	5	0
Oat	7	9	1
Triticale	11	0	0
Oilseed Rape	36	67	0
Corn (maize)	4	0	100
Grass seed	45	0	0
Flax	7	0	0
Beets	5	2	0
Potatoes	9	7	0
Peas	13	21	0
Linseed	0	14	0
Beans	0	28	0
Soybean	0	2	87
Grain sorghum	0	0	10
Other[1]	10	2	9

[1] Crops where 5 percent or less precision practices are used in the three countries include: Seed corn, grass, herbage seed, edible beans and alfalfa.

Source: Pedersen *et al.*, 2009

There are differences among countries in how long farmers have used precision practices, with farmers in the UK tending to have used them longer than farmers in the US or Denmark (figure 4). The majority of Danish and US respondents have used one or more precision practice between two and four years, while the majority of UK respondents have used a precision practice between five and seven years.

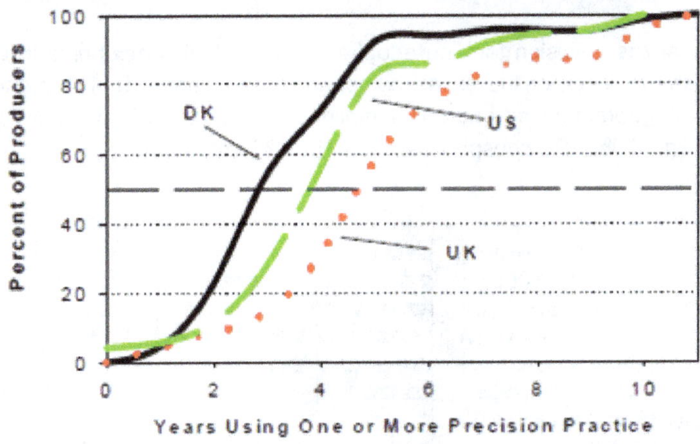

Source: Pedersen et al., 2009

Figure 4. Length of time producers have used a precision practice (cumulative distribution)

Yield mapping with GPS is the most common practice used in all three countries. Yield monitoring without GPS is used significantly only in the US. Directed soil sampling, according to yield maps or other spatial information, is a practice of increasing popularity in the US and UK, given the high investment required for grid sampling. A larger percentage of producers have tried aerial photography or remote sensing in the US relative to the UK and Denmark. More producers in the UK and Denmark have mapped soil conductivity on their farms than in the US. A significant number of producers in the UK indicated they had used conventional soil surveys on their farm (this question was not asked of producers in the US or Denmark). In the US, a significant percentage of producers had had their fields mapped for topography, primarily in relation to leveling for irrigation.

Table 2. Use of precision practices among countries (Percent of respondents in each country who indicate they have used the practice)

Practice	DK	UK	US
		%	
Information Gathering			
Grid soil sampling	49	56	50
Directed soil sampling	14	27	39
Yield monitoring (no GPS)	7	17	29
Yield mapping (GPS)	80	100	76
Aerial photography	3	27	40
Remote sensing	3	17	19
GPS-pest monitoring	1	2	3
Soil conductivity mapping	14	19	6
Soil topography mapping	1	4	22
Conventional soil mapping		17	
Mean	19	29	32
Taking Action			
VRT fertilization	26	58	47
VRT lime	37	33	26
VRT pesticide	10	10	3
Mean	24	34	25
Other	3	6	14

Source: Pedersen et al., 2009

Table 2 shows that, on average, US and UK producers are highest in adoption of evaluative, or information gathering, practices (32% and 29% average adoption, respectively, compared to an average adoption of 19% in Denmark). Producers in the UK are more likely than Danish or US producers to take action on the information they have gathered, with 34%, on average, adoption of prescriptive practices, compared to 24% and 25% in Denmark and the US.

Figure 5 summarizes precision farming adoption by individual components in Central Ohio, Japan. The number of responses for each of the seven components was about 50. The highest adoption percentage, 84.6%, occurred with georeferenced grid soil sampling, followed by 78% adoption of georeferenced variable rate application of lime, 70% VRT phosphorus, and 64% VRT potassium.

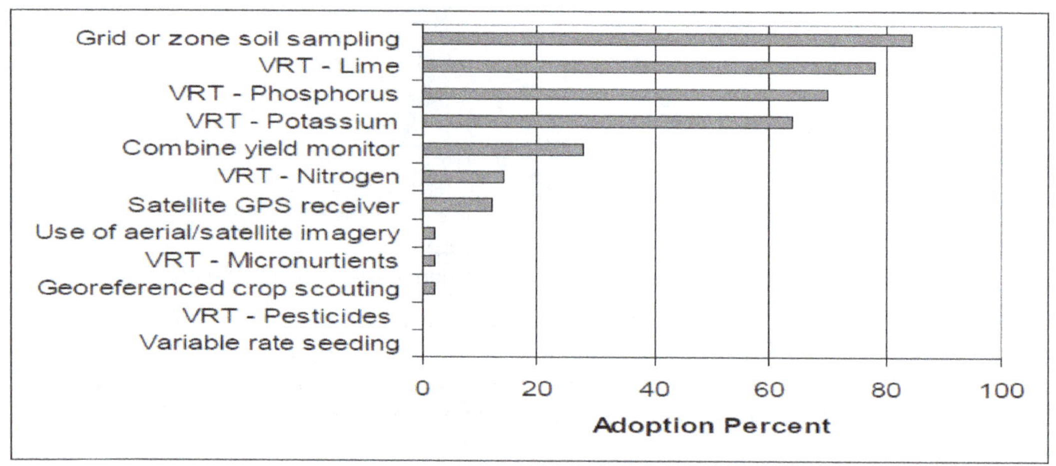

Source: Arnholt et al., 2007

Figure 5. Adoption percentages for various precision farming component technologies and practices in Central Ohio, Japan

When farmers are asked to list the top three practices they felt would have the most potential economically for their farm, producer responses varied somewhat by country (table 3). The most commonly listed practice among all three countries was variable rate fertilization. The second most commonly listed practice in the UK and US was yield mapping, while in Denmark the second most commonly listed practice was variable rate lime application. The third most commonly listed practice in the US and UK was grid soil sampling, while the third most commonly listed practice in Denmark was variable rate pesticide application.Two-thirds of Danish farmers believe either variable rate fertilization, liming, or pesticide application will be economically beneficial, compared to 34% in the UK and 25% in the US. This is likely related to restrictions on N rates below the economic optimum in Denmark, as well as substantial taxation on the use of pesticides (Langkilde, 1999).

In case of variable rate application producers in different countries sometimes had different opinions regarding the impact of adopting precision agriculture practices on the total amounts of inputs, such as fertilizer and seed, for crop production (table 4). Table 4 Impact of adoption of precision practices on total inputs (↑ indicates an increase, ↓ indicates a decrease, and → indicates no change in the overall total of each input; where two arrows are shown, there were significant numbers of farmers who held each opinion – the first arrow shown is the predominate opinion)

Table 3. Precision farming practices that farmers believe to have the greatest potential economic impact on their farm

Practice	DK	UK	US
		%	
Information Gathering			
Grid soil sampling	16	28	33
Directed soil sampling	2	13	14
Yield monitoring (no GPS)	5	0	20
Yield mapping (GPS)	28	58	46
Aerial photography	0	5	3
Remote sensing	3	10	10
GPS-pest monitoring	10	5	0
Soil conductivity mapping	3	5	3
Soil topography mapping	2	5	1
Conventional soil mapping		3	
Mean	8	13	14
Taking Action			
VRT fertilization	81	60	54
VRT lime	66	18	20
VRT pesticide	53	25	1
Mean	67	34	25
Other	9	38	39

Note: The respondents were asked to mention up to three practices they believed would be beneficial on their farm.

Source: Pedersen et al., 2009

Table 4. Impact of adoption of precision practices on total inputs

Input	DK	UK	US
Nitrogen	→	→	→
Phosphorous	→↓	→↓	→↑
Potassium	→↓	→↓	→
Other Fertilisers	→	→	→
Lime	↓	→	→↑
Herbicides	↓	→	→
Insecticides	→↓	→	→
Fungicides	↓→	→	→
Seed	→↓	↓→	→
Growth regulators	↓→	→	→

Source: Pedersen *et al.*, 2009

Only nitrogen and other fertilizers showed no difference in opinion among countries, with all farmers feeling there would be no change in these total inputs. Farmers in Denmark and the UK felt precision practices would leave unchanged or decrease their total use of phosphorous fertilizers, while farmers in the US felt if anything it would increase total P use. Farmers in Denmark and the UK felt potassium use would remain the same or decrease, while farmers in the US believe it would remain unchanged.

Producers in Denmark felt lime use would decrease, while US farmers felt it would increase. Danish farmers also felt herbicide, insecticide, and fungicide use would decrease or remain unchanged, while producers in the other two countries felt they would be unchanged. US farmers felt that total seed and growth regulator use would remain unchanged, while farmers in Denmark and the UK felt they would remain the same or decrease. In Denmark and the UK, where agriculture is often more intensive than in Nebraska, soil phosphorus levels are generally higher. Soil potassium levels generally are higher in Nebraska than in the UK or Denmark due to relatively younger, less weathered soils. Also, the use of fungicides and growth regulators is more common in Denmark and the UK for intensively grown cereals. The expectation that precision practices will decrease overall herbicide use in Denmark may be related to more stringent regulations on the use of herbicides in Denmark, compared to the UK or US.

An interesting observation is that respondents in Denmark and the UK felt precision practices would leave unchanged or reduce total inputs.The relative benefits and costs associated with the use of each PF component is evaluated in Table 5. Farmers were asked to respond to the statement, "On my farm the benefits of (named component) clearly exceed its costs". The scale used was 1 (strongly disagree) to 5 (strongly agree). The combine yield monitor ranked the highest (4.2), suggesting that this is the most profitable single PF component. No producer disagreed that benefits of the yield monitor exceeded its costs. Similarly, all adopters of Satellite GPS receivers agreed that benefits of this component clearly exceeded its costs. Nearly 77 percent agreed or strongly agreed that the benefits of VRT of fertilizers or lime exceeded their costs, and 59 percent felt that grid soil sampling had positive net benefits.

Table 5. Farmer evaluation of relative benefits and costs for precision farming technology

		Percent [a]					
Precision Farming Component [b]	Adoption Percent	Strongly Disagree	Disagree	Neutral	Agree	Strongly Agree	Average Score
Georeferenced grid or zone soil sampling	84.6	0.0	4.9	36.6	43.9	14.6	3.7
Combine yield monitor	27.5	0.0	0.0	15.4	46.2	38.5	4.2
Satellite GPS receiver	12.0	0.0	0.0	0.0	100.0	0.0	4.0
VRT of Fertilizers or Lime	73.3	0.0	2.6	20.5	64.1	12.8	3.9

Source: Arnholt*et al.*, 2007

a: this is based on a scale of 1 to 5, with 1 meaning strongly disagree, 2 meaning disagree, 3 meaning neutral, 4 meaning agree, and 5 meaning strongly agree; b: farmers were asked "On my farm, the benefits of (precision farming component) clearly exceed its costs"

Table 6. Farmer evaluations of the costs associated with precision farming adoption and use

	Cost Level [a]						
Costs	Not a cost	Very Low Cost	Low Cost	Medium Cost	High Cost	Very High Cost	Average Score
Service charges for Variable Rate Application of fertilizer/lime	0.00	0.00	13.64	45.45	31.82	9.09	3.36
Soil testing fees	2.27	15.91	20.45	36.36	22.73	2.27	2.68
Soil sample collection costs	8.89	15.56	20.00	31.11	20.00	4.44	2.51
Manger time required (including your time)	30.23	23.26	27.91	11.63	4.65	2.33	1.44
Consulting fees paid	39.53	20.93	16.28	16.28	4.65	2.33	1.33
GPS equipment and differential correction subscription	66.67	4.76	0.00	9.52	14.29	4.76	1.14
Input application equipment (spreaders/sprayers/planters)	76.19	2.38	0.00	7.14	4.76	9.52	0.90
GIS and mapping software costs	71.43	4.76	2.38	11.9	7.14	2.38	0.86
Computer hardware	76.19	2.38	4.76	9.52	2.38	4.76	0.74
Service charges for Variable Rate Application of herbicides, insecticides, and fungicides	80.49	0.00	2.44	9.76	2.44	4.88	0.68
Service charges for Variable Rate Planting	87.80	4.88	0.00	2.44	2.44	2.44	0.34
Remote sensing data fees	88.10	4.76	0.00	2.38	2.38	2.38	0.33

a - this scale is rated 0 to 5, where 0 means NOT a cost, 1 means very low cost, and 5 means very high cost.

Source: Arnholtet al., 2007

The various costs associated with the adoption and use of precision farming is identified in table 6. Farmers were asked to evaluate these cost sources using a scale of 0 to 5, where 0 is not a cost, 1 is a low level of cost and 5 is a very high cost. It is clear that these farmers perceive the highest cost associated with the adoption and use of PF to be the service charges for variable application of fertilizers and lime (3.36). The next two highest costs are soil testing fees (2.68) and soil sample collection costs (2.51). Nearly all of the mean cost scores (9 of 12) are below 1.50. This suggests that nine of the cost descriptions listed is considered low to very low cost factors associated with the adoption of PF according to the sample surveyed. The lowest cost scores are associated with remote sensing data fees (not a cost for 88% of the farmers)), service fees for variable rate planting (also not a cost for 88% of the farmers) and service fees for VRT application of herbicides, and pesticides (not a cost for 80% of the farmers).

A benefits gained over the cost associated with adoption of precision farming is listed in Table 7. The same 0 to 5 scale was used to score each possible benefit a farmer may have gained by using PF. The highest rated benefit is precise knowledge of soil pH levels in grids and or management zones (4.07) followed by precise knowledge of soil nutrient levels in grids and or management zones (3.76) and reduction in lime usage (3.25). The lowest ranked benefits include reduction in insecticide or fungicide usage (0.25), reduction in herbicide usage (0.43), and precise knowledge of weed problem areas (0.89). Most of the farmers surveyed (84.6%) have adopted grid or management zone soil sampling.

Table 7. Farmer evaluations of the benefits gained over the cost from precision farming adoption and use

	Benefit Level [a]						
Benefits	Not a Benefit	Very Low Benefit	Low Benefit	Medium Benefit	High Benefit	Very High Benefit	Average Score
Precise knowledge of soil pH levels in grids/management zones	2.33	0.00	0.00	13.95	53.49	30.23	4.07
Precise knowledge of soil nutrient levels in grids/management zones	4.44	0.00	4.44	22.22	44.44	24.44	3.76
Reduction in lime usage	9.09	6.82	11.36	15.91	36.36	20.45	3.25
Increased farm average yield	9.09	6.82	4.55	31.82	31.82	15.91	3.18
Reduction in fertilizer usage	11.11	8.89	6.67	28.89	31.11	13.33	3.00
Decreased variability (risk) in yields	25.00	6.82	13.64	15.91	31.82	6.82	2.43
Better knowledge for future selection of hybrids and varieties	32.56	9.30	11.63	11.63	23.26	11.63	2.19
Quantified and precise knowledge of areas of high/low yields	46.51	6.98	4.65	11.63	20.93	9.30	1.81
Better environmental records	41.86	6.98	11.63	18.60	18.60	2.33	1.72
Improved information for crop rotation management	44.19	6.98	25.58	13.95	6.98	2.33	1.40
Identification of drainage problems	57.78	8.89	0.00	17.78	11.11	4.44	1.29
Increased ability to compete/negotiate for leased land	55.81	6.98	6.98	18.60	9.30	2.33	1.26
Reduction in soil compaction	54.76	7.14	11.9	14.29	11.90	0.00	1.21
Better information for crop insurance claims	65.12	6.98	6.98	4.65	9.30	6.98	1.07
Knowledge of where equipment failure may have occurred	65.12	4.65	11.63	11.63	4.65	2.33	0.93
Precise knowledge of weed problem areas	68.18	4.55	4.55	18.18	2.27	2.27	0.89
Reduction in herbicide usage	77.5	12.50	0.00	10.00	0.00	0.00	0.43
Reduction in insecticide or fungicide usage	86.36	6.82	2.27	4.55	0.00	0.00	0.25

a - this scale is rated 0 to 5, where 0 means NOT a value/benefit, 1 means very low value/benefit, and 5 means very high value/benefit.

Source: Arnholt *et al.*, 2007

From the quantitative data, it can be seen in the figure 6 below, that the major benefits were the improved understanding of field characteristics, explaining field variations and making farming more interesting and simulating.

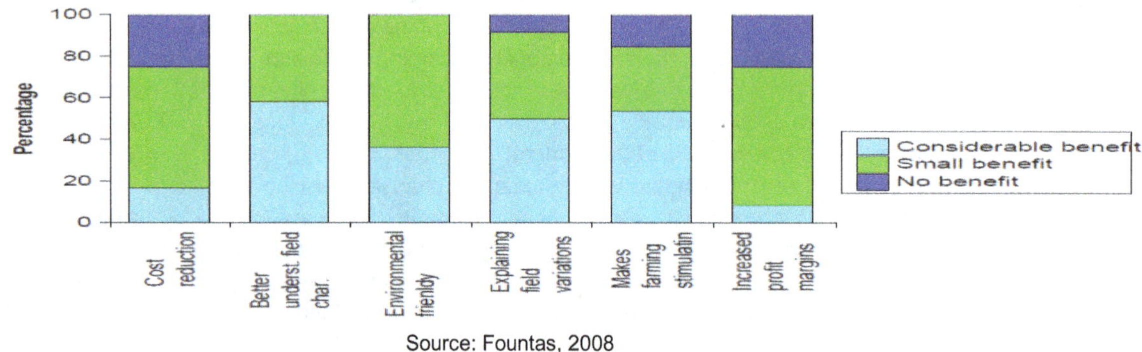

Source: Fountas, 2008

Figure 6. Major benefits of using precision farming

Efficient use of land through precision agriculture

Precision agriculture helps to use the lands in efficient manner through:

I. GPS helps to select appropriate the lands for appropriate crops.
II. GIS helps to analyse the data.
III. VART helps to provide the appropriate amount of pesticide in the crops whichhelps to keep the soil fertile.

Prospect of precision agriculture in Bangladesh

- Bangladesh is a over populated country and by precision agriculture we can produce more by using available resources to feed these population not only in quantity but also can provide them nutritious food.
- In Bangladesh, we can use precision agriculture by using the data (rainfall, temperature, humidity, sunshine, wind speed etc.) available in different institutions.
- Precision agriculture helps to produce and improve crops at minimum cost which is very essential for Bangladesh as it is a developing country where money or investment is a very big problem.
- In Bangladesh, precision agriculture has great prospect as our country is highly natural calamity sensitive country and through it we can easily take measure to prevent our agricultural products from damage caused by natural calamities.
- Though, precision agriculture is very costly but the benefit from it is more than its cost for most of the developing country.

Disadvantages of Precision Farming

The most common disadvantage listed in all three countries was the cost of using the technology, and the apparent lack of economic return (table 8). (An interesting concern given the generally optimistic outlook producers have for the eventual positive economic impact of adopting precision agriculture). The second most commonly listed disadvantage was the time required – for making equipment work and analyzing and summarizing data.

Table 8. Proportion of disadvantages listed among all producers (%)

Cost vs. Return	Equipment Problems	Potential Government Regulation	Lack of Research and Advice	Time Spent on Precision Agriculture
58	5	4	12	21

Source: Blackmore, 2008

Reasons for not using any Precision Farming techniques

The main important reasons are 'too expensive' 65% and 'not sufficient information' 41%. Less importance the farmers consider that precision farming is 'too big a change' 38% and 'too technical' 37%.

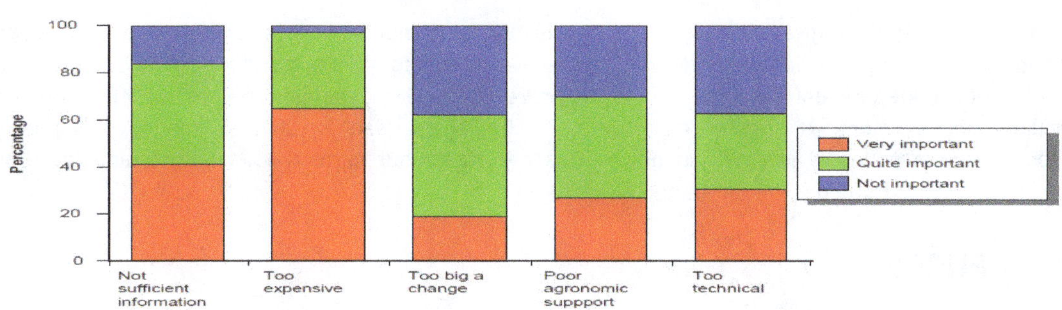

Source: Fountas, 2008

Figure 7. Reasons for not using precision farming techniques

Ways of encouraging the adoption Precision Farming technology

Key reasons are 'lower set-up cost' 75%, and 'more convincing benefits' 69%. Not important are 'endorsement by agricultural organizations' 40% and 'endorsement by other farmer' 38%. Quite important are 'advice from agronomist' 51%, 'A more developed integration with current systems' 56% and 'more information' 54%.

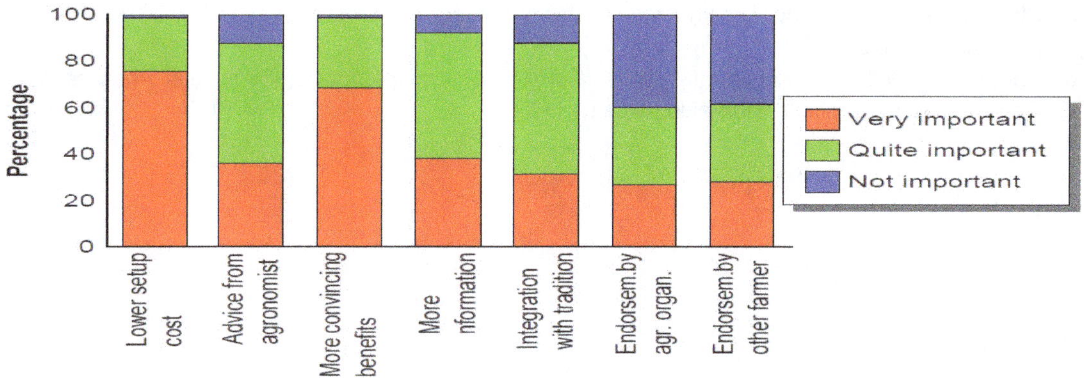

Source: Fountas, 2008

Figure 8. Ways of encouraging the adoption of precision farming

Obstacles

There are many obstacles to adoption of precision farming in developing countries in general. Some are as follows.

- Culture and perceptions of the users
- Small farm size
- High cost
- Lack of success stories
- Heterogeneity of cropping systems and market imperfections
- Land ownership, infrastructure and institutional constraints
- Lack of local technical expertise

CONCLUSION

Precision agriculture gives farmers the ability to use crop inputs more effectively including fertilizers, pesticides, tillage and irrigation water. More effective use of inputs means greater crop yield and/or quality, without polluting the environment. However, it has proven difficult to determine the cost benefits of precision agriculture management. At present, many of the technologies used are in their infancy, and pricing of equipment and services is hard to pin down. This can make our current economic statements about a particular technology dated.

REFERENCES

1. Arnholt M, MT Batte and S Prochaska, 2007. Adoption and Use of Precision Farming Technologies: A Survey of Central Ohio Precision Farmers, OSU AED Economics (AEDE-RP-OO11-O1), December 5.
2. Blackmore BS, 2008. Interview, Centre for Precision Farming, KVL, Taastrup.
3. Lowenberg –DeBoer J and M Boehlje, 1996. Revolution, evolution or dead-end: economic perspectives on precision agriculture. InProceedings of the Third International Conference on Precision Agriculture. P.C. Robert RH Rustand W.E. Larson, eds. American Society of Agronomy, Crops Science Society of America, Soil Science Society of America. Madison, WI.
4. Pedersen SM RB Ferguson and M Lark, 2009. A Comparison of Producer Adoption of Precision Agricultural Practices in Denmark, the United Kingdom and the United States, SJFI – Working Paper no. 2.
5. Stephen WS, 2009. Precision Farming: A New Approach to Crop Management, Texas Agricultural Extension Service.
6. Srinivasan, 2001. Precision farming in Asia: Progress and prospects, Geospatial Analysis Center, Regional Science Institute, Hokkaido, Japan.

7

INCREASING RICE PRODUCTIVITY BY MANIPULATION OF CALCIUM FERTILIZER IN USTIC ENDOAQUERT

Fauzan Zakaria[1]* and Nurdin[2]

Faculty of Agriculture, State University of Gorontalo, Indonesia

***Corresponding author:** Fauzan Zakaria; E-mail: fauzandza@gmail.com

ARTICLE INFO | **ABSTRACT**

Key words
Calcium fertilizer
Potassium
Rice Productivity
Ustic endoaquert

National rice production needs to be improved and maintained to meet the demands of fast growing population. One of the ways to meet this demand is through cultivating the rain fed land in many areas which its physical characteristics are challenging factor. This research aims at finding out the feedback of the rice production on the calcium fertilizer following the administration of river sand, beach sand, coco peat, and banana peat in ustic endoaquert. This research is implemented in rain fed field composed of vertisol soil in Sidomukti village of Mootilango Gorontalo, Indonesia. The subjects are randomly chosen and the treatments are separately implemented in two sub-group of vertisol soil. There are five treatments that were repeated three times, thus, there are 15 pieces of trials in each sub-vertisol groups. This research reveals that the administration of K fertilizer following the administration of river sand, beach sands, coco peat, and banana trunks fiber has significant effect on the number of grain, the weight of 1000 grains and the total weight of the grains. Meanwhile, the administration of K fertilizer following the administration of beach sand, coco peat and banana peat has significantly influenced the number of stalk, the length of stalk, and the total weight of the grains.

INTRODUCTION

The 2% population growth rate per year has caused the increased demands on rice. Up to 2006, the national rice consumption was 36,350,000 tons (BPS RI, 2007), thus Indonesia has to import the rice because our national production was only 57,157,435 tons of grains or equal to 32,304,029 tons (Deptan RI, 2007). From that number 54,199,693 tons of the grains (94,83%) comes from the irrigated rice fields and the rest are the product of the dry land farming. Although our current rice production is sufficient, considering our population growth rate, this rice production needs to be maintained and increased. Rain fed rice field is a rice field ecosystem that water source rely dominantly on rain water and is the second biggest producer of rice after irrigated field. This rain fed rice field amount to 2.1 million ha (Toha and Pringadi, 2004). The areal of rain fed rice field in Paguyaman, province of Gorontalo are dominated by vertisol soil that developed from lacustrine sediments (Hikmatullah *et al.* 2002; Prasetyo 2007; Nurdin 2010). Chemically, this vertisol soil is rich with high nutrition (Deckers *et al.,* 2001). However, its physical characteristics are challenging factors for the development of the crop and the crops ability to yield more. The characteristics of vertisols soil are have a high content of clay mineral, easy to shrink and swell, low water permeability, and slow draining (Mukanda and Mapiki 2001). Consequently, the growth and yield of the plants are obstructed. Soil ameliorant is needed to improve these soil characteristics. Sand is one of the ameliorant in high clay soil. In Ravina and Magier report (1984); Narka and Wiyanti (1999) showed that administration of sand had significant positive influence in lowering the value of cole and plasticity index, increasing the soil permeability, and reducing the moisture level. However, the rice farming in rain fed field need medium permeability with sufficient water available, thus, another soil ameliorant is needed to improve both characteristics, and the ameliorants needed for these are coco peat and banana peat.

Coco peat has been used as water storage medium in farming (Subiyanto *et al.* 2003). Meanwhile, the banana peat is still rarely used regardless to this dry banana peat has interrelated pores (Indrawati, 2009). The administration of those three ameliorants is suspected to be able to improve the physical characteristics of the vertisols soil in rice farming at rain fed field. Hence, the productivity of the rain fed farming as the second biggest producer of rice can be increased. This research aims at finding out the feedback of the rice product on calcium fertilizer following the administration of river sand, beach sand, and coco peat and banana peat in endoaquert ustic.

METHODOLOGY

This research is conducted in rain fed field composed of vertisols soil in Sidomukti village of Mootilang sub-district, District of Gorontalo, Gorontalo province. The object of this research is vertisols soil that has been previously treated with river sand, beach sand, coco peat and banana peat as ameliorants.

This research uses random group design method administered separately in two sub-group of vertisols soil. There are five levels of treatments. Each treatment is repeated three times, thus, there are 15 experiment plots for each sub-vertisols group and in total, there are 30 plots trial. (Table 1).

Table 1. Treatment of each Kalium fertilization in vertisols soil

Treatments		Ustic Endoaquerts	
Symbol	KCl Fertilizer Levels	0 DAP	30 DAP
	(kg ha^{-1})		
K0	0	0	0
K1	50	25	25
K2	100	50	50
K3	150	75	75
K4	200	100	100

DAP = day after plantings

Before the planting, the basic fertilizers are weighed. The lists of the fertilizers are in Table 2 below.

Table 2. Basic fertilizer, source, and day of fertilizer for ages after planting

Fertilizers as Starter	Source of Fertilizer	Recommendation of Fertilizer	Ages/Level of Fertlizing	
			0 DAP	60 DAP
		(kg ha^{-1})		
N	Urea (46% of N)	125	62,5	62,5
P	Phonska (15% of P2O5)	100	50,0	50,0

The farmland uses as the plot trials are the lots used in the first phase of the research. The Mekonga is the rice variety used in this research and it has been previously seeded for 21 days and planted with 25 cm x 25 cm spacing and 3 seeds are planted in one planting hole. The N, P and K fertilizers are given twice, half dosage in day 0 after the planting (HST) and on the 60th day after the planting. The irrigation is first done when the plants are ± 5 cm high up to when the plants are 10 days old. The next irrigation is regulated based on the growth and development of the crops. The weeding is done manually when the crops are 15 days old, and next weeding is determined by the weed condition in the field. Viruses, diseases, harmful insects are managed through understanding the relationship among environments, pests, natural enemies, host plants to help determine what action necessary. The harvesting time is when the crops are ± 115 days old. Physical appearance of ready to harvest crops is when > 95% of the crops have turned yellow.

The harvesting is done manually by cutting the upper half of the crop that contains the rice stalk. The rice then dried under the sun for 3-5 days to reach the 15% moisture level. After that, the rice then weighed per trial plot to gain the parameter of the rice crops yield. Those parameters are:

1. Number of stalk

This parameter is calculated per bunch in each treatment. The number then add together to find the mean of the number of stalks per bunch of crops for each treatment.

2. Length of stalk (cm)

This parameter is calculated in cm per bunch in each treatment. The result of this measurement then add together to find the mean of stalk length per bunch for each treatment.

3. Number of grain

This parameter is calculated per stalk in each treatment to find out the mean of number of grain per stalk for each treatment.

4. The weight for 100 dried grains (kg)

This parameter is calculated by weighing 100 dried grains using the digital scales for each treatment. The result then added to find the mean weight for 100 dried grains for each treatment.

5. The weight of dried grains (kg ha -1)

This parameter is obtained by weighing the dried grains using the digital scales for each treatment. The result then added up to find the mean of dried grains weight for each treatment. The weight then converted to weight of dried grains per ha.

All the obtained data, whether from calculation, measurement, and weighing processed and analyzed statistically. The presentation on the data on the influence of ameliorants administration on the crops yield is presented in tables and graphs. The data are further analyzed using the variance randomized block design analysis. If there is a significant difference, then the least significance difference test is conducted with the 5% level test.

FINDINGS AND DISCUSSION

Rice Crops Yields with K Fertilizer in Endoaquert Ustic following the administration of river sand, coco peat, and banana peat. The variance result shows that the K fertilizer does not give significant influence on the number of stalk and the length of stalk, however, it gives significant influence on the number of grain, the weight of 1000 grains and the total weight of the grains in Endoaquert ustic. The average yield of rice crops using K fertilizer in endoaquert ustic with Least Significant difference (P>0.05) is presented in Table 3 below.

Table 3. The mean of rice crops yields component using K fertilizers in endoaquert ustic following the administration of river sand, coco peat, and banana peat

Treatments	Number of stalk	Length of stalk (cm)	Number of grain	The weight for 100 dried grains (g)	The weight of dried grains (g)
0 kg ha^{-1} (K0)	15.83^{ns}	24.47^{ns}	103.42a	18.00a	356.70a
50 kg ha^{-1} (K1)	13.16	24.73	140.00b	20.66a	652.30ab
100 kg ha^{-1} (K2)	16.58	24.44	135.83b	25.00b	690.30ab
150 kg ha^{-1} (K3)	17.16	23.34	167.42c	25.33b	478.30ab
200 kg ha^{-1} (K4)	16.58	23.98	177.67c	26.66b	758.00b
LSD (0,05)			23.50	2.68	368.44
CV (%)	14.67	4.50	8.61	6.16	33.32

Note: Number that following by same latter in same column did not significantly different at LSD level of 0.05; ns=not significant effect at F level test 0.05; LSD=least significant different; CV=coefficient of variant.

The highest amount of stalks (17.16 stalks) are obtained in 150 kg ha-1 administration of K fertilizer (K3) and the least amount of stalks are obtained in 50 kg ha-1 administration of K fertilizer (K3). The longest stalk is obtained in K1 where the longest stalk is 24.73 cm and the shortest stalk is found in the administration of 150 kg ha-1 of K fertilizer (K3). It appears that the variety of numbers and length of stalks tend to fluctuate. It is assumed due to K fertilizer may not play a role in the growth and development of stalks. The number of grains in the administration of 200 kg ha-1 of K fertilizer (K4) significantly yields more grains (177.67 grains) than any other treatments. This is due to the Calcium (K) nutrient that is widely available in treatment K4, thus, the development and grain filling processes are not obstructed. Calcium (K) is one of important macro nutrients for the crop due to this nutrient plays direct roles in some physiological processes such as, (1) biophysical aspect of the Calcium plays important role in managing the osmotic and turgor pressure of the cell and to stabilize the pH, and (2) biochemical aspect, calcium plays a role in enzyme activities in carbohydrate and protein, and increasing the translocation of photosynthesis out of the leaves (Marschener, 195).

The heaviest weight of 1000 grains is shown in the administration of 200 kg ha-1 of K fertilizer (K4) and the significant difference of K fertilizer administration in K0 and K1 but there are no significant difference in K2 and K3. This shows that the 100-200 kg ha-1 of K fertilizer (K2-K4) have shown significant weight difference in 1000 grains. The more the K fertilizer given, the heaviest the 1000 grains would be. Further, this highest total weight of grains is shown by the administration of 200 kg ha-1 K fertilizer (K4) and only significantly differs with K0 treatment. It appears that the variety of total grains weight have fluctuate pattern.

The result of the regression analysis shows that there is a positive and linier correlation between the numbers of stalks with all the applied treatments, meanwhile, the length of stalks shows a reversed pattern, however, both have positive correlation with all treatments (Figure 1). It appears that the increase in the dosage of K fertilizer administration would be followed by the increase of stalks number, but the reverse happened with the length of stalks. Meanwhile, the number and weight of 1000 grains tend to be positively linier with strong positive correlation (Figure 2).

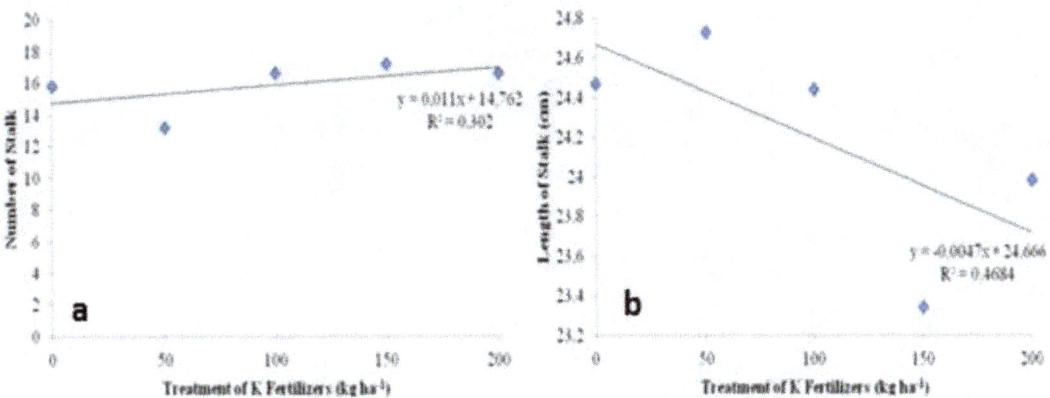

Figure 1. Regression between number of rice stalks **(a)** and length of stalks with the administration of K fertilizer in endoaquert ustic **(b)**

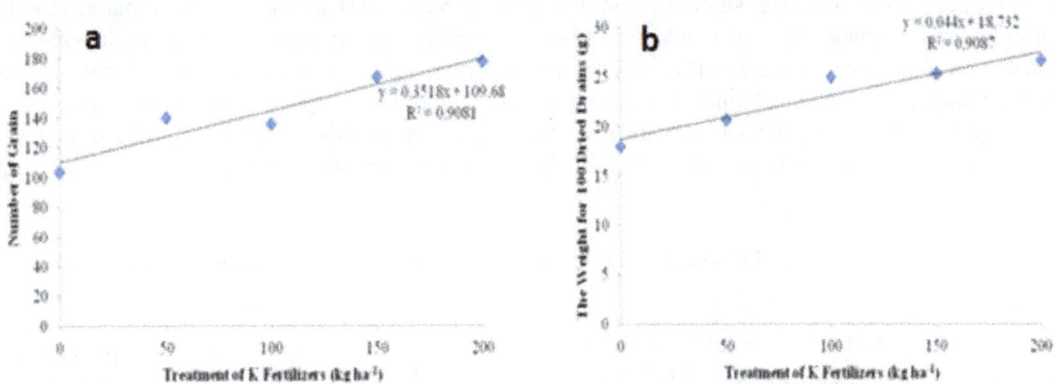

Figure 2. Regression correlation between number of grains **(a)** and weight of 1000 grains **(b)** in administration of K fertilizer in Endoaquert Ustic

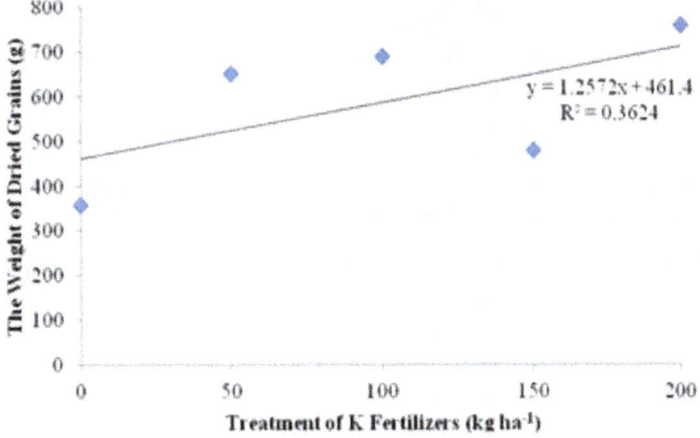

Figure 3. Regression correlation between the total weight and the K fertilizer administration in Endoaquert Ustic

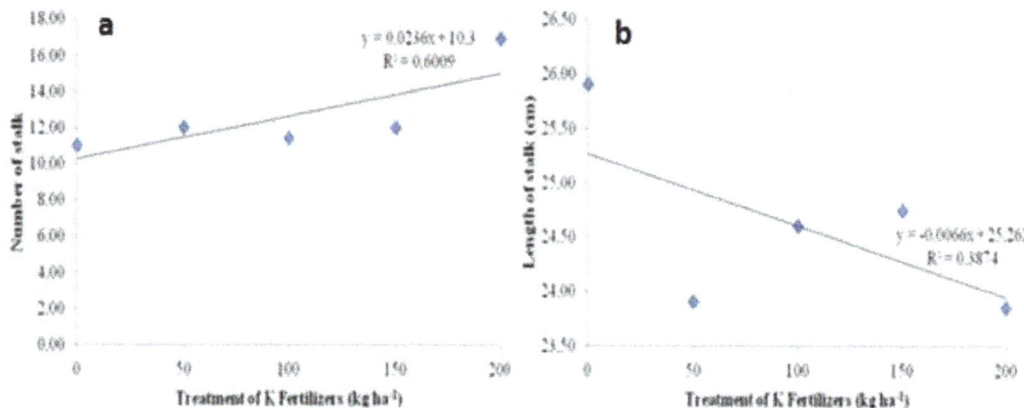

Figure 4. Regression correlation between number of rice stalk **(a)** and length of rice stalks **(b)** on administration of K fertilizer in Endoaquert Ustic

Correlation between administration of K fertilizer and the total weight shows positive and linier pattern (Figure 3). It appears that the increase in one unit of K fertilizer administration will yield 36 grams increase in the total weight of the grains. The rest is influenced by other factors such as washing, dissolved in water and fixated in the crystalized clay mineral). The variance result shows that administration of K fertilizer only has significant difference on number of stalk and in length of stalk and total weight of the grains in endoaquert ustic. The rest do not have significant difference on the length of stalk and the weight of 1000 grains. The average rice crop yield in endoaquert ustic with Least significant difference test ($P > 0.05$) is presented in table 4 below.

Table 4. Average rice crops yield of K fertilizer in Endoaquert Ustic following the administration of beach sand, coco peat and banana peat

Treatments	Number of stalk	Length of stalk (cm)	Number of grain	The weight for 100 dried grains (g)	The weight of dried grains (g)
0 kg ha^{-1} (K0)	11.00a	25.91a	127.16^{ns}	24.00^{ns}	418.00a
50 kg ha^{-1} (K1)	12.00a	23.91b	125.50	23.66	634.67b
100 kg ha^{-1} (K2)	11.41a	24.61ab	118.58	24.33	677.00b
150 kg ha^{-1} (K3)	12.00a	24.75ab	116.75	24.00	699.00b
200 kg ha^{-1} (K4)	16.91b	23.85b	111.50	25.00	615.00b
LSD (0,05)	1.51	1.72			137.24
CV (%)	10.55	3.71	7.75	5.08	12.81

Note: Number that following by same latter in same column did not significantly different at LSD level of 0.05; ns=not significant effect at F level test 0.05; LSD=least significant different; CV=coefficient of variant.

Administration of 200 kg ha-1 K fertilizer shows the most stalks and the most significant difference than the other treatments. The length of stalk in no fertilizer treatment (K0) shows the longest stalks and significantly differs from K2 and K4 treatments. Meanwhile, the total weight in K3 treatments significantly heavier than other treatments and significantly different from K0 treatment.

The result of regression analysis shows that there is a positive and linier correlation between number of rice stalks and all the applied treatments. However, the length of stalks have a reversed correlation with the treatments, regardless, both length and number of stalks have positive correlation with all treatments (Figure 4). It appears that the increase of K fertilizer dosage will be followed by the increase of number of stalks but not the length of stalks. on the other hand, the number of grains and the weight of 1000 grains have a reversed pattern compared to the number of stalks and the length of stalks (Figure 5).

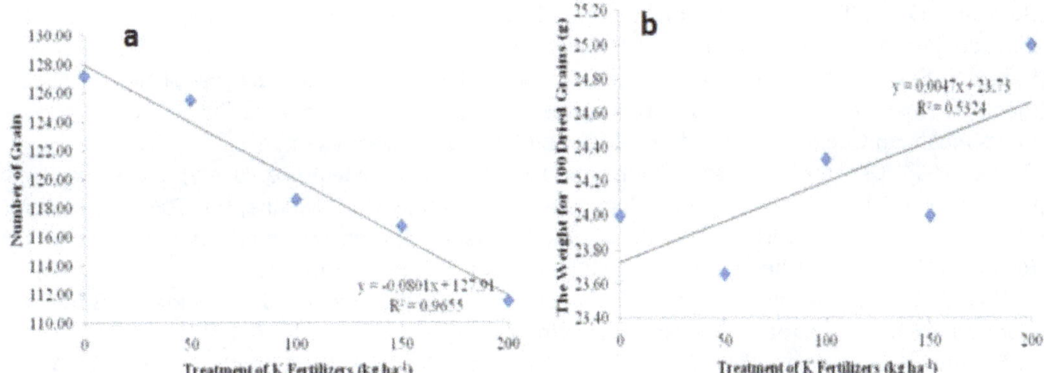

Figure 5. Regression correlation between number of rice grains **(a)** and the weight of 1000 rice grains on administration of K fertilizer **(b)** in Endoaquert Ustic

It appears that the administration of one unit of K fertilizer would decrease as much as 96.5 grains and administration of K fertilizer can increase the total weight of 1000 rice grains into 25 grams. Therefore, the dosage of K fertilizer administration should be adjusted to the need of K nutrient in rice crops, thus it would not decrease the number of grains into an extreme amount.

In addition, the correlation between the total weight and the administration of K fertilizer shows a strong positive and linier correlation (Figure 20). This shows that the administration of one unit of K fertilizer will would increase the 42.1 grams of the total weight. However, 200 kg ha-1 dosage of K fertilizer (K4) have a decreasing effect on the total rice grains weight.

CONCLUSION

The study concluded the following: (1) Administration of K fertilizer after the administration of river sand, coco peat and banana peat has significant influence on the number of rice grains, the weight of 1000 grains, and the total weight of the grains and (2) Administration of K fertilizer after the administration of beach sand, coco peat and banana peat has significant influence on the number of stalks, the length of stalks, and the total weight of the grains.

REFERENCES

1. Alwi M and D Nazemi, 2000. Pemberian brangkasan kedelai dan pupuk N untuk meningkatkan hasil jagung di lahan gambut (Administration of soya peat and N fertilizer to increase the corn yields in peat moss). Prosiding Simposium Nasional dan Kongres VII Peragi (Proceeding of National Symposium and VII Congress of Peragi) , Bogor. pp 253-259.
2. APCC. 2003. Coconut statistical yearbook 2002. Asia Pacific Coconut Community.
3. Agustian A, S Friyatno, Supadi and A Askin, 2003. Analisis pengembangan agroindustri komoditas perkebunan rakyat (kopi dan kelapa) dalam mendukung peningkatan daya saing sektor pertanian (Agro industry development analysis of community plantations commodity (coffee and coconut) in supporting the competitiveness of agricultural sector). *Makalah* Seminar Hasil Penelitian Pusat Penelitian dan Pengembangan Sosial Ekonomi Pertanian Bogor (Paper, Research Seminar in Agricultural Research and Socio-Economic Development Bogor). 38 pages.
4. Adam FP, J Moenandir, and M Santoso, 2008. Pengaruh pencampuran herbisida dan persiapan lahan terhadap pertumbuhan dan hasil padi sawah (the influence of herbicide mix and land preparation on the growth and rice crops yield). J. Agritek 16: 1601-1615.
5. Arabia T, 2009. Karakteristik tanah sawah pada toposekuen berbahan induk volkan di daerah Bogor-Jakarta [disertasi] (Characteristics of rice field land in toposequence composed mainly of volcanic materials in Gogor-Jakarta [dissertation] . Bogor: Sekolah Pascasarjana Institut Pertanian Bogor (Postgraduate school of Institut Pertanian Bogor).

6. Borchardt G. 1989. Smectites. p675-727 *in* Minerals in Soil Environments. Second Edition. Soil Science Society of America Madison, Wisconsin, USA.
7. Bahcri S, Sukido, and Ratman N. 1993. Peta geologi lembar tilamuta, Sulawesi Skala 1 : 250.000 (geologic map of Tilamuta, Sulawesi Scale 1: 250,000). Bandung: Pusat Penelitian dan Pengembangan Geologi (Center for research and Geologic Development).
8. Balitpa. 2004. Deskripsi varitas unggul padi (description of prominent rice variety). *Dikompilasi* oleh OS Lesmana, HM Toha, I Las, dan B Suprihatno (compiled by OS Lesmana, HM Toha, I Las, and B Suprihatno). Balai penelitian tanaman padi, Badan penelitian dan pengembangan pertanian (Rice Research Center, Agricultural Research and Development Center). 68 pages.
9. BPS RI. 2007. Statistik Indonesia tahun 2007 (Statistics of Indonesia in 2007). Jakarta: BPS Republik Indonesia (Statistics Bureau of the Republic of Indonesia).
10. BPS Provinsi Gorontalo. 2010. Gorontalo dalam angka tahun 2010 (Gorontalo in Figures 2010). Gorontalo: BPS Provinsi Gorontalo (Statistical Bureau of Gorontalo Province).
11. BPS Kabupaten Gorontalo. 2010. Kabupaten gorontalo dalam angka tahun 2010 (Gorontalo district in Figures 2010). Limboto: BPS Kabupaten Gorontalo (Statistical Bureau of Gorontalo District).
12. Driessen PM and R Dudal (Eds). 1989. Lecture notes on the geography, formation, properties, and use of the major soils of the world. Agricultural University, Wageningen.
13. Dudal R and H Eswaran. 1988. Distribution, properties and classification of Vertisols. *In* LP Wilding and R Puentes (Eds), Vertisol: Their distribution, properties, classification and management . SMSS Technical Monograph 18, Texas A&M University, College station.
14. Djusar D, 1996. Aplikasi polimer hidroksi aluminium sebagai alternatif perbaikan beberapa sifat fisik tanah vertisol [Skripsi] (Application of aluminum hydroxide polymer as alternative to improve some physical characteristics of vertisols [thesis]. Bogor: Jurusan Tanah Fakultas Pertanian IPB (school of soil science, faculty of agriculture, Institut Pertanian Bogor).
15. Deckers J, O Spaargaren, and F Nachtergaele, 2001. Vertisols: Genesis properties and soilscape management for sustainable development. p. 3-20. *In* Syers JK, FWT Penning De Vries, and P Nyamudeza (Eds): The Sustainable Management of Vertisols. IBSRAM Proceeding No. 20.
16. Dharmawan AH, 2004. Sistem pengendalian konversi lahan pertanian: Perspektif sosiologi pertanian. Makalah pada Round Table Pengendalian Konversi dan Pengembangan Lahan Pertanian (management system of agricultural land conversion: agricultural sociology perspective, a paper in round table discussion on conversion management and development of agricultural land). Jakarta, 14 December 2004. Departemen Pertanian RI (Department of Agriculture of the Republic of Indonesia). 2008. Produksi padi nasional (national rice production). Jakarta: Departemen Pertanian RI Department of Agriculture of the Republic of Indonesia).
17. Eswaran H and T Cook, 1988. Classification and management- related properties of Vertisols. p. 431. *In* Jutzi S, I Haque, J McIntire, and J Stares (Eds): Proceeding of a Conference held at ILCA, Addis Ababa, Ethiopia, 31 August to 4 September 1987
18. FAO, 2000. Vertisol. http://www.fao.org/ag/agl/prosoil/verti.htm. last update 21 August 2000.
19. Firmansyah MA, 2011. Arang Sumber Amelioran Tanah Yang Ramah Lingkungan (Charcoal as the environmental friendly soil ameliorant). http://www.sinartani.com/bumiair/arang-sumber-amelioran-tanah-ramah-lingkungan-1272881571.htm. Last update 24/03/2011.
20. Havlin JL, JD Beaton, SL Tisdale, WL Nelson, 1999. Soil Fertility and Fertilizer. An Introduction to Nutrient Manegement. [New Jersey] Prentice Hall, Upper Saddle River. p. 198 – 216.
21. Hikmatullah, BH Prasetyo, and M Hendrisman, 2002. Vertisol dari daerah Gorontalo: Sifat-sifat fisik-kimia dan komposisi mineralnya (vertisols from Gorontalo: physic-chemical characteristics. Journal of Land and Water, 3: 21-32.
22. Hidayat P, 2008. Teknologi pemanfaatan serat daun nanas sebagai alternatif bahan baku tekstil (the usage of pineapple leaves fiber as alternative textile material). *J. Teknoin (teknoin journal)* 13(2):31-35.
23. Ismangil and A Maas, 2006. Potensi batuan belu sebagai amelioran pada tanah mineral masam (belu soils potentials as ameliorant in acid minerals soil). Tanah Tropika Journal, 11: 81-88.
24. Indrawati E, 2009. Koefisien penyerapan bunyi bahan akustik dari pelepah pisang dengan kerapatan yang berbeda [Skripsi] (sound absorption coefficient of the acoustic material made from different densities of banana midribs) [final paper]. Malang: Jurusan Fisika Fakultas Sains Dan Teknologi

Universitas Islam Negeri Maliki (department of Physics, Faculty of Science and Technology of Universitas Islam Negeri Maliki).

25. Kumarawarman B. 2008. Lingkungan pengendapan lakustrin [thesis] (Lacustrine sediments environment) . Yogyakarta: Program Pascasarjana Universitas Gadjah Mada (postgraduate program of Universitas Gajah Mada).
26. Kamus Besar Bahasa Indonesia Edisi III (3rd edition of Bahasa Dictionary). 2010. Jakarta: Pusat Bahasa Kementrian Pendidikan Nasional RI (center for language ministry of national education of the republic of Indonesia).
27. Lopulisa C. 2005. Studi karakteristik lahan sawah dan budidaya padi di Kabupaten Maros (A study on the characteristics of rice field and rice cultivation in district of Maros). *Journal of Sains & Teknologi* 5(1):1-11.
28. Marschner, H. 1995. Measurement and assessment of soil potassium. Int. Potash Inst. IPI Res. Topics No.4. Mukanda N and A Mapiki. 2001. Vertisols Management in Zambia. p. 129-127. *In* Syers JK, FWT Penning De Vries, and P Nyamudeza (*Eds*): The Sustainable Management of Vertisols. *IBSRAM Proceedings No.* 20.
29. Mulyanto D, M Nurcholis, and Triyanto. 2001. Minertalogi Vertisol dari bahan induk tuf, napal dan batupasir (minertalogy of vertisols mainly composed of tuff, napalm and sandstones . *Journal of Tanah dan Air (land and water)* 2(1):38-46.
30. Mahmud Z and Y Ferry. 2005. Prospek pengolahan hasil samping buah kelapa (the prospect of coconut side products processing). *Prospektif* 4(2):55-63
31. Muchtar and Y Soelaeman. 2010. Effects of green manure and clay on the soil characteristics, growth and yield of peanut at the coastal sandy soil. *J. Trop. Soils* 15(2):139-146.
32. Nelson, L.A., L. Anderson. 1977. Partitioning of soil test-crop respon probability. In Stelly et al. (Eds). Soil Testing : Correlating and Interpreting the Analytical Result. ASA Special Publication No. 29.
33. Nuryani SHU and T Notohadiprawiro. 1994. Pengaruh sari kering limbah pabrik kulit atas populasi mikrobia dan susunannya pada berbagai jenis tanah (the influence of dry sewage from leather manufacturer on microbial population and its composition in different kinds of soils). *J. Manusia dan Lingkungan (journal of human and environment)* 1(2):1-8.
34. Narka IW and Wiyanti. 1999. Pengaruh pemberian pasir dan bahan organik terhadao sifat fisik tanah Vertisol (the influence of sand and organic materials administration of physical characteristics of vertisol soil). *J. Agritrop* 18(1):11-15.
35. Nursyamsi D, D Setyorini, and JS Adiningsing. 1996. Pengelolaan hara dan pengaturan drainase untuk menanggulangi kendala produktifitas sawah baru (nutrient management and drainage management as a solution to new rice field productivity problem. pp 113-128. *Dalam* Prosiding Pertemuan Pembahasan dan Komunikasi Hasil Penelitian Tanah dan Agroklimat. Buku III Bidang Kesuburan dan Produktifitas Tanah (in the proceeding of discussion and communication meeting on soils and agro climate research). Cisarua Bogor, 26-28 September 1995.
36. Pusat Penelitian Tanah dan Agroklimat, Bogor. Nursyamsi D ans Suprihati. 2005. Sifat-sifat kimia dan mineralogi tanah serta kaitannya dengan kebutuhan pupuk untuk padi (*Oryza sativa*), jagung (*Zea mays*), dan kedelai (*Glycine max*) (chemical and mineralogy characteristics of soils and their relations to fertilizers need for rice, corn, and soya bean). *Bull. Agron.* 33(3):40-47.
37. Nursyamsi D, K Idris, S Sabiham, DA Rachim, and A Sofyan. 2008. Pengaruh asam oksalat, Na+, NH4+, dan Fe3+ terhadap ketersediaan K tanah, serapan N, P, dan K tanaman, serta produksi jagung pada tanah-tanah yang didominasi smektit (the influence of oxalate acid,, Na^+, NH4+, and Fe3+, on the availability of K soil, N, P and K absorption in plants, and corn production in smectit dominated soils). Journal of soil and climate, 28: 69-82.
38. Nursyamsi D, 2009. Pengaruh kalium dan varietas jagung terhadap eksudat asam organik dari akar, serapan N, P, dan K tanaman dan produksi brangkasan jagung (*Zea mays* L.). (the influence of calcium and corn variety on organic acid exudate from the roots, absorption of N, P, and K of the plants and the production of corn straw) *J. Agron. Indonesia* 37(2):107-114.
39. Nur II, Kardiyono, Umar, ans A Aris, 2003. Pemanfaatan limbah debu sabut kelapa dalam usahatani padi pasang surut. Kelembagaan Perkelapaan di Era Otanomi Daerah (the usage of dust coco peat sewage in tidal rice fields). Prosiding Konferensi Nasional Kelapa V (proceeding of 5th national conference of coconut). Tembilahan 22-24 October 2002. pp160-165.

40. Nasution LI. 2004. Review peraturan perundangan dalam mengendalikan konversi lahan (regulations review on land conversion). *Makalah* pada Round Table Pengendalian Konversi dan Pengembangan Lahan Pertanian (paper in round table discussion of conversion management and agricultural land development), Jakarta, 14 December 2004.
41. Noor M, A Maas, and T Notohadikusumo, 2005. Pengaruh pelindian dan ameliorasi terhadap pertumbuhan padi (*Oryza sativa*) di tanah sulfat masam Kalimantan (the influence of leachability and ameliorants on rice growth in acid sulfate soil in Kalimatan) . *J. Ilmu Tanah dan Lingkungan (journal of soils science and environment)* 5(2):38-54.
42. Nurdin, 2010. Perkembangan, klasifikasi dan potensi tanah sawah tadah hujan dari bahan lakustrin di Paguyaman, Gorontalo [disertasi] (development, classification and potentials of rain fed rice field composed of lacustrine material in Paguyaman, Gorontalo). Bogor: Sekolah Pascasarjana Institut Pertanian Bogor (postgraduate program of Institut Pertanian Bogor).
43. Pusat Penelitian Tanah, 1983. Terms of reference survei kapabilitas tanah no 22/1983 (term of reference of survey on soils capability). Bogor: Proyek Penelitian Pertanian Menunjang Transmigrasi (P3MT) Badan Penelitian dan Pengembangan Pertanian Departemen Pertanian RI (agricultural research project to support transmigration in agricultural research and development of the agricultural development of the republic of Indonesia).
44. Prasetyo BH, H Sosiawan, and S Ritung. 2000. Soil of Pametikarata, East Sumba: Its suitability and constraints for food crop development. Indonesian Journal Agricultural Science, 1: 1-9.
45. Prasetyo BH, M Soekardi, and Subagjo H. 1996. Tanah-tanah sawah intensifikasi di Jawa: Susunan mineral, sifat-sifat kimia dan klasifikasinya (intensification of rice field soils in Java: mineral composition, chemical characteristics and its classification). Research on Soils and Fertilizers News, 14: 12-24.
46. Prasetyo BH. 2007. Perbedaan sifat-sifat tanah vertisol dari berbagai bahan induk (the characteristics differences of vertisols from other soils composition. *Journ*al of Agricultural Science, 9: 20-31.
47. Prasetyo BH, D Setyorini, 2008. Karakteristik tanah sawah dari endapan aluvial dan pengelolaannya (characteristics of rice field made from alluvial sediments and its management). Journal of land Resource, 2: 1-14.
48. Partohardjono S, JS Adiningsih, and IG Ismai l, 1990. Peningkatan produktivitas lahan kering beriklim basah melalui teknologi sistem usahatani (improvement of wet climate dry lands' productivity through farming system technology . *in* M. Syam *et al.* (Eds). Risalah lokakarya penelitian sistem usahatani di lima agroekosistem (proceeding of agricultural business system research in five agro ecosystems). Pusat Penelitian dan Pengembangan Tanaman Pangan (Centers for Crops Research and Development. pp 47-62.
49. Prihatin DSH, 2000. Pertumbuhan stek pucuk dan stek batang kepuh (*Sterculia foetida* Linn) pada berbagai media dan zat pengatur tumbuh rootone-F [Skripsi] (the growth of shoot cuttings and cuttings of kepuh (sterculia foetida Linn) in many medium and rootone-F growth regulator substance) [final paper] , . Bogor: Fakultas Kehutanan IPB (forestry faculty of IPB).
50. Pawirosemedi and Marsadi. 2000. Pengaruh pemberian belerang (S) dan inokulasi Rhizobium pada vertisol terhadap pertumbuhan dan produksi kedelai (*Glycine max* L. Merr) serta kadar hara N dan S daun indeks (the influence of sulfate (S) and inoculation of Rhizobium in vertisols on the growth and production of soya bean (Glycine mas L. Merr) and the N and S nutrients of the leaves index). J. Agrivita, 22: 58-63.
51. Pramono J. 2004. Kajian penggunaan bahan organik pada padi sawah (the study on the usage of organics materials on rice field). J. Agrosains, 6: 11-14.
52. Permadi K and HM Toha. 1996. Kultivar padi pada lingkungan gogo rancah dan sawah di lahan sawah tadah hujan (cultivation of rice in dry land and rice field in the rain fed land). J. Kultura, 137: 14-19.
53. Permadi K, I Nurhati, and Y Haryati. 2005. Penampilan padi gogo rancah varietas Singkil dan Ciherang melalui model teknologi pengelolaan tanaman dan sumberdaya terpadu di sawah tadah hujan (visibility of the gogo rancah rice of singkil and ciherang variety through model of plant management technology and integrated resources in rain fed rice field). J. Agrivigor, 4: 227-233.

54. Pusat Analisis Sosial Ekonomi dan Kebijakan Pertanian. 2008. Kebijakan untuk Menciptakan Lahan Pertanian Pangan Abadi (policy to create sustainable agriculture field). Bogor: Pusat Analisis Sosial Ekonomi dan Kebijakan Pertanian Departemen Pertanian RI (center of socio-economic and agriculture policy analysis, Department of Agriculture of the Republic of Indonesia).
55. Ravina I and J Magier, 1984. Hydraulic conductivity and water retention of clay soil containing coarse fragments. Soil Science Society of America Journal, 48: 738-740.
56. Ristori GG, E Sparvalie, M deNobili, and LP D'Aqui, 1992. Characterization of organic matter in particle size fractions of Vertisols. Geoderma, 54: 295-305.
57. Rindengan B, A Lay, H Novarianto, H Kembuan and Z Mahmud. 1995. Karakterisasi daging buah kelapa hibrida untuk bahan baku industri makanan (characteristics of the hybrid coconut fruit for the food industry material). *Laporan* Hasil Penelitian Kerjasama Proyek Pembinaan Kelembagaan Penelitian Pertanian Nasional (research report on cooperation project for institutional development for national agriculture research)
58. Badan Litbang Departemen Pertanian RI. 49 pp. Ruskandi. 2006. Teknik pembuatan kompos limbah kebun pertanaman kelapa polikultur (technique of producing copost from the poly-culture coconut plantation) . *Bulletin of Teknik Pertanian* 11(1):33-36.
59. Subagjo H, 1983. Pedogenesis dua pedon Grumosol (Vertisols) dari bahan volkanik gunung Lawu dekat Ngawi dan Karanganyar (pedogenesis of two pedon grumosol (vertisols) of volcanic materials from Lawu mountain near Ngawi and Karanganyar) . *Pemberitaan Pen. Tanah dan Pupuk* 2:8-18.
60. Subagyo H, N Suharta and AB Siswanto, 2004. Tanah-tanah pertanian di Indonesia (agricultural lands in Indonesia) pp 21-66. in A Adimihardja *et al* (Eds). Sumberdaya Lahan Indonesia dan Pengelolaannya (Indonesian land resources and its management) . Puslitbangtanak. 2nd edition.
61. Smith C, 1995. Coir: a viable alternative to peat for potting. *J. Horticulturist* 4(3): 25-28.
62. Suhartatik E and R Sismiyati, 2000. Pemanfaatan pupuk organik dan agen hayati pada padi sawah (the usage of organics fertilizer and natural agents in rice field). in Suwarno *et al.* (*Eds*): Tonggak kemajuan teknologi produksi tanaman pangan, paket dan komponen teknologi produksi padi (the milestone of crops production technology development, package and components of rice production technology). Bogor: Pusat Penelitian dan Pengembangan Tanaman Pangan (center for crops research and development).
63. Saparso. 2001. Kajian serapan N dan pertumbuhan tanaman kubis pada berbagai kombinasi mulsa dan dosis pupuk N di lahan pasir pantai (a study on the absorption of N nutrient and the development of cabbage plants in many mulch combination and dosage of N fertilizer in beach sand field) [Thesis]. Yogyakarta: Program Pasca Sarjana UGM (postgraduate program of UGM).
64. Saparso. 2010. Teknologi efisiensi pemanfaatan air dalam peningkatan produktivitas bawang merah di lahan pasir pantai (water efficiency technology in increasing the roductivity of onion in beach sand field). http://www.lontar.ui.ac.id//opac/themes/libri2/detail.jsp?id =133748&lokasi=lokal. accessed on 24 March 2011.
65. Subiyanto B, R Saragih and E Husin, 2003. Pemanfaatan serbuk sabut kelapa sebagai bahan penyerap air dan oli berupa panel papan partikel (usage of coco peat as water and oil absorption agent in particle board panel). Journal of Ilmu dan Teknologi Kayu Tropis, 1: 26-34.
66. Sudadi, YN Hidayati and Sumani. 2006. Ketersediaan K dan hasil kedelai (*Glycine max* L. Merril) pada tanah vertisol yang diberi mulsa dan pupuk kandang (availability of K and soya beans yield (*Glycine max* L. Merril) in vertisols soils given the mulch and manure). Journal of Ilmu Tanah dan Lingkungan, 7: 8-12.
67. Suyamto, Toha HM, P Hamdan, MY Sumaullah, TS Kadir, and F Agus. 2008. Petunjuk teknis pengelolaan tanaman terpadu (PTT) padi sawah tadah hujan (technical guideline for integrated plants management of rice in rain fed field). Jakarta: Badan Penelitian dan Pengembangan Departemen Pertanian RI (research and development agency in department of agriculture of the republic of indonesia)
68. Silalahi MD, C Shiallagan, and E Monica. 2007. Penyisihan Mn2+ dalam air sumur dengan memanfaatkan sabut kelapa (filtering the Mn2+ in well water using the coco peat). Journal of Teknologi Lingkungan, 4: 44-49.

69. Shiddieq Dj, Tohari, Saparso and B Setiadi. 2008. Karakterisasi berbagai jenis bahan lapisan kedap, ketebalan dan nisbah bentonit dengan pasir pada pengelolaan lahan pasir pantai (characteristics of many proof layers, the thickness and the bentonite ratio in management of beach sand land). Journal of Ilmu Tanah dan Lingkungan, 8: 93-101.
70. Syafiisab AA, 2010. Pengaruh komposit core berbasis limbah kertas, dengan pencampur sekam padi, dan serabut kelapa terhadap kekuatan bending panel [Skripsi] (the influence of paper sewage core composite with the rice husks mix and coco peat on the panel bending strength) [final paper]. Surakarta: Fakultas Teknik Universitas Sebelas Maret.
71. Soil Survey Staff, 2010. Key of soil taxonomy. Ed ke-11. Washington DC: USDA-Natural Resources Concervation Service.
72. Taufik M, 2003. Pengaruh komposisi serat sabut kelapa dan batu kapur terhadap tegangan lentur pada eternity (the influence of coco peat composition and lime stone on the bending stress of the eternity).http://elib.unikom.ac.id/gdl.php?mod=browse&op=read&id=jiptumm -gdl-s1-2003-mohamad-8838-2003. Diakses tanggal 24 Maret 2011.
73. Toha HM and K Pirngadi, 2004. Pengaruh kerapatan tanaman dan pengendalian gulma terhadap hasil beberapa varitas padi sistem tabela pada lahan sawah tadah hujan (the influence of plants density and the weed management on some wet seeded rice crops yields in rain fed fields) . Journal of Agrivigor, 3: 170-177.
74. Van Bemmelen RW, 1949. The geology of Indonesia; general geology of indonesia and adjacent archipelagoes. Vol ke-1A. Hague: Goverment Printing Office.
75. Wiqoyah Q, 2006. Pengaruh kadar kapur, waktu perawatan dan perendaman terhadap kuat dukung tanah lempung (the influence of lime leve, treatment time, and soaking time toward the strength of clay stone). Journal of Teknik Sipil, 6: 16-24. W
76. ihardjaka A and S Abdurachman, 2007. Dampak pemupukan jangka panjang padi sawah tadah hujan terhadap emisi gas metana (the effect of long term administration of fertilizer in rain fed field on methane gas emission). Journal of Penelitian Pertanian Tanaman Pangan, 26: 199-205.
77. Widiawati D, Z Rais, A Haryudant, and ES Amanah, 2007. Pemanfaatan limbah sabut kelapa sebagai bahan baku alternatif tekstil (the usage of coco peat as alternative material for textile). Journal of Ilmu Desain, 2: 57.
78. Wuryaningsih S, T Sutater and B Tjia, 2008. Pertumbuhan tanaman hias pot *Anthurium andraeanum* pada media curah sabut kelapa (growth of potted plants Anthurium andraeanum in coco peat media). Journal of Penelitian Pertanian, 18: 31-38.

8

FARMERS' PERCEPTION AND KNOWLEDGE OF CLIMATE CHANGE IN BANGLADESH – AN EMPIRICAL ANALYSIS

Muntaha Rakib[1*] and Shah Mohammad Hamza Anwar[2]

[1]Department of Economics, Shahjalal University of Science and Technology, Sylhet, Bangladesh; [2]Department of Economics, Metropolitan University, Sylhet, Bangladesh

***Corresponding author:** Muntaha Rakib; E-mail: muntaha_rakib@yahoo.com

ARTICLE INFO

Key words

Climate change, Agriculture, Farmers, Perception, Bangladesh

ABSTRACT

The lack of sufficient knowledge about climate changes and the impact on agricultural production is an impediment to long term sustainable agriculture in most developing countries, including Bangladesh. This paper presents the results of an investigation to determine perception of farmers about changes in climate in Bangladesh. The study finds the determinants of farmers' perception on climate variability in different specifications of household characteristics. The sample was adult farmers with at least 20 years of farming experience in the area. Data was collected on perceptions about temperature changes and variability in precipitation over a 20 year period. The results indicated that more than 80% of farmers believe that temperature in the district had become warmer and over 90% were of the opinion that rainfall timing had changed, resulting in increased frequency of drought.

INTRODUCTION

Determining farmers' decision to adapt to and cope with shocks in one hand and for improving existing policies and to formulate new policies and supportive programs on the other hand; which types of farmers perceive that climate is changing is imperative to understand. Perception refers to the process of acquisition and understanding of information from one's environment (Maddox, 1995). Farmers have to perceive first that the climate has changed, and then identify useful adaptations and implement them (Maddison, 2006).Among the few researches on perception of farmers in Bangladesh, most of the studies focus on coastal area and very few of them on hill tracts. None of the studies analyzed the influencing factors of perception of agricultural farmers using econometric tools which is what the paper contributes. This paper however, covers all the geopolitical and agro-ecological zones in Bangladesh specially focusing on agricultural households, and is more country representative and is useful for policy implications.

Climate change is a major challenge to agricultural development and general livelihood conditions in Bangladesh. Living in a developing, densely populated flat land area, the poor people of Bangladesh are at severe risk due to climate change. Despite of the expanding service sector, agriculture which is heavily affected by climatic shocks, is the employment source of 46 percent of the total labor force (Labor Force Survey 2010) and approximately 17 percent of the country's GDP (Gross domestic product) (BBS 2013).It is forecasted that a rise of 1 meter sea level will inundate a 29,846 square kilometers area and will displace around 15 million people and to lose 15-17percent of its land in the coming decades (Akter 2009, IPCC 2001). These geographic and demographic characteristics make the country one of the most vulnerable due to climate change and other shocks.

Literature review

Most of the papers analyzing farmers' perception in Bangladesh revealed that farmers have the perception that climate is changing (Kamruzzaman, 2015, Rashid et al. 2014, Syeda and Nasser 2012) among which Kamruzzaman (2015) looked for the influencing factors of perception in Sylhet hilly region using tabular analysis. Level of education and access to extension services had significant association with their perceived cause of climate change (Kamruzzaman, 2015). Akanda and Howlader (2015) in a study on coastal farmers of Patuakhali District analyzed the correlates of farmers' perception where, education, farm size, annual family income, communication exposure and agricultural knowledge were positively correlated to perception while fatalism was negatively significant (Akanda and Howlader, 2015). Information of uncertainty of consequences of climate change influence risk perception (Moniruzzaman2013).

The studies find that both rainy season and cold season delays to start but ends early while mean duration of both seasons has been significantly reduced (Kamruzzaman, 2015, Rashid et al. 2014, Rakib et al. 2014). Farmers found areduction in overall rainfall and variation in wind speed, duration of strong wind. They felt the incidence of drought has been increased and flood has been decreased (Kamruzzaman, 2015, Rashid et al. 2014, Rakib et al. 2014). The extended summer periods with increasing average temperatures have resulted in decreased growth and yield of crops and increased pest infestations. Farmers also have perception that severe cold wave along with dense fog has been observed in Bangladesh, particularly in central and northern part, over couple of years (Sealand Baten 2011). Moniruzzaman (2013) in a different study revealed gradually decreasing knowledge and confidence levels of climate change from adult group to young groups. The results of the study support the claim that both knowledge and confidence levels will increase when people learn more about climate change. Changing of climate affects crop yield and cultivation especially the production of wheat (Syeda and Nasser 2012).

In a study on southern Khulna Rashid et al. (2014) found that, local people perceived changes in rainfall patterns and water-logging, resulting in delayed rice planting, decreased yield and damaged crops. The increased period during which river water is saline limits the scope of irrigation with river water (Rashid et al. 2014).Among several consequences of climate change, Seal and Baten (2011) found drainage congestion resulting from flood as the most dominating natural shock. In this study, the respondents of Sirajgonj identify riverbank erosion while inhabitant's of Shariatpur find salinity intrusion as the effect of climate change which is adversely affecting agriculture.

Most of the other existing studies find that farmers are perceived about temperature increase and rainfall decrease (Gbetibouo 2009, Maddison 2006, Nhemachena and Hassan 2007). However, delayed rainfall, prolonged drought and extreme temperature are the most severe threats to agriculture in Bangladesh due to climate change (Bangladesh MoEF, 2005). BCAS (2009) finds erratic behavior of temperature and rainfall, extreme weather events and salinity are the most prominent indicators of climate change in Bangladesh. Adger et al. (2003) analyzed the annual average temperature and find that the country has experienced an increase of 0.40C to 0.80C temperature from 1900 to 2000 and until 2050 it is expected to increase by 1-20C more. Yu et al. (2010) analyzed sixteen global climate change model and simulated a positive trend of temperature increase from 2030s in Bangladesh while a higher intensity of precipitation in wet season and dryer winter which will intensify more droughts.

Survey data

The study utilized panel data collected through primary surveys of 800 agricultural households in rural Bangladesh in 2010 and 2012 that were administered by the International Food Policy Research Institute (IFPRI) and Data Analysis and Technical Assistance Limited (DATA), with additional inputs from the Center for Development Research (ZEF) in 31 out of 64 districts from all divisions across the seven agro-ecological zones which are categorized by Bangladesh Centre for Advanced Studies (BCAS).Therefore it is more representative with the whole set of major climatic shocks then the very few empirical past surveys in Bangladesh. Information was collected on demographic characteristics, physical asset, livestock and land ownership, crop management practices, access to credit and extension services, prior experience with climatic and non-climatic shocks, and perceptions about climate change. In 2012, additional information was collected on social and political capital and participation in groups by main male and female member of the household, coping mechanisms, adaptation strategies related to livestock; hence, the analysis involving these variables is for 2012 only. Note that, of the responding households used in the analysis, 89 percent were headed by men, and 11 percent were headed by women.

Perception and knowledge of climate change

Households were asked about their perceptions of temperature and rainfall changes and about overall climate change in the previous 20 years. Overall 90 percent households noticed that climate is changing from the last twenty years. Table 1 shows almost 88 percent of households reported being aware that rainfall was decreasing and approximately 86 percent noticed that temperatures were increasing.The three most cited changes were - more erratic rainfall, longer periods of drought and later onset of rains. This is consistent with the analysis of Thomas et al. (2013) using the baseline data, which is unsurprising given that the sample was a resurvey of Thomas's survey sites and that the follow-up survey was conducted only two years after the baseline.

Table 1. Households' perceptions of changes in rainfall and temperature in the past 20 years

Households' Perception	Rainfall	Temperature
Share of households (%) that …		
perceived an increase	8.5	86.0
perceived a decrease	88.4	8.9
perceived no change	2.8	4.6
did not know	0.3	0.5
Total	100.0	100.0

Source: Calculated by author based on survey datain 2012.

The farmers were asked about whether they perceive that climate is changing and if so, to mention the three most important changes they perceived. The most important changes they noticed and ranked as first are summarized in table 2.

Table 2. Farmers' beliefs about the likely response to more extreme weather due to climate change

Most important change farmers perceived	Percent
Rains have become more erratic	71.25
Rains come later	8.42
Longer periods of drought	15.43
Rains are heavier	3.09
Rains come earlier	1.82
Total	100

Source: Calculated by author based on survey data in 2012.

Among the other important indicators, longer periods of drought and more floods were noticed largely. They were also asked about whether they noticed any other changes and the result is shown in table 3. Other than rain and drought, change of starting time of season, level of water down, frequently cold wave and fogging and more floods etc. were noticed by the farmers.

Table 3. Noticed any other climatic changes

Any other changes farmers perceived	Percent
Rains have become more erratic	61.57
Longer periods of drought	19.02
Average temperature changed	9
Rains are heavier	3.21
Starting time of season changed	3.21
Level of water is down	2.83
Frequently cold wave/fogging	0.51
More floods	0.26
Others	0.39
Total	100

Source: Calculated by author based on survey data in 2012

Empirical approach

In this section the factors associated with the perception that climate is changing by male and female heads are investigated. The paper starts with simple probit model estimated by maximum likelihood method to identify the correlates and subsequently check whether the strategies are taken as complement or substitute to each other by applying bivariate probit in appropriate cases.

A set of independent variables are chosen according to the relevance and on the basis of theory and previous works. The independent variables are gender of household head, years of schooling, work experience and age of household head, households' land ownership, physical assets measured in index using principal component analysis (PCA), livestock ownership in tropical livestock unit (TLU), access to extension and credit, group participation and information on social and political capitalof husbands and wives, experiences of climate change shocks and access to ICT by husbands and spouses. Even though the existing studies include physical capital, natural capital and livestock variables in their analysis (Deressa et al., 2010,

Gbetibouo 2009, Nhemachena and Hassan 2007), access to these assets might be influenced by higher level of adaptation strategies and perception, which brings forth the endogeneity issue. Therefore, the asset variables, for instance physical capital, livestock, social and political capital are calculated as leave out village mean by considering that the people from same village are more similar to each other in their asset endowment than households of other villages. In calculating leave out village mean of a household's asset, the average of the rest of the households from the same village are taken into account dropping the household in question. Land ownership is defined as a binary variable with whether the household own more than 50 decimals of land which reflects functional land ownership in the perspective of Bangladesh (Quisumbing 2011, Hossain et al. 2007). Robustness of the results is checked by both including and excluding the asset variables in the model to minimize the simultaneity bias. Instead of using the social and political capital index, merely group participation of husbands and wives are also used to check the robustness of the results. To address possible heteroscedasticity, robust standard error is used throughout the paper. Besides, correlations among different independent variables are checked to avoid multicollinearity in the model.

Farmers' perception of climate change

It is interesting to know which types of farmers are likely to observe the climate change - an important issue to understand for practicing adaptation strategies. For this study, temperature increase and rainfall decrease are considered as the two measures of perceptions. To identify the correlates of farmers' perception of change in climate, the dependent variable is a binary variable that takes the value 1 if the head of household perceives that temperature is increasing or rainfall is decreasing from last twenty years and the value 0 otherwise.

Table 4 summarizes the marginal effects of probit regression in first two columns with asset variables calculated as leave out village mean. Results from Table 4 shows, female heads are more likely to notice rainfall decrease probably because it affects domestic water management, which is mainly women's duty in rural Bangladesh. Households which functionally own land and with physical and livestock assets especially when measured with leave out village mean, are more likely to be perceived about rainfall decrease probably because adequate rainfall is necessary for preparing land for cultivation, crop production and fodder for livestock. Gbetibouo (2009) found similar result for South Africa.

However, the perception of increasing temperature and decreasing rainfall are likely to be correlated to each other which demands seemingly unrelated biprobit model, the results of which are presented in table 5. Column (1) and (2) investigates whether farmers have perception of climate change with asset variables calculated as leave out village mean while column (3) and (4) look at the same without leave out village mean.

Results from both of the tables (table 4 and 5) are consistent in terms of signs and statistical significance for most of the variables. Important finding is that, access to information and technology, political capital by male and group participation by female are positively and significantly correlated with the perception of both temperature and rainfall change. Probably female groups more emphasize the issue or alternatively they come in contact with different people which help them to develop their outlook about perception of climate change while, male who actively take part in politics and are more concerned about citizen rights and duty have better perception that the climate is changing. Surprisingly, access to credit and social capital of male and female are negatively correlated to perception either because the groups do not emphasize on this issue, or the households which have access to credit and other groups might already manage to overcome the problem of rainfall decrease by more irrigation for example. Surprisingly, education has no effect in influencing perception probably because farmers of survey sites has on an average 3 years of schooling which is very low to give rise the notion of climate change and variability.

Table 4. Results of marginal effects of farmers' perception of change in the climate

Variable	Probit			
	Temperature increase (1)	Rainfall decrease (2)	Temperature increase (3)	Rainfall decrease (4)
Male headed household	-0.012	-0.050*	-0.011	-0.059**
	(0.040)	(0.028)	(0.043)	(0.028)
Age of HH head	0.002	0.000	0.001	0.000
	(0.001)	(0.001)	(0.001)	(0.001)
Years of schooling of HH head	0.003	-0.001	0.003	-0.001
	(0.003)	(0.003)	(0.003)	(0.003)
Experience of HH head	0.000	0.000	0.001	0.000
	(0.001)	(0.001)	(0.001)	(0.001)
Whether HH functionally own land	-0.029	0.022	-0.006	0.038*
	(0.027)	(0.023)	(0.026)	(0.024)
Physical asset index (leave out village mean)	0.283	0.472**	-	-
	(0.247)	(0.238)		
Total livestock in TLU (leave out village mean)	0.041	0.102***	-	-
	(0.037)	(0.033)		
Physical asset index	-	-	0.024	-0.047
			(0.071)	(0.070)
Total livestock in TLU	-	-	-0.022	0.002
			(0.014)	(0.014)
Access to credit	-0.048*	-0.004	-0.062**	-0.024
	(0.025)	(0.024)	(0.025)	(0.023)
Access to extension	-0.026	0.035	-0.017	0.040*
	(0.030)	(0.024)	(0.029)	(0.024)
Affected by climatic shocks	0.024	0.022	0.049*	0.025
	(0.026)	(0.024)	(0.027)	(0.024)
Access to ICT by male	0.128***	0.076**	0.163***	0.099***
	(0.037)	(0.036)	(0.039)	(0.036)
Access to ICT by female	0.104	0.078	0.020	0.071
	(0.073)	(0.073)	(0.076)	(0.073)
Group participation by male	-	-	-0.036	0.016
			(0.031)	(0.025)
Group participation by female	-	-	0.064**	0.070***
			(0.028)	(0.023)
Social capital of male (leave out village mean)	-0.514**	-0.909***	-	-
	(0.231)	(0.197)		
Social capital of female (leave out village mean)	-1.176**	-0.136	-	-
	(0.499)	(0.471)		
Political capital of male (leave out village mean)	0.954***	0.065	-	-
	(0.242)	(0.202)		
Political capital of female (leave out village mean)	-0.495	-0.597	-	-
	(0.744)	(0.612)		
Pseudo R-squared	0.084	0.084	0.061	0.050
Total observations	740	740	740	740

Source: Calculated by authorfrom The Economics of Adaptation to Climate Change in Bangladesh follow-up Survey 2012.
Notes: Robust standard errors are given in parentheses; ***p<0.01, **p<0.05, *p<0.1.

Table 5. Results of marginal effects of farmers' perception of change in the climate

Variable	Seemingly unrelated biprobit			
	Temperature increase (1)	Rainfall decrease (2)	Temperature increase (3)	Rainfall decrease (4)
Male headed household	-0.058 (0.207)	-0.352 (0.234)	-0.055 (0.217)	-0.389* (0.236)
Age of HH head	0.008 (0.006)	0.000 (0.006)	0.005 (0.005)	-0.002 (0.006)
Years of schooling of HH head	0.017 (0.016)	-0.010 (0.018)	0.016 (0.016)	-0.009 (0.018)
Experience of HH head	0.002 (0.006)	-0.003 (0.005)	0.005 (0.006)	0.000 (0.005)
Whether HH functionally own land	-0.144 (0.136)	0.135 (0.135)	-0.025 (0.127)	0.215* (0.133)
Physical asset index (leave out village mean)	1.386 (1.251)	2.748** (1.414)	-	-
Total livestock in TLU (leave out village mean)	0.197 (0.187)	0.585*** (0.197)	-	-
Physical asset index	-	-	0.106 (0.340)	-0.260 (0.390)
Total livestock in TLU	-	-	-0.102 (0.065)	0.010 (0.080)
Access to credit	-0.258* (0.144)	-0.017 (0.141)	-0.315** (0.141)	-0.132 (0.136)
Access to extension	-0.119 (0.144)	0.220 (0.154)	-0.079 (0.138)	0.247* (0.145)
Affected by climatic shocks	0.118 (0.124)	0.124 (0.130)	0.232* (0.122)	0.140 (0.124)
Access to ICT by male	0.651*** (0.190)	0.455** (0.209)	0.803*** (0.188)	0.566*** (0.199)
Access to ICT by female	0.516 (0.358)	0.469 (0.432)	0.083 (0.367)	0.397 (0.405)
Group participation by male	-	-	-0.184 (0.141)	0.077 (0.140)
Group participation by female	-	-	0.341* (0.178)	0.449** (0.188)
Social capital of male (leave out village mean)	-2.585** (1.170)	-5.316*** (1.148)	-	-
Social capital of female (leave out village mean)	-6.000** (2.529)	-0.724 (2.738)	-	-
Political capital of male (leave out village mean)	4.947*** (1.263)	0.418 (1.150)	-	-
Political capital of female (leave out village mean)	-2.412 (3.752)	-3.380 (3.475)	-	-
Wald chi2(32) 80.62***				
Total observations 740				

Source: Calculated by authorfrom The Economics of Adaptation to Climate Change in Bangladesh follow-up Survey 2012.
Note: Robust standard errors are given in parentheses; ***p<0.01, **p<0.05, *p<0.1.

CONCLUSIONS

The study explores the detail empirical picture of farmers' perception of climate change and the level of their perception in agricultural households in Bangladesh. Results find that farmers of Bangladesh especially those with assets, access to credit, extension services and ICT, greater female participation in groups and more exposed to climate change shocks; are already perceived that climate is changing.Participation in social groups by women is particularly important in enhancing their perceptions of climate change which should be encouraged by Government with appropriate policy intake. Government policies should be initiated to improve household access to extension services and access to credit and information, which would improve and diversify farmers' knowledge of climate change and perception and thereby to improve their adaptation strategies. Improving opportunities for households to generate off-farm income could provide a further strategy in response to negative shocks. Political capital of male is important for farmers' perception rather than social capital which show the importance of knowledge diversification and dissemination. The study does not consist of gender disaggregated data of adaptation and coping. Future research using panel data and disaggregated by gender will increase the understanding of existing knowledge.

ACKNOWLEDGEMENTS

This work is supported by the Federal Ministry for Economic Cooperation and Development (BMZ), Germany under the project "Enhancing Women's Assets to Manage Risk under Climate Change: Potential for Group–Based Approaches". We express our sincerest gratitude to Joachim von Braun and Julia Anna Matz for comments and to ZEF, IFPRI, and DATA for collecting the data. Furthermore, we are grateful for comments and discussions of participants at various workshops and conferences. All errors are ours.

REFERENCES

1. Adger WN, S Huq, K Brown, D Conway and M Hulme, 2003. Adaptation to climate change in the developing world, Progress in Development Studies, 3: 179–195.
2. Akanda MGR and MS Howlader, 2015. Coastal Farmers' Perception of Climate Change Effects on Agriculture at Galachipaupazila under Patuakhali District of Bangladesh, Global Journal of Science Frontier Research: D, Agriculture and Veterinary, 15: Version 1.0.
3. Akter T, 2009. Climate Change and Flow of Environmental Displacement in Bangladesh, Unnayan Onneshan-The Innovators, www.unnayan.org
4. BBS, 2013. Bangladesh Bureau of Statistics, Statistics Division, Ministry of Planning, Government of the People's Republic of Bangladesh.
5. Bangladesh MoEF (Ministry of Environment and Forest), 2005. Bangladesh National Adaptation Program of Action (NAPA), Dhaka, Bangladesh.
6. BCAS (Bangladesh Center for Advanced Studies), 2009. Policy Study on the Probable Impacts of Climate Change on Poverty and Economic Growth and the Options of Coping with Adverse Effect of Climate Change in Bangladesh. Dhaka, Bangladesh: General Economics Division, Planning Commission; United Nations Development Program, Bangladesh.
7. Davis P and S Ali, 2014.Exploring Local Perceptions of Climate-Change Impact and Adaptation in Rural Bangladesh, IFPRI discussion paper 01322, January.
8. Deressa TT, C Ringler and RM Hassan, 2010. Factors Affecting the Choices of Coping Strategies for Climate Extremes: The Case of Farmers in the Nile Basin of Ethiopia, IFPRI Discussion Paper 01032, November, International Food Policy Research Institute, Washington, DC, USA.
9. Fankhauser S, 1998. The Costs of Adapting to Climate Change, GEF Working paper 16, Global Environment Facility, Washington, DC, USA.
10. Gbetibouo GA, 2009. Understanding farmers' perceptions and adaptations to climate change and variability: the case of the Limpopo Basin, South Africa. IFPRI Discussion Paper, 00849, International Food Policy Research Institute (IFPRI), Washington, DC, February.

11. Hossain M, D Lewis,M LBose and A Chowdhury, 2007. Rice Research, Technological Progress, and Poverty: The Bangladesh Case. In Agricultural Research, Livelihoods, and Poverty: Studies of Economic and Social Impacts in Six Countries, edited by M. Adato and R. Meinzen-Dick. Baltimore: Johns Hopkins University Press.
12. IPCC (Intergovernmental Panel on Climate Change), 2001. Climate Change 2001.Overview of Impacts, Adaptation, andVulnerability to Climate Change. IPCC Third Assessment Report, Cambridge University Press. Downloaded from www.ipcc.ch.
13. Kamruzzaman M, 2015. Farmers' Perceptions on Climate Change: A Step toward Climate Change Adaptation in Sylhet Hilly Region, Universal Journal of Agricultural Research 3: 53-58.
14. Labor Force Survey, 2010. Bangladesh Bureau of Statistics, Statistics Division, Ministry of Planning, Government of the People's Republic of Bangladesh.
15. Maddison D, 2006. The perception of an adaptation to climate change in Africa. World Bank Policy Research Working Paper, 4308. The World Bank, Washington, DC.
16. Maddox GL, 1995. The Encyclopedia of Aginy.2nd edn. New York: Springer Publishing Company,Inc.
17. Moniruzzaman M, 2013. People's Perception on Climate Change and Variability: A Study of Sabrang Union, Teknaf, Cox's bazar, Bangladesh, ASA University Review 7 (2), July–December.
18. Nhemachena, C., R. M. Hassan, 2007. Micro-level analysis of farmers' adaptation to climate change in Southern Africa. IFPRI Discussion Paper, 714. International Food Policy Research Institute (IFPRI), Washington, DC: 30.
19. Quisumbing A, 2011. Do Men and Women Accumulate Assets in Different Ways? Evidence from Rural Bangladesh. International Food Policy Research Institute (IFPRI) Discussion Paper No. 1096.
20. Rakib MA, M A Rahman, M S Akterand, MAH Bhuiyan, 2014. Climate change: Farmers Perception and Agricultural Activities. Herald Journal of Geography and Regional Planning, 3: 115 - 123 June.
21. Rashid MH, S Afroz, D Gaydon, A Muttaleb, P Poulton, C Rothand and Z Abedin, 2014. Climate Change Perception and Adaptation Options for Agriculture in Southern Khulna of Bangladesh, Applied Ecology and Environmental Sciences, 2: 25-31.
22. Santos I, I Sharif, HZ Rahman and H Zaman, 2011. How do the poor cope with shocks in Bangladesh? Evidence from survey data. Policy Research Working Paper 5810. The World Bank, South Asia Region, Social Protection Unit, September.
23. Seal L and MA Baten, 2011. Reckoning Climate Change: Local Peoples' Perception on the Impacts of Climate Change in South-Central and Northern Bangladesh, The Unnayan Onneshan – The Innovators, Annual Report.
24. Syeda J. A. and M. Nasser2012. Farmers' Perception Regarding Climate Change and Crop Production, Especially for Wheat in Dinajpur District, Journal of Environmental Science & Natural Resources, 5: 129 – 136.
25. Thomas TS, K Mainuddin , C Chiang, A Rahman, A Haque, N Islam, S Quasem and Y Sun, 2013. Agriculture and Adaptation in Bangladesh: Current and Projected Impacts of Climate Change. IFPRI Discussion Paper 01281, International Food Policy Research Institute, Washington, DC.
26. Yu WH, M Alam, A Hassan, A S.Khan, A C.Ruane,C Rosenzweig, DC Major and J Thurlow, 2010. Bangladesh - Climate change risks and food security in Bangladesh. World Bank, Washington DC.

9

EFFECT OF POLLUTED RIVER WATER ON GROWTH, YIELD AND HEAVY METAL ACCUMULATION OF RED AMARANTH

KM Mohiuddin*, Md. Mehediul Alam, Md. Shahinur Rahman, Md. Shafiqul Islam and Istiaq Ahmed

Department of Agricultural Chemistry, Faculty of Agriculture, Bangladesh Agricultural University, Mymensingh-2202, Bangladesh

***Corresponding author:** KM Mohiuddin; E-mail: mohiagchem@gmail.com

ARTICLE INFO **ABSTRACT**

Key words
Heavy metal,
Vegetable,
River water,
Irrigation

The present study was carried out to assess the levels of different heavy metals like chromium (Cr), lead (Pb), cadmium (Cd), iron (Fe), copper (Cu), zinc (Zn) and manganese (Mn) on red amaranth vegetable irrigated with polluted river water. Atomic Absorption Spectrometer was used for analyzing the heavy metals in the samples. The results indicated a substantial build-up of heavy metals accumulation in red amaranth irrigated with polluted river water. The ranges of various metals in red amaranth samples irrigated with polluted river water were 0.45–0.93, 0.147–0.175, 42.33–479.73, 1.31–12.04, 3.71–35.11 and 10.9–142.9 μg g^{-1} for Cr, Pb, Fe, Cu, Zn and Mn, respectively. Cadmium concentration was below the detection limit (0.01 μg g^{-1}) of the method used in the analysis. However, the regular monitoring of levels of these metals from effluents and sewage, in vegetables and in other food materials is essential to prevent excessive build-up of these metals in the food chain. In general, our results indicated that using polluted river water had no significant variation in growth and yield of red amaranth from the crops irrigated with fresh water.

INTRODUCTION

River water is becoming contaminated for the disposal of sewages and municipal waste, medical, dying, tannery waste and from free flowing of pesticide and fertilizer and most of these effluents contain various hazardous metals like As, Cr, Cd, Pb, Cu, Ni, Fe, Zn, Mn, Hg etc which offer potential threats for both human being and environment even at trace levels. Heavy metals are very harmful because of their non-biodegradable nature, long biological half-lives and their potential to accumulate in different body parts. Most of the heavy metals are extremely toxic because of their solubility in water. Now-a-days heavy metals are ubiquitous because of their excessive use in industrial applications. Wastewater contains substantial amounts of toxic heavy metals, which create problems (Chen, Wang, & Wang, 2005; Singh, Mohan, Sinha, & Dalwani, 2004). Excessive accumulation of heavy metals in agricultural soils through wastewater irrigation may not only result in soil contamination, but also affect food quality and safety (Muchuweti *et al.,* 2006). Heavy metals are easily accumulated in the edible parts of leafy vegetables, as compared to grain or fruit crops (Mapanda, Mangwayana, Nyamangara, & Giller, 2005). Vegetables take up heavy metals and accumulate them in their edible (Bahemuka & Mubofu, 1991) and non-edible parts in quantities high enough to cause clinical problems both to animals and human beings consuming these metal-rich plants (Alam, Snow, & Tanaka, 2003). A number of serious health problems can develop as a result of excessive uptake of dietary heavy metals.

Rivers around Dhaka city, Bangladesh are economically very important. River water is used for irrigation during dry season. Not only in Dhaka city, irrigation of crops with river water though it is contaminated, is a very common practice in Bangladesh due to its easy availability and scarcity of fresh water. Hundreds of industries had already been established along the rivers and are frequently used for dumping of industrial waste. As a result, rivers in Dhaka city and it's surrounding are under serious threat of toxic heavy metal contamination (Alam, 2008). Thus, when those rivers water used as irrigation purposes, it gets an easy way to alter the soil physical and chemical properties which may offer a potential risk for agricultural activities and become harmful for growth and yield of different agricultural crops especially on leafy vegetables. Vegetables constitute an important part of the human diet. Some vegetables are very rich with carbohydrates, proteins, vitamins and minerals. But, recently different heavy metals have been found on the surface and in the tissue of fresh vegetables with a mentionable concentration (Bigdeli and Seilsepour, 2008). Some experimental report suggested that, in addition to human health hazard about 6 μg mL^{-1} of Cu, Hg and Ni in water also seriously decrease germination of seed, shoot length, root length, leaf area, fresh and dry mass of leaves of cowpea (Joshi *et al.,* 1999).

The previous studies on the river water were focused on the river water chemistry and physicochemical properties in the river water (e.g. Ali *et al.,* 2008; Moniruzzaman *et al.,* 2009) and some studies on seasonal and spatial distribution of heavy metals (Alam, 2003; Mohiuddin *et al.,* 2011). But, no detailed study on growth, yield and quality of different leafy vegetables has so far been conducted. The present study was conducted to observe the concentration of accumulated metals and also effects of irrigation with polluted river water on the growth, yield and quality of winter vegetable such as Red amaranth *(Amaranthus tricolor* L.).

MATERIALS AND METHODS

Collection and preparation of water sample

About 16 river water samples were collected from 16 different sites around the Dhaka city (Table 1) following the method as described by APHA (2005). About 20 L water samples from each location were collected in plastic container for irritation. About 100 mL of water samples from each location were collected separately for chemical analysis, in which 0.5 mL conc. nitric acid were added to avoid any microbial growth. The chemical analyses of water were done as quickly as possible on arrival at the Laboratory of Agricultural Chemistry, BAU, Mymensingh-2202.

Test crop used for the study

Red amaranth (*Amaranthus tricolor L.*) was used as test crop to determine the effects of chemical constituents of river water on the growth, yield and quality of a winter vegetable. The seeds of Red amaranth were collected from the Bangladesh Agricultural Development Corporation (BADC), Mymensingh before 5 days of planting. The fresh soils were thoroughly mixed and air dried separately in a clean room to avoid contamination and ground to pass through 2 mm sieve. 10 kg of soil was taken into each of the plastic pot labeled earlier. Individual pot size was 30 cm diameter at the top and 20 cm diameter at the bottom with 25 cm depth. The top surface area of each pot was 0.07071 m^2. All the pots were filled with processed soil leaving 5 cm from the top. The seeds were sown by hand into the pot at 24 February, 2014 keeping uniform distance as far as possible and then the seeds were covered by soils. The seeds were germinated uniformly after 6 to 8 days of sowing. The seedlings were irrigated with the water samples collected from 16 different sampling sites around the Dhaka city. Irrigation with distilled water was considered as control treatment. The experimental design was randomized complete block design (RCBD) with three replications

Table 1. Name of the locations of different sampling sites of the river around the Dhaka city, Bangladesh

Sample No.	Location
1	Tongi Bazar Boat Mooring
2	Sluice Gate Area
3	Noyanichala Bailey Bridge
4	Asulia Highway Bridge
5	Uttara Sector 16
6	Sinnirtak
7	Gabtoli
8	Aminbazar
9	Hazaribag – ZH Sikder MC
10	Nobabgonj Bara Masjid
11	Shohid Nagar Beribadh
12	Kellarmor Truck Stand
13	Raghunathpur
14	Swarighat
15	Nowab Barir Ghat
16	Mererbag

Data collection

Plant height (cm), shoot length (cm), root length (cm), no. of leaves plant^{-1}, total no. of plant pot^{-1}, fresh and dry weight of plant were recorded and their mean values were calculated from the sample plants after harvesting. The plant was harvested after 50 days of sowing on 14 April, 2014.

Digestion of the vegetable samples

All the collected vegetable samples were thoroughly washed with double distilled water to get rid of air born pollutants. All the samples were then oven-dried in a hot air oven at 70–80 °C for 24 h, to remove all moisture. Dried samples were powdered using a pestle and mortar and sieved. All reagents were of analytical reagent grade, 69–72% HNO_3, 30% H_2O_2, and 70% $HClO_4$ were used for digestion of samples. Double deionized water was used for all dilutions. During the experiments, all glasswares and equipment were carefully cleaned starting with 2% HNO_3 and ending with repeated rinsing distilled deionized water to prevent contamination. Samples (0.6 g) were digested with 10 mL of HNO_3, 2 mL of H_2O_2, and 4 mL of $HClO_4$ in block digester system maintained temperature 180-200°C until white fumes are evolved and finally diluted to 50 mL with 2% nitric acid. All sample solutions were clear. A blank digest was carried out in the same way.

Determination of metals

Titration using 0.02N EDTA was applied to analyze Ca and Mg (Page *et al.,* 1982). Heavy metals (Cr, Pb, Cd, Fe, Cu, Zn and Mn) concentration in the extract were determined by atomic absorption spectrophotometer (SHIMADZU AA-7000), using a deuterium background correction. Sodium and potassium were determined by flame emission spectrophotometer (JENWAY, PFP7).

Determination of protein

The protein content of the samples was estimated by the micro Kjeldhal method, in which the sample was digested with a known quantity of concentrated sulphuric acid in the block digester. The digested material was distilled after the addition of alkali. The released ammonia was collected in 4% boric acid in the distillation unit. The resultant boric acid, now contained the ammonia released from the digested material, was then titrated against 0.1 N HCl, manually. The nitrogen content thus determined was multiplied by a factor of 6.25 to arrive at the amount of protein.

Statistical analysis

Data obtained were subjected to statistical analysis by using the software package SPSS 12 (SPSS Inc., Chicago, IL, USA). The statistical tests performed were analysis of variance (ANOVA; P = 0.05) and Duncan's multiple-range test (P = 0.05).

RESULT

The analysis of river water is given in Table 1 and 2. The application of polluted river water generally led to changes in the physicochemical characteristics of soil and consequently heavy metal uptake by red amaranth. The table reports a relative enrichment of that water by Cr, Fe, Cu, Zn and Mn,

Table 2. Concentration of major nutrients in water samples collected from different sites of the river around the Dhaka city, Bangladesh

Sample no.	**Ca (μg mL^{-1})**	**Mg (μg mL^{-1})**	**Na (μg mL^{-1})**	**K (μg mL^{-1})**
1	0.64	1.07	29.51	135.8
2	0.32	0.68	21.14	132.7
3	0.48	0.58	20.30	282.6
4	0.64	0.68	21.98	263.7
5	0.48	0.78	21.98	62.4
6	0.64	0.68	29.51	172.1
7	0.32	0.87	14.44	165.0
8	0.48	1.17	18.63	153.2
9	0.80	0.97	30.3	80.6
10	0.64	0.78	38.7	142.1
11	0.96	1.07	28.7	61.6
12	0.96	0.87	35.4	101.9
13	0.96	0.87	22.8	135.8
14	0.80	0.58	27.8	151.6
15	0.64	1.07	20.3	152.4
16	0.96	1.17	17.0	146.1
Mean	0.67	0.86	24.9	146.2
Range	**0.32-0.96**	**0.58-1.17**	**14.44-38.7**	**61.6-282.6**

Effect of polluted river water on growth and yield of red amaranth

Plant growth parameters such as plant height (cm), shoot length (cm), root length (cm), no. of leaves plant^{-1}, total no. of plant pot^{-1}, fresh and dry weight of plant were not significantly affected by the river water application (Table 4). Total no. of plants among the all treatments varied from 16 to 45 with an average of 30. On the other hand, the maximum no. of plant (58.0) was observed in pot where normal water used for irrigation. These observed results might be due to use of river water for irrigation which contaminated with different heavy metals. The plant height of the red amaranth were ranging from 10.7 to 24 cm and the average height of the control plants were 22.2 cm and average height of the 16 treatments were 18.6 cm. Root and shoot length of red amaranth were decreased due to application of polluted river water ranging from 1.9 to 5.2 cm 8.8 to 18.8 cm, respectively. The average root and shoot length of 16 treatments were 3.6 cm and 14.7 cm, respectively but 4.72 cm and 17.48 cm were in control treatment, respectively. The average leaf no. of five plants per pot ranged from 4 to 9.4 with an average of 6.8.The average no. of leaf in control of five plants per pot was 7.6.Fresh and dry weights of 10 plants per pot were observed among all treatments ranging from 5.60 to 18.40 g and 0.54 to 1.50 g, respectively with an average of 10.11 and 1.01 g, respectively and 12.20 g and 1.50 g, respectively, in control. Similar result was also observed in tomato and cucumber plants (Qaryouti *et al.*, 2015).Subsequently, these authors suggested that the reclaimed river polluted water can replace fresh water in irrigation of vegetables during dry season but the water quality is continuously monitored to avoid heavy metal accumulation and microbial contamination.

Nutritive quality of red amaranth

Protein and total nutrient uptake of Ca, Mg, K and Na are summarized in Table 5 and were not significantly affected by the river water application except Na. The maximum protein content was (13%) observed in sample no. 15 and the minimum value of N was (6.5 %) in treatment no. 10. Mean value of protein content was found 9.0 % of 16 treatments. Control treatment contains (9.8%) in which normal water was used. The content of potassium varied from 1.1% to 2.6%. The maximum value of K content was (2.63%) observed in control treatment and minimum (1.1%) in treatment no. 14. The average content of K was 2.0 %. Aykroyd (1963) observed that K in red amaranth was 341 mg l00g^{-1} in edible portion. The content of Ca varied from 0.3% to 1.0%. The maximum value of Ca content was (1.0%) observed in sample no. 1, 5, 12, and 15; and minimum (0.3%) was observed in sample no. 2 and 7. The average content of Ca was (0.7%) but in control it was 0.96%. Begum (2006) reported that the Ca content of red amaranth was 374 mg 100g^{-1} of the edible portion. The content of Mg was ranging from 0.1% to 0.27%. The maximum and minimum value of Mg was observed 0.27% and 0.1%, respectively with the average value of 0.21% and control treatment contains 0.17% Mg. Aykroyd (1963) observed that Mg in red amaranth was 247 mg 100g^{-1} of the edible portion. Statistical analysis showed that Na content was increased significantly in all treatments except sample no. 2, 6, 7, 8 and 11 as compared with control treatment. The highest (182.2 ppm) Na was found in sample no. 3 and the lowest (69.3 ppm) in sample no. 2 and 8. The average content of Na was observed 126.7 ppm and in control treatment where 81.64 ppm Na was found.

Table 3. Concentration of heavy metals in water samples collected from different sites of the river around the Dhaka city, Bangladesh

Sample no.	Heavy metal concentration (µg mL^{-1})						
	Cr	**Pb**	**Cd**	**Fe**	**Cu**	**Zn**	**Mn**
1	0.022	BDL	BDL	0.832	0.135	0.243	0.125
2	0.014	BDL	BDL	0.921	0.172	0.262	0.201
3	0.031	BDL	BDL	0.812	0.211	0.252	0.163
4	0.123	BDL	BDL	0.235	0.166	0.392	0.017
5	0.089	BDL	BDL	0.364	0.182	0.390	0.066
6	0.101	BDL	BDL	0.275	0.183	0.278	0.012
7	0.173	BDL	BDL	0.285	0.189	0.204	0.109
8	0.166	BDL	BDL	0.492	0.215	0.334	0.103
9	0.135	BDL	BDL	1.446	0.203	0.243	0.283
10	0.142	BDL	BDL	0.912	0.209	0.591	0.167
11	0.139	BDL	BDL	1.312	0.217	0.821	0.241
12	0.152	BDL	BDL	0.605	0.219	0.575	0.229
13	0.169	BDL	BDL	0.639	0.215	0.399	0.192
14	0.163	BDL	BDL	0.372	0.220	0.572	0.171
15	0.192	0.025	BDL	0.321	0.217	0.483	0.165
16	0.179	0.082	BDL	0.373	0.242	0.342	0.132
Range	0.014-0.192	BDL-0.082	-	0.235-1.446	0.135-0.242	0.204-0.821	0.012-0.283
Mean	0.124	0.0535	BDL	0.637	0.200	0.399	0.149
DWGV*		**2[a]**		**0.3[b]**	**0.05[a]**		**0.4[a]**
IWGV*		**0.011[d]**		**0.75[c]**	**0.2[c]**		**2[c]**

BDL: Below Detectable Limit; DWGV: Drinking Water Guideline Value; IWGV: Irrigation Water Guideline Value
*a - WHO (2008); b - USEPA (2009); c - Ayers and Westcot (1985); d- ADB (1994)

Table 4. Effect of Waste Water on Morphological and Physiological attributes of red amaranth

Sample no.	Total No. of plant	Average plant height (cm) of 05 plant per pot	Average shoot length (cm) of 05 plant per pot	Average root length (cm) of 05 plant per pot	Average of 05 plant leaf No. per pot	Fresh weight (g) of 10 plant per pot	Dry weight (g) of 10 plant per pot	Moisture content (%) of 10 plant
1	45.0	24.0	18.8	5.2	9.4	13.10	1.40	89.3
2	17.0	13.6	11.2	2.4	6.2	8.18	1.21	85.2
3	52.0	21.4	16.9	4.5	8.6	10.10	1.32	86.9
4	26.0	18.2	14.6	3.6	6.9	8.50	0.89	89.5
5	41.0	18.8	15.8	3.0	7.2	11.10	1.00	91.0
6	43.0	18.9	15.1	3.8	7.2	13.00	1.10	91.5
7	24.0	10.7	8.8	1.9	4.7	7.43	0.60	91.9
8	31.0	21.8	17.9	3.9	6.4	11.60	1.25	89.2
9	32.0	18.4	14.8	3.6	7.0	10.80	1.13	89.5
10	29.0	21.0	17.1	3.9	6.8	7.30	1.00	86.3
11	26.0	21.2	16.4	4.8	7.4	18.40	1.50	91.8
12	16.0	18.0	14.3	3.7	6.4	5.60	0.54	90.4
13	31.0	18.8	12.3	2.0	4.0	8.51	0.81	90.5
14	29.0	23.6	18.0	5.6	8.2	12.40	1.40	88.7
15	21.0	14.9	12.5	2.4	6.8	7.35	0.54	92.7
16	17.0	13.8	11.1	2.7	6.4	8.31	0.41	95.1
Average of (16) sample	**30.0**	**18.6**	**14.7**	**3.6**	**6.8**	**10.11**	**1.01**	**90.0**
Range	**16.0-45.0**	**10.7-24**	**8.8-18.8**	**1.9-5.2**	**4-9.4**	**5.60-18.40**	**0.54-1.50**	**85.2-91.9**
Control	58.0	22.20	17.48	4.72	7.6	12.20	1.50	87.7

Table 5. Concentration of protein and major nutrients in red amaranth

Sample No.	Protein (%)	Ca (%)	Mg (%)	K (%)	Na (ppm)
1	8.2	1.0	0.17	2.6	170.4
2	9.2	0.3	0.23	2.6	69.3
3	8.0	0.5	0.17	2.4	182.8
4	9.0	0.7	0.20	2.1	124.0
5	10.2	1.0	0.25	2.0	131.0
6	8.2	0.6	0.25	2.0	71.8
7	9.3	0.3	0.12	2.6	81.6
8	8.7	0.8	0.10	2.6	69.3
9	8.8	0.5	0.27	1.6	145.8
10	6.5	0.6	0.17	2.0	145.8
11	8.0	0.6	0.10	2.1	98.9
12	10.5	1.0	0.23	1.6	150.7
13	7.8	0.8	0.27	1.6	135.9
14	8.2	0.8	0.21	1.1	131.0
15	13.0	1.0	0.25	1.3	170.4
16	10.2	0.6	0.27	2.4	148.2
Average of (16) sample	**9.0**	**0.70**	**0.21**	**2.0**	**126.7**
Range	**6.5-13.0**	**0.3-1.0**	**0.10-0.27**	**1.1-2.6**	**69.3-182.8**
Control	9.8	0.96	0.17	2.63	81.64

Heavy metal accumulation in red amaranth

The application of polluted river water generally led to changes in the physicochemical characteristics of soil and consequently heavy metal uptake by vegetable. The heavy metals concentrations in edible part of red amaranth are shown in Fig. 1. It can be clearly observed that the concentration of all the heavy metals is higher in polluted river water-irrigated than freshwater-irrigated except Pb and Zn. Heavy metals concentration was in the order of Fe > Mn > Zn > Cu > Cr > Pb > Cd in the polluted river water irrigated vegetable.

Table 6. Concentration of heavy metal (μg g^{-1}) in red amaranth samples

No. of Sample	Concentration (μg g^{-1})						
	Cr	Pb	Cd	Fe	Cu	Zn	Mn
1	0.519	0.0147	BDL	139.99	9.47	11.60	84.0
2	0.448	0.0161	BDL	81.92	1.31	5.24	16.7
3	0.489	0.0153	BDL	272.32	6.29	9.04	93.5
4	0.532	0.0159	BDL	245.66	7.83	3.81	89.6
5	0.485	0.0156	BDL	176.53	7.01	9.48	86.7
6	0.487	0.0150	BDL	128.16	8.75	9.76	70.6
7	0.769	0.0170	BDL	42.33	7.11	3.71	10.9
8	0.837	0.0152	BDL	142.11	8.55	8.98	72.0
9	0.918	0.0153	BDL	298.50	11.32	10.26	104.1
10	0.899	0.0159	BDL	368.92	6.08	10.45	114.1
11	0.899	0.0160	BDL	199.91	12.65	8.95	78.2
12	0.907	0.0159	BDL	388.85	12.04	9.45	140.5
13	0.893	0.0175	BDL	344.87	9.47	9.17	103.0
14	0.933	0.0164	BDL	341.81	11.93	8.64	124.7
15	0.880	0.0166	BDL	326.67	12.04	7.61	117.4
16	0.933	0.0163	BDL	479.73	6.08	35.11	142.9
Average	**0.74**	**0.02**	BDL	**248.64**	**8.62**	**10.08**	**90.56**
Range	**0.448-0.933**	**0.147-0.175**		**42.33-479.73**	**1.31-12.04**	**3.71-35.11**	**10.9-142.9**
Control	0.10	0.0159	BDL	197.92	1.92	13.63	74.51

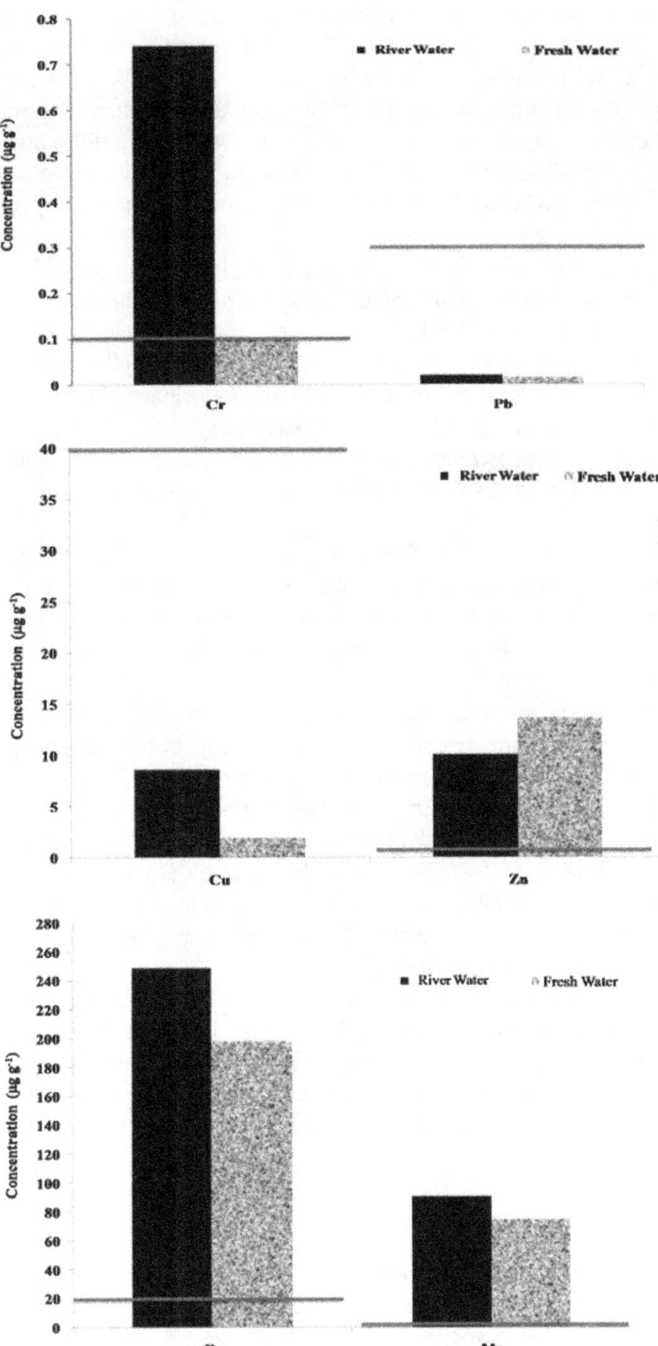

Figure 2. Heavy metal concentration in different red amaranth samples irrigated with contaminated river water and fresh water [Red line showing the standard acceptable limit for consumption proposed by FAO/WHO]

The concentration of Cr in red amaranth samples ranged from 0.45 to 0.93 µg $g^{-1}$with a mean value of 0.74µg g^{-1} (Table 6) of 16 samples and control treatment contains 0.10 µg g^{-1} Cr which is much below with the previous report, where the concentration of this metal was found to be 6.4 µg g^{-1} (Salvatore, Carratù, and Carafa, 2009). However this value is higher than 0.1 µg g^{-1} limit set by FAO/WHO (1993) and SEPA (2005), and 0.002 µg g^{-1} by Bose and Bhattacharyya (2008).

The level of Pb in red amaranth ranged from 0.0147 to 0.0175 µg g^{-1}, having an average value of 0.02 µg g^{-1} (Table 6) of 16 treatments. Control treatment contains 0.02 µg g^{-1} Pb. The permissible limit in plants recommended by FAO/WHO (1993) is 0.3 µg g^{-1} and SEPA (2005) limit is 9.0 µg g^{-1} but in all the plant were below these limit. Lead is a serious cumulative body position which enters into the body system through air, water and food and cannot be removed by washing fruits and vegetables (Kumar Sharma *et al.,* 2007).

The permissible limit of Cd in vegetables, recommended by FAO/WHO (1993) and SEPA (2005), is 0.02 µg g^{-1} and our values of Cd in all samples were below 0.01 µg g^{-1} (below detectable limit by flame AAS method). Cadmium is highly toxic metal not known to have any beneficial effects for plants and animals. Many Cd compounds are also believed to be carcinogenic.

From the results it was observed that the concentration of Fe in red amaranth samples varied from 42.33 to 479.73 µg g^{-1} having an average value of 248.64 µg g^{-1} (Table 6) and this value is greater than the concentration of Fe in control treatment which is 197.92 µg g^{-1} and it is much higher than the recommended value by FAO/WHO (20 µg g^{-1}). Excess amount of Fe (more than 10 mg kg^{-1}) causes rapid increase in pulse rate and coagulation of blood in blood vessels, hypertension and drowsiness. Fe toxicity in plants occurs when they accumulate greater than 300 µg g^{-1} of Fe, at soil pH less than 5.0 (Li, Wang, Gou, Su, & Wang, 2006). However, high concentration of iron reported from all vegetables analyzed, coupled with its lowest percentage loss from plants can be attributed to its role in chlorophyll synthesis in plants in addition to its relative abundance in the earth crust.

The concentration of Cu in red amaranth ranged from 1.31 to 12.04 µg g^{-1}, having an average value of 8.62 µg g^{-1} (Table 6). Out of 16 plant samples, 04 samples exceed the mean value. In control treatment where normal water was used for irrigation purpose contain 1.92 µg g^{-1} Cu which was less than the mean value. The permissible limit of copper for plants is 40 µg g^{-1} recommended by FAO/WHO. All plant samples concentration was recorded below the permissible limit.

Among all the heavy metals, Zn is the least toxic and an essential element in human diet as it is required to maintain the functioning of the immune system. Zinc deficiency in diet may be highly detrimental to human health than too much Zn in diet. The recommended dietary allowance for Zn is 15 mg/day for men and 12 mg/day for women but high concentration of Zn in vegetables may cause vomiting, renal damage, cramps (Alexander, Alloway, & Dourado, 2006). According to this study, the level of Zn in vegetable was found in the range of 3.71 to 35.11 µg g^{-1}, having an average value of 10.08 µg g^{-1} (Table 6), which is in close agreement to 22.7 µg g^{-1} reported by Zakir *et al.* (2009). The value of zinc found in this study is far higher than 0.3 µg g^{-1} (FAO/WHO, 1993). The value of Zn in control treatment was 13.63 µg g^{-1}(Table 6). However, Zn concentration was substantially lower than the SEPA limit (100 µg g^{-1}).

The concentration of Mn in vegetable samples varied from 10.9 to 142.9 µg g^{-1} with a mean value of 90.56 µg g^{-1} (Table 6). Control treatment contains74.51µg g^{-1}which was less than from mean value. This value is higher than 0.2 µg g^{-1} (Pennington *et al.,* 1995) which is a safe limit according to WHO/EU (1990). Sridhara Chary, Kamala, and Raj (2008) reported the accumulation of Mn in *Allium cepa* to a level of 5.39 µg g^{-1}. High concentration of Mn causes hazardous effects on lungs and brains of humans.

CONCLUSIONS

It may be concluded that irrigation by untreated river water are the main reasons for accumulation of heavy metals in vegetables around the Dhaka city. In addition, crop parameters were not significantly influenced by river water with the replacement of fresh water. Hence, it is imperative to treat sewage water and industrial effluents before their discharge into water bodies. Awareness should be created among the farmers of the area regarding the serious consequences of using polluted river water for irrigation purpose. Therefore, continuous monitoring of wastewater irrigated crops is needed to control the dangerous accumulation of various toxic metals in plant parts and to prevent the possible health hazards due to the consumption of toxic metal contaminated agricultural products. Further research is needed on the accumulation of heavy metals in different vegetable crops cultivated in different types of soils irrigated with wastewater and their health risk after consumption.

REFERENCES

1. ADB (Asian Development Bank), 1994. Training Manual for Environmental Monitoring. USA: Engineering Science Inc., 2–16.
2. Alam AMS, MA Islam, MA Rahman, MN Siddique, MA Matin, 2003. Comparative study of the toxic metals and non-metal status in the major river system of Bangladesh. Dhaka University Journal of Science, 51: 201–208.
3. Alam K, 2008. Cost–Benefit analysis of restoring buriganga river, Bangladesh. Water Resources Development, 24: 593–607.
4. Alam MGM, ET Snow and A Tanaka, 2003. Arsenic and heavy metal contamination of vegetables grown in Santa village, Bangladesh. Science of the Total Environment, 308: 83–96.
5. Alexander PD, BJ Alloway and AM Dourado, 2006. Genotypic variations in the accumulation of Cd, Cu, Pb and Zn exhibited by six commonly grown vegetables. Environmental Pollution, 144: 736–745.
6. Ali MY, MN Amin and K Alam, 2008. Ecological Health Risk of Buriganga River, Dhaka, Bangladesh. Hydro Nepal: Journal of Water, Energy and Environment, 3: 1–4.
7. APHA (American Public Health Association), 2005. Standard Methods for the Examination of Water and Wastewater. 21st Edition. AWWA and WEF, Washington, USA.
8. Ayers RS and DW Westcot, 1985. Water Quality for Agriculture. FAO Irrigation and Drainage Paper, 29: 8–96.
9. Aykroyd WR, 1963. ICMR Special Report. Series. No.42.
10. Bahemuka TE and EB Mubofu, 1991. Heavy metals in edible green vegetables grown along the sites of the Sinza and Msimbazi rivers in Dar-es-Salaam, Tanzania. Food Chemistry, 66: 63–66.
11. Begum RA, 2006. Assessment of water and soil pollution and its effect on rice and red amaranth. Ph.D. Thesis, Department of Agricultural Chemistry, Bangladesh Agricultural University, Mymensingh.
12. Bigdeli M and M Seilsepour, 2008. Investigation of metals accumulation in some vegetables irrigated with waste water in Shahre Rey-Iran and toxicological implications. American - Eurasian Journal of Agricultural & Environmental Sciences, 4: 86–92.
13. Bose S and AK Bhattacharyya, 2008. Heavy metal accumulation in wheat plant grown in soil amended with industrial sludge. Chemosphere, 70: 1264–1272.
14. Chen Y, C Wang and Z Wang, 2005. Residues and source identification of persistent organic pollutants in farmland soils irrigated by effluents from biological treatment plants. Environment International, 31: 778–783.
15. FAO/WHO, 1993. Food Additives and Contaminants. Joints FAO/WHO Food Standard Programme ALINORM 01/12A.
16. Joshi UN, SS Rathore and SK Arora, 1999.Effect of effluents on growth and development of cow pea (*Vigna unguiculata* L.). Indian Journal of environmental Research,19: 157–162.
17. Kumar SR, M Agrawal and F Marshall, 2007. Heavy metal contamination of soil and vegetables in suburban areas of Varanasi, India. Ecotoxicology and Environmental Safety, 66: 258–266.
18. Li Y, Y Wang, X Gou,Y Su, and G Wang, 2006. Risk assessment of heavy metals in soils and vegetables around non-ferrous metals mining and smelting sites, Baiyin, China. Journal of Environmental Sciences, 18: 1124–1134.
19. Mapanda F, EN Mangwayana, J Nyamangara and KE Giller, 2005. The effects of long-term irrigation using water on heavy metal contents of soils under vegetables. Agriculture, Ecosystem and Environment, 107: 151–156.
20. Mohiuddin KM, Y Ogawa, HM Zakir,K Otomo and N Shikazono, 2011. Heavy metals contamination in the water and sediments of an urban river in a developing country. International Journal of Environmental Science and Technology, 8: 723–736.

21. Moniruzzaman M, SF Elahi andMAA Jahangir, 2009. Study on Temporal Variation of Physico-chemical Parameters of Buriganga River Water through GIS (Geographical Information System) Technology. Bangladesh Journal of Science and Industrial Research, 44: 327–334.
22. Muchuweti M, JW Birkett, E Chinyanga, R Zvauya, MD Scrimshaw and JN Lester, 2006. Heavy metal content of vegetables irrigated with mixture of wastewater and sewage sludge in Zimbabwe: implications for human health. Agriculture, Ecosystem and Environment, 112: 41–48.
23. Page AL, RH Millerand DR Kenny, 1982. Methods of Soil Analysis. Part II. Second edition, American Society of Agronomy, Inc. Madison, Wisconsis, USA.
24. Pennington JAT, SA Schoen, GD Salmon, B Young, RD JohnsonandRW Marts, 1995. Composition of core foods of the US food supply, 1982–1991 III. Copper, manganese, selenium, and iodine. Journal of Food Composition and Analysis, 8: 171–217.
25. Qaryouti M, N Bani-Hani, TM Abu-Sharar, I Shnikat, M Hiari andM Radiadeh, 2015. Effect of using raw waste water from food industry on soil fertility, cucumber and tomato growth, yield and fruit quality. Scientia Horticulturae, 193: 99–104
26. Salvatore M, G Carratù and AM Carafa, 2009. Assessment of heavy metals transfer from a moderately polluted soil into the edible parts of vegetables. Journal of Food, Agriculture and Environment, 7: 683–688.
27. SEPA, 2005. The Limits of Pollutants in Food. State Environmental Protection Administration, China.
28. Singh KP, D Mohan, S Sinha and R Dalwani, 2004. Impact assessment of treated/ untreated wastewater toxicants discharged by sewage treatment plants on health, agricultural, and environmental quality in the wastewater disposal area. Chemosphere, 55: 227–255.
29. Sridhara CN, CT Kamala and SSD Raj, 2008. Assessing risk of heavy metals from consuming food grown on sewage irrigated soils and food chain transfer. Ecotoxicology and Environmental Safety, 69: 513–524.
30. USEPA (U.S. Environmental Protection Agency), 2009. National Primary Drinking Water Regulations. EPA 816-F-09-004, Washington, D.C. 20460.
31. WHO (World Health Organization), 2008. Guidelines for Drinking Water Quality. Third edition.
32. WHO, 1990. Cadmium Environmental Health Criteria, Geneva, World Health Organization.
33. Zakir SN, I Ihsanullah, MT Shah, Z Iqbal and A Ahmad, 2009. Comparison of heavy and trace metals levels in soil of Peshawar basin at different time intervals. Journal of the Chemical Society of Pakistan, 31: 246-252.

10

EFFECTS OF ROOTING MEDIA AND VARIETIES ON ROOTING PERFORMANCE OF DRAGON FRUIT CUTTINGS (*Hylocereu sundatus* Haw.)

MAB Khalil Rahad, M. Ashraful Islam*, M Abdur Rahim and S Monira

Department of Horticulture, Faculty of Agriculture, Bangladesh Agricultural University, Mymensingh-2202, Bangladesh

***Corresponding author:** M. Ashraful Islam; E-mail: ashrafulmi@bau.edu.bd

ARTICLE INFO | **ABSTRACT**

The experiment was carried out to investigate the effects of different rooting media on rooting performance, plant growth and development of two varieties of Pitahaya (dragon fruit) cuttings at the Fruit Tree Improvement Project (FTIP), Germplasm Centre (GPC) of Bangladesh Agricultural University (BAU), Mymensingh during the period from May 2014 to August 2014. The two factors experiment consisted with of two different varieties of dragon fruit viz., BAU Dragon fruit-1 (V_1) and BAU Dragon fruit-2 (V_2) with 14 rooting media. viz. control (soil 100%) (T_0), 50% cow dung + 50% soil (T_1), Saw dust (100%) (T_2), Compost (100%) (T_3), 50% soil + 50% sand (T_4), 50% soil + 50% saw dust (T_5), Sand (100%) (T_6), Indole-3 Acetic Acid (IAA) 500 ppm solution + soil (100%) (T_7), 300 ppm solution + soil (100%) (T_8), 200 ppm solution of IAA + soil (100%) (T_9), 500 ppm solution of Indole-3 Butyric Acid (IBA) + soil (100%) (T_{10}), 300 ppm solution IBA + soil (100%) (T_{11}), 200 ppm solution IBA + soil (100%) (T_{12}), (IBA+IAA) 200 ppm solution of each + soil (100%) (T_{13}). All the parameter showed significant effect except number of branches per plant. In case of variety less time was needed for first root initiation (22.33 days) with the longest plant height (34.02 cm). The better result regarding number of roots per cutting (6.00) was found in case of BAU dragon fruit-2 (V_2) where the highest root length (15.22 cm). In case of different treatment, the better result regarding number of roots per cutting (8.17) at 100 DAP was counted from the combination of (IAA+IBA) 200 ppm solution + soil (100%) (T_{13}). The highest root length (25.38 cm) was observed in IAA 200 ppm solution + soil (100%) (T_9). In case of combined effect, the minimum time was required for first root initiation (20.78 days) with IAA 300 ppm solution + soil (100%) (T_8). The highest root length (25.87 cm) was observed in BAU Dragon fruit-1 with IAA 200 ppm solution + soil (100%) (V_1T_9). Number of roots per cutting was noticed (9.67) in BAU Dragon fruit-2 200 ppm solution of each IAA and IBA + soil (100%) (V_2T_{13}).

Key words
Compost,
Cowdung,
Dragon fruit,
Indole-3 Acetic Acid (IAA),
Indole-3 Butyric Acid (IBA),
Saw dust

INTRODUCTION

Pitahayas (*Hylocereus* spp.) (Commonly known as dragon fruit) are perennial climbing cactus plants native to tropical areas of North, Central, and South America (Morton 1987). Major pitahaya fruit growing countries are Vietnam, Colombia, Mexico, Costa Rica, and Nicaragua and, to a lesser degree, cultivation occurs in Australia, Israel, and Reunion Island. The European Union and Asia, especially China, are the largest import markets (Le Bellec et al. 2006). Different types of dragon fruit like most *Hylocereus* species have red-purple pigmented skin, while the pulp color ranges from white (in *H. undatus*) to red and purple (in *H. polyrhizus* and *H. costaricensis*) (Esquivel et al. 2007). Dragon fruits is an excellent source of vitamin C and therefore are abundant with minerals, particularly calcium supplement as well as phosphorus. Also, it is a good source of of natural pigments in food processing, due to their high content of betalains. Pitahaya is considered a promising crop to be grown commercially in dry regions (Vaillant et al., 2005). This species is found to have high water-use efficiency. One of the pitahaya mechanisms to secure water requirement is developing aerial roots from the sides of the stem to collect water from the surroundings (Nobel et al., 2004). Pitaya fruit has red or pink thornless skins, while its juicy flesh can range from white to magenta. The skin is covered with bracts or scales. The small seeds are consumed with the fruit.

The most advantage of this crop is that once it is planted, it will grow for about 20 years, and one hectare could accomodate 1000 to 1200 Dragon fruit plants.More importantly, it is a fast return perennial fruit crop with production in the second year after planting and full production within five years. Usually, pitahaya fruit is propagated sexually by seed and asexually by grafting and stem cutting. The easiest, cheapest and convenient method of propagating dragon fruit is by stem cutting. After 1 week, 100% of the big cuttings developed roots. However, (87 and 65%) of the medium and small cuttings, respectively, developed roots after the same time (Obeidy 2006). On the other hand, there are some reports of success of micropropagation in dragon fruit. According to Vinas et al. 2012, central region of new joint as explant showed about 100 per-cent survival rates and the growth was higher compare to basal and distal joints.

Dragon fruit is suitable for everyone to eat. Flesh and seeds are edible parts and they are eaten altogether. It supplies fiber which is digestive and helpful for healthy liver. Dragon fruits consist of phytoalbumins, which may have anti-oxidant qualities which help to stop the development of cancer cells. Dragon fruit also reported to have health benefits including prevention of memory losses, control of blood glucose level in diabetic patients, prevention of oxidation, aiding in healing of wounds etc. In addition, it has the ability to promote the growth of probiotics in the intestinal tract (Zainoldin and Baba, 2012). In Bangladesh, it is newly extending in all the area. It is important to select the soil media where it can grow well which will help to provide the planting materials of dragon fruit to the whole country of Bangladesh. Considering the present experiment was therefore undertaken to examine the root and plant growth, to evaluate the varietal effect and to find out the suitable condition of rooting media and variety.

MATERIALS AND METHODS

The present experiment was conducted at the Fruit Tree Improvement Project (FTIP), BAU Germplasm Centre (GPC), Bangladesh Agricultural University (BAU), Mymensingh during the period from May, 2014 to August, 2014. The site is situated between 24.6°N and 90.5°E latitude and at latitude 18 m from sea level. The soil of the experimental area is sandy loam type and belongs to the Old Brahmaputra Flood Plain Alluvial Tract (UNDP, 1988) of AEZ 9 having non-calcareous dark grey flood plain soil. The matured Dragon fruit plants were established by cuttings and planted in May, 2014. The experiment consisted with following treatments:

Factor A: Variety

(a) BAU Dragon fruit-1, (V_1)
(b) BAU Dragon fruit-2, (V_2)

Two varieties of Dragon fruit bear different characteristics. BAU Dragon fruit-1 bears white fleshed fruit and BAU Dragon fruit-2 bear red fleshed fruit. Generally, red fleshed one is the sweater than white fleshed one.

Factor B: Rooting media:

T_0=	Control (soil 100%)	T_7=	Dipped in IAA (500 ppm solution) + soil (100%)
T_1=	50% cow dung + 50% soil	T_8=	Dipped in IAA (300 ppm solution) + soil (100%)
T_2=	Saw dust (100%)	T_9=	Dipped in IAA (200 ppm solution) + soil (100%)
T_3=	Compost (100%)	T_{10}=	Dipped in IBA (500 ppm solution) + soil (100%)
T_4=	50% soil+50% sand	T_{11}=	Dipped in IBA (300 ppm solution) + soil (100%)
T_5=	50% soil + 50% saw dust	T_{12}=	Dipped in IBA (200 ppm solution) + soil (100%)
T_6=	Sand (100%)	T_{13}=	Dipped in IBA (200 ppm solution) + IAA (200 ppm solution) + soil(100)

The 2 factors experiment consisting of two varieties, and 14 treatment combinations with 3 replications was laid out in Randomized Complete Block Design (RCBD).

Application of rooting hormone

A. Rooting media as different organic fertilizer

Different types and amount of organic fertilizer like cow dung, sawdust, compost and sand were applied in the field before planting.

B. Media Solution preparation

Different concentration of solution of IAA and IBA were prepared where powder of IAA and IBA was mixed into the water according to the different concentration (200, 300, 500 ppm). Before planting, cutting were dipped into targeted solution for 10 minutes and then placed into the 100% soil.

Data collection

Data on Plant height, Number of branches per cuttings were recorded at 25 day's interval after planting the cutting in the bed. Days to first root initiation were recorded at different treated cutting. During final harvest (after 100 DAP) number of roots per cutting (at final harvest), Root length, Fresh weight roots and shoots, Fresh weight of shoots, Fresh weight of roots, Dry weight of roots and Survival rate of plant (%) were recorded.

Statistical analyses

The recorded data for each parameter from the present experiment was analyzed statistically to find out the variation resulting from experimental treatments using MSTAT-C package program developed by Russel (1986). The means for all treatments were calculated and analyses of variances of the parameter under study were performed by F variance test at 5% and 1% levels of significance. The means of the parameter were compared by least significant difference test (LSD) (Gomez and Gomez, 1984).

RESULTS AND DISCUSSION

A statistically significant difference was found in case of varieties, treatments and combined effect in terms of days to first root initiation. In case of variety BAU dragon fruit-1 (V_1) required shorter time for root initiation (22.33 days) comparing with BAU dragon fruit-2 (V_2). V_2requiredlonger time (24.03 days) compare to V_1. Among the different rooting media (treatments), Indole-3 Butyric Acid (IBA) 200 ppm solution + soil (100%) (T_{12}) required longer time for root initiation (24.00 days) comparing with Indole-3 Acetic Acid (IAA) 200 ppm solution + soil (100%)(T_9) where shorter time (22.08 days) was needed. In case of combined effect of variety and treatment the maximum time (25.11 days) was required for BAU Dragon-2 with IAA 300 ppm solution + soil (100%) (V_2T_8) and minimum time (20.78 days) was needed for first root initiation in BAU Dragon-1 with IAA 300 ppm solution + soil (100%) (V_1T_8) (Table 1, 3 and 5).

Plant height was significantly influenced by different treatments. The plant height was recorded at different dates after planting (DAP) i.e. at 25, 50, 75 and 100 days after planting. In case of variety at 100 DAP; the longer plant (34.02 cm) was noticed in BAU Dragon fruit-1 (V_1), whereas the shorter (32.89 cm) was in BAU Dragon fruit-2 (V_2). In case of treatment at 100 DAP; the longer plant (37.5 cm) was noticed in 50% soil + 50% saw dust (T_5) whereas the shorter (30.33 cm) was in IBA 300 ppm solution + soil (100%) (T_{11}). In case of combined effect at 100 DAP the longest plant (41.88 cm) was noticed in BAU Dragon fruit-1 with 50% soil + 50% sand (V_1T_4) and shortest plant (30.75 cm) was found in BAU Dragon fruit-2 with 50% soil + 50% saw dust (V_2T_5) (Figure 1, Table 2, Table 4).

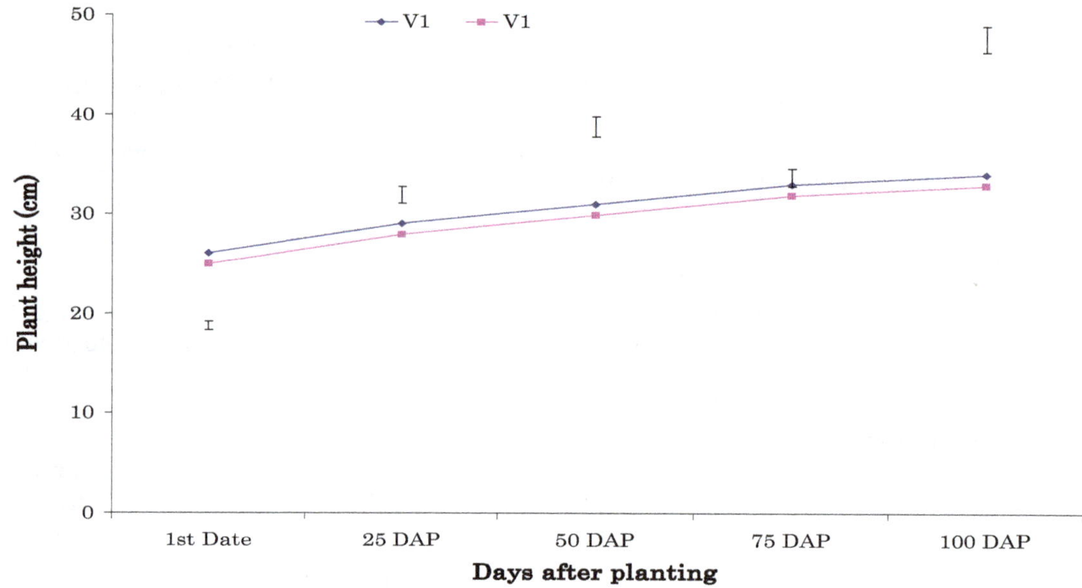

Figure 1. Effect of variety on plant height.

[BAU Dragon fruit-1, (V_1), BAU Dragon fruit-2, (V_2) and DAP= Days after Planting]

Moreover, statistically non-significant difference was found in case of varieties and significant difference was found in case of treatments and combined effect in terms of number of branches. In case of variety the number of branches per cutting (.984) was noticed in BAU Dragon-1 (V_1) and (.960) was noticed in BAU Dragon fruit-1 (V_1). In case of treatment the highest number of branches per cutting (1.44) was noticed in control (soil 100%) and IAA 200 ppm solution + soil (100%) (T_9) and the lowest number of branches (0.22) was noticed in IAA 500 ppm solution + soil (100%) (T_7). In case of combined effect the maximum number of branches (2.00) was noticed in BAU Dragon fruit-1 with IBA 200 ppm solution + soil (100%) (V_1T_{12}) and minimum number of branches (0.11) was found in BAU Dragon fruit-1 with 50% soil + 50% saw dust (V_1T_5) and IAA 500 ppm solution + soil (100%)(V_1T_7) respectively (Table 1, Table 3 and Table 5).

On the other hand, statistically significant difference was found in case of varieties, treatments and combined effect in terms of number of roots per cutting. In case of variety the higher numbers of roots (6.00) at 100 DAP was counted in the BAU Dragon fruit-1 (V_1) but the lower number of roots (5.36) was in BAU Dragon fruit-2 (V_2). In case of treatment the higher numbers of roots (8.17) at 100 DAP was counted in (IAA+IBA) 200 ppm solution + soil (100%) (T_{13}) but the lower number of roots (4.39) was found in control (soil 100%) condition. In case of combined effect the maximum number of roots per cutting was noticed (9.67) in BAU Dragon fruit-2 with (IAA+IBA) 200 ppm solution + soil (100%) (V_1T_{13}) and minimum number of branch was found (3.44) in BAU Dragon fruit-2 with 50% soil + 50% saw dust (V_2T_5) (Table 1, Table 3 and Table 5).

In addition, statistically significant difference was found in case of varieties, treatments and combined effect in terms of root length. In case of variety the highest root length was measured in case of BAU Dragon fruit-2 (V_2) (15.22 cm) but lower length was found in BAU Dragon fruit-2 (V_2) (14.44 cm). In case of treatment the highest root length was measured (25.38 cm) in case of IAA 200 ppm solution + soil (100%) (T_9) but lower length (8.39 cm) was found in sand (100%) (T_6). In case of combined effect the longest (25.87 cm) was observed in BAU Dragon fruit-1 in case of IAA 200 ppm solution + soil (100%) (V_1T_9)

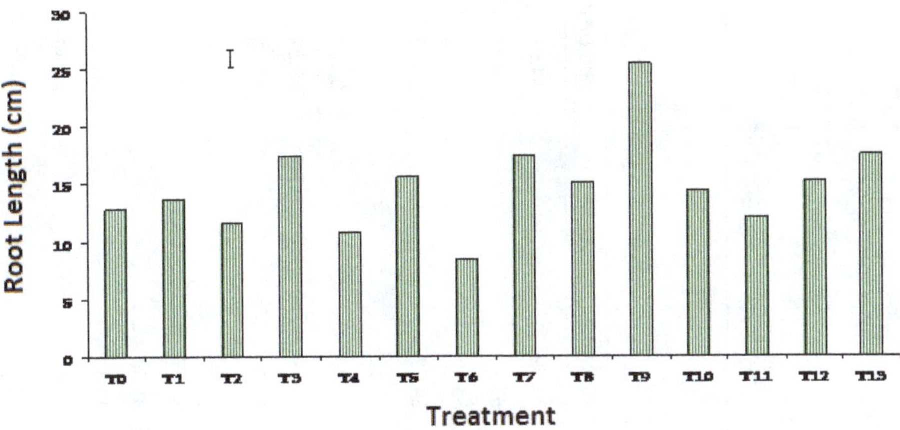

Figure 2. Effect of Treatment on root length (cm)

[T_0= Control (soil 100%), T_1= 50% cow dung + 50% soil, T_2= Saw dust (100%), T_3= Compost (100%), T_4= 50% soil+50% sand, T_5= 50% soil + 50% saw dust, T_6= Sand (100%), T_7= Dipped in IAA (500 ppm solution) + soil (100%), T_8= Dipped in IAA (300 ppm solution) + soil (100%), T_9= Dipped in IAA (200 ppm solution) + soil (100%), T_{10}= Dipped in IBA (500 ppm solution) + soil (100%), T_{11}= Dipped in IBA (300 ppm solution) + soil (100%), T_{12}= Dipped in IBA (200 ppm solution) + soil (100%), T_{13}= Dipped in IBA (200 ppm solution + IAA (200 ppm solution) + soil (100)] and the shortest (7.78 cm) was recorded from BAU Dragon fruit-2 with sand (100%) (V_2T_6) (Table 1 and 5).

Again, statistically significant difference was found in case of varieties, treatments and combined effect in terms of fresh weight of roots and shoots. In case of variety the higher fresh weight of roots and shoots (108.25 g) was measured in case of BAU Dragon fruit-2 (V_2) whereas the lower (104.58 g) was in BAU Dragon fruit-1 (V_1). In case of treatment the higher fresh weight of roots and shoots (128.61 g) was measured in case of saw dust (100%) (T_2) and the lower (75.69 g) were in sawdust (100%) (T_2). In case of combined effect the maximum fresh weight of roots and shoots (144.17 g) was found in BAU Dragon-2 with IBA 300 ppm solution + soil (100%) (V_2T_{11}) and minimum (66.54 g) was observed in BAU Dragon-2 with IBA + IAA 200 ppm solution + soil (100%) (V_2T_{13}) (Table 1, 3 and 5).

Moreover, statistically significant difference was found in case of varieties, treatments and combined effect in terms of fresh weight of shoots. In case of variety the higher fresh weight of shoots (106.4 g) was measured in case of BAU Dragon fruit-2 (V_2) whereas the lower (103.12 g) was in BAU Dragon fruit 1 (V_1). In case of treatment the higher fresh weight of shoots (125.96 g) was measured in case of saw dust (100%) (T_2) and the lower (74.27 g) were in (IAA+IBA) 200 ppm solution + soil (100%) (T_{13}). In case of combined effect the maximum fresh weight of shoots (142.69 g) was found in BAU Dragon-2 with IBA 300 ppm solution + soil (100%) (V_2T_{11}) and minimum (65.03 g) was observed in BAU Dragon-2 with IBA + IAA 200 ppm solution + soil (100%) (V_2T_{13}) (Table 1, 3 and 6).

Moreover, statistically significant difference was found in case of varieties, treatments and combined effect in terms of fresh weight of roots. In case of variety the higher fresh weight of roots (1.85 g) was measured in case of BAU Dragon fruit-2 (V_2) whereas the lower (1.46 g) was in BAU Dragon fruit-1(V_1). In case of treatment the higher fresh weight of roots (2.66 g) was measured in case of compost (100%) (T_3) whereas the lower (1.18 g) was in 50% soil + 50% saw dust (T_5) and IBA 200 ppm solution + soil (100%) (T_{12}). In case of combined effect the maximum fresh weight of roots (142.69 g) was found in BAU Dragon-2 with IBA 300 ppm solution + soil (100%) (V_2T_{11}) and minimum (65.03 g) was observed in BAU Dragon-2 with IBA + IAA 200 ppm solution + soil (100%) (V_2T_{13}) (Table 1, 3 and 6).

Moreover, statistically significant difference was found in case of varieties, treatments and combined effect in terms of dry weight of roots. In case of variety the higher dry weight of roots (0.433 g) was measured in case of BAU Dragon fruit-2 (V_2) whereas the lower (0.353 g) was in BAU Dragon fruit-1 (V_1). In case of treatment the higher dry weight of roots (0.679 g) was measured in case of saw dust (100%) (T_2) and the lower (0.253 g) were in case of IBA 200 ppm solution + soil (100%) (T_{12}). In case of combined effect the maximum dry weight of roots (0.992 g) was found in BAU Dragon-2 with saw dust (100%) (V_2T_2) and minimum (0.218 g) was observed in BAU Dragon-2 with IBA 200 ppm solution + soil (100%) (V_2T_{12}) (Table 1,3 and 6).

Table 1. Effect of varieties on plant growth and development of Dragon fruit

Variety	Days to first root initiation	Root length (cm)	Number of roots per plant	Fresh weight of Roots and Shoots/plant (g)	Fresh weight of Roots/plant (g)	Fresh weight of shoots (g)	Dry weight of roots/plant (g)	Survival rate of plant (%)
V_1	22.33	14.441	5.36	104.58	1.46	103.12	0.353	86.01
V_2	24.03	15.219	6.00	108.25	1.85	106.40	0.433	84.41
LSD at 5%	0.132	0.327	0.065	1.145	0.028	1.206	0.014	0.245
LSD at 1%	0.176	0.435	0.087	1.526	0.037	1.607	0.018	0.327
Level of sig.	**	**	**	**	**	**	**	**

** = Significant at 1% level of probability; * = Significant at 5% level of probability

Table 2. Effect of different rooting media (treatments) on plant height

Treatments	Plant height (cm)				
	1st Date	25 DAP	50 DAP	75 DAP	100 DAP
T_0	26.60	29.60	31.59	33.59	34.59
T_1	25.55	28.55	30.41	32.41	33.41
T_2	26.75	29.75	31.46	33.46	34.46
T_3	27.76	30.76	32.72	34.72	35.72
T_4	29.53	32.53	34.50	36.50	37.50
T_5	25.11	28.11	29.80	31.80	32.80
T_6	27.30	30.30	32.54	34.54	35.54
T_7	23.57	26.57	28.98	30.98	31.98
T_8	24.07	27.07	28.90	30.90	31.90
T_9	25.91	28.91	31.29	33.29	34.29
T_{10}	24.63	27.63	29.56	31.56	32.56
T_{11}	22.56	25.56	27.33	29.33	30.33
T_{12}	24.07	27.07	28.70	30.70	31.70
T_{13}	23.87	26.87	28.56	30.56	31.56
LSD at 5%	0.463	0.881	1.095	0.934	1.377
LSD at 1%	0.617	1.174	1.460	1.245	1.836
Level of sig.	**	**	**	**	**

** = Significant at 1% level of probability; * = Significant at 5% level of probability; DAP = Days after planting

Table 3. Effect of different rooting media (treatments) on plant growth and development of Dragon fruit

Treatment	Number of roots per plant	Fresh Weight of Roots and Shoots/plant (g)	Fresh Weight of Roots/plant (g)	Fresh weight of hoots/plant (g)	Dry weight of roots/plant (g)	Survival rate of plant %	Days to first root initiation	Survival rate of plant %
T_0	4.39	101.67	1.59	100.08	0.521	71.2	23.83	71.2
T_1	5.17	100.83	1.58	99.25	0.36	74.81	23.56	74.81
T_2	6.17	128.61	2.65	125.96	0.679	94.25	23.33	94.25
T_3	5.89	118.06	2.66	115.4	0.511	79.77	23.56	79.77
T_4	5.22	93.06	1.26	91.8	0.398	75.43	23.61	75.43
T_5	6	106.11	1.18	104.93	0.292	91.58	22.94	91.58
T_6	4.69	116.81	1.19	115.62	0.264	89.68	23.06	89.68
T_7	5.28	95.5	1.75	93.75	0.418	85.01	22.50	85.01
T_8	5.56	110.42	1.78	108.64	0.398	92.39	22.94	92.39
T_9	5.5	103.75	2.11	101.64	0.382	89.04	22.28	89.04
T_{10}	5.56	102.67	1.36	101.31	0.392	99.25	22.78	99.25
T_{11}	6.5	124.17	1.45	122.72	0.297	84.55	23.06	84.55
T_{12}	5.42	112.5	1.18	111.32	0.253	78.82	24.00	78.82
T_{13}	8.17	75.69	1.42	74.27	0.341	87.22	23.06	87.22
LSD at 5%	0.172	3.026	0.073	3.186	0.037	0.648	0.349	0.648
LSD at 1%	0.229	4.034	0.098	4.248	0.049	0.863	0.466	0.863
Level of sig.	**	**	**	**	**	**	**	**

Table 4: Combined effect of varieties and rooting media (treatments) on plant height

Variety and Treatment	Plant height (cm)				
	1st Date	25 DAP	50 DAP	75 DAP	100 DAP
V_1T_0	27.04	30.04	32.23	34.23	35.23
V_2T_0	26.16	29.16	30.94	32.94	33.94
V_1T_1	25.97	28.97	30.64	32.64	33.64
V_2T_1	25.13	28.13	30.18	32.18	33.18
V_1T_2	26.32	29.32	31.16	33.16	34.16
V_2T_2	27.18	30.18	31.76	33.76	34.76
V_1T_3	28.24	31.24	33.39	35.39	36.39
V_2T_3	27.27	30.27	32.04	34.04	35.04
V_1T_4	33.87	36.87	38.88	40.88	41.88
V_2T_4	25.18	28.18	30.11	32.11	33.11
V_1T_5	27.07	30.07	31.85	33.85	34.85
V_2T_5	23.15	26.15	27.75	29.75	30.75
V_1T_6	29.97	32.97	35.6	37.6	38.6
V_2T_6	24.63	27.63	29.47	31.47	32.47
V_1T_7	23.21	26.21	28.8	30.8	31.8
V_2T_7	23.92	26.92	29.17	31.17	32.17
V_1T_8	23.09	26.09	28.17	30.17	31.17
V_2T_8	25.04	28.04	29.63	31.63	32.63
V_1T_9	25.52	28.52	30.51	32.51	33.51
V_2T_9	26.3	29.3	32.07	34.07	35.07
V_1T_{10}	26.15	29.15	31.04	33.04	34.04
V_2T_{10}	23.1	26.1	28.08	30.08	31.08
V_1T_{11}	20.87	23.87	25.22	27.22	28.22
V_2T_{11}	24.24	27.24	29.43	31.43	32.43
V_1T_{12}	24.01	27.01	28.58	30.58	31.58
V_2T_{12}	24.13	27.13	28.81	30.81	31.81
V_1T_{13}	23.43	26.43	28.16	30.16	31.16
V_2T_{13}	24.3	27.3	28.95	30.95	31.95
LSD at 5%	0.656	1.248	1.551	1.323	1.95
LSD at 1%	0.872	1.659	2.063	1.759	2.593
Level of sig.	**	**	**	**	**

** = Significant at 1% level of probability; * = Significant at 5% level of probability

Table 5.Combined effect of varieties and rooting media (treatments) on plant growth and development of Dragon fruit

Variety and Treatment	Number of branches per plant	Root length (cm)	Number of roots per plant	Fresh Weight of Roots and Shoots (g)	Days to first root initiation
V_1T_0	1.11	13.33	3.78	103.33	23.33
V_2T_0	1.78	12.33	5	100	22.67
V_1T_1	0.44	13.06	4.44	101.67	22.67
V_2T_1	1.78	14.22	5.89	100	23.00
V_1T_2	1.67	12.33	5.22	115	22.56
V_2T_2	0.22	11	7.11	142.22	22.67
V_1T_3	0.38	15	4.89	117.78	23.11
V_2T_3	1.22	19.78	6.89	118.33	21.11
V_1T_4	1.22	11.11	4.78	97.22	20.78
V_2T_4	1.22	10.56	5.67	88.89	21.33
V_1T_5	0.11	18.33	8.56	113.89	22.78
V_2T_5	0.89	12.83	3.44	98.33	21.22
V_1T_6	1.44	9	4.56	122.78	23.56
V_2T_6	0.33	7.78	4.83	110.83	21.78
V_1T_7	0.11	15.72	5.89	87.67	24.33
V_2T_7	0.33	19.17	4.67	103.33	24.44
V_1T_8	0.75	12.56	5.11	106.67	24.00
V_2T_8	1.11	17.67	6	114.17	24.11
V_1T_9	1.56	25.87	5.67	104.17	24.67
V_2T_9	1.33	24.89	5.33	103.33	23.22
V_1T_{10}	0.56	17.44	6.11	113.33	23.00
V_2T_{10}	1.22	11.33	5	92	23.89
V_1T_{11}	1.33	13.5	5.33	104.17	25.11
V_2T_{11}	1.22	10.5	7.67	144.17	23.22
V_1T_{12}	2	13.67	4	91.67	22.78
V_2T_{12}	0.67	16.83	6.83	133.33	24.89
V_1T_{13}	0.78	10.83	6.67	84.83	24.44
V_2T_{13}	0.44	24.17	9.67	66.54	24.33
LSD at 5%	0.073	1.22	0.243	4.285	0.495
LSD at 1%	0.098	1.63	0.323	5.698	0.658
Level of sig.	**	**	**	**	**

** = Significant at 1% level of probability; * = Significant at 5% level of probability

Table 6.Combined effect of varieties and rooting media (treatments) on plant growth and development of Dragon fruit

Variety and Treatment	Fresh Weight of Roots (g)	Fresh weight of shoots (g)	Dry weight of roots (g)	Survival rate of plant %
V_1T_0	1.31	102.02	0.351	71.5
V_2T_0	1.86	98.14	0.691	75.36
V_1T_1	1.54	100.13	0.299	95.36
V_2T_1	1.62	98.38	0.421	80.42
V_1T_2	1.57	113.43	0.367	76.75
V_2T_2	3.72	138.5	0.992	92.45
V_1T_3	1.64	116.14	0.45	90.47
V_2T_3	3.68	114.65	0.572	86.31
V_1T_4	1.21	96.01	0.427	93.41
V_2T_4	1.31	87.58	0.37	89.4
V_1T_5	1.5	112.39	0.35	100
V_2T_5	0.86	97.47	0.233	85.14
V_1T_6	1.19	121.59	0.261	79.31
V_2T_6	1.2	109.63	0.267	88.27
V_1T_7	1.7	85.97	0.433	70.89
V_2T_7	1.8	101.53	0.402	74.25
V_1T_8	1.69	104.98	0.34	93.14
V_2T_8	1.88	112.29	0.456	79.12
V_1T_9	1.79	102.38	0.363	74.11
V_2T_9	2.43	100.9	0.4	90.71
V_1T_{10}	1.49	111.84	0.396	88.89
V_2T_{10}	1.24	90.76	0.388	83.71
V_1T_{11}	1.43	102.74	0.31	91.36
V_2T_{11}	1.48	142.69	0.283	88.67
V_1T_{12}	1.04	90.63	0.288	98.49
V_2T_{12}	1.33	132	0.218	83.95
V_1T_{13}	1.33	83.5	0.308	78.33
V_2T_{13}	1.51	65.03	0.373	86.17
LSD at 5%	0.104	4.513	0.052	0.917
LSD at 1%	0.138	6.001	0.069	1.22
Level of sig.	**	**	**	**

** = Significant at 1% level of probability; * = Significant at 5% level of probability

Again, statistically significant difference was found in case of varieties, treatments and combined effect in terms of survival rate of plant. In case of variety the higher survival rate of plant (86.01%) was observed in case of BAU Dragon fruit-1 (V_1) whereas the lower (84.41%) was in BAU Dragon fruit-2 (V_2). In case of treatment the higher survival rate of plant (99.25%) was observed in case of IBA 500 ppm solution + soil (100%) (T_{10}) and the lower (71.2%) were in case of control (soil 100%). In case of combined effect the highest survival rate of plant (100%) was observed in BAU Dragon fruit-1 with IBA 500 ppm solution + soil (100%) (V_1T_{10}) and the lowest survival rate of plants (70.89%) in BAU Dragon fruit-1 with control (soil 100%) (V_1T_0) (Table 1, 3 and 6).So in our experiment, it is observed that all the parameter were significant at 1% level in case of varieties, treatments and combined effect except the number of branches which was non- significant in case of variety.

CONCLUSION

In this experiment there were significant effects of rooting media and varieties of dragon fruit cuttings. Among the observations the best result was found in case of variety for the highest root length (15.22 cm) was measured in case of BAU Dragon fruit-2 (V_2). In case of treatment for the highest root length (25.38 cm) was measured in case of Indole-3 Acetic Acid (IAA) 200 ppm solution + soil (100%) (T_9) and in case of combined effect the highest root length (25.87 cm) was measured in case of BAU Dragon fruit-1 with (IAA) 200 ppm solution + soil (100%) (V_1T_9). On the other hand, in case of variety the longest plant height (34.02 cm) was noticed in BAU Dragon fruit-1 (V_1). Saw dust (100%) (T_2) also gave better result for another parameter in case treatment. So, the results obtained from this investigation that among the rooting media IAA 200 ppm solution + soil (100%)(T_9), IBA 300 ppm solution + soil (100%)(T_{11}), 50% soil + 50% saw dust (T_5) and 50% soil +50% sand (T_4) gave the better result for maximum parameter of Dragon fruit cultivation.

REFERENCES

1. Esquivel P, Stintzing FC and R Carle, 2007. Comparison of morphological and chemical fruit traits fromdifferent pitaya genotypes (*Hylocereus*sp.) grown in Costa Rica. Journal of Applied Botany and Food Quality, 81: 7–14.
2. Gomez KA and AA Gomez, 1984: Statistical Produce for Agricultural Research. In: John Willy and Sons (Editors), New York. pp. 272-279.
3. Le Bellec F, Vaillant F and E Imbert, 2006.Pitahaya (*Hylocereus* spp.): a new fruit crop, a market with a future. Fruits, 61: 237–250.
4. Morton J, 1987. Strawberry pear. In: Morton J. F. (ed) Fruits of warm climates. Miami, FL, pp 347–348.
5. Nobel P and de la Barrera E, 2004. CO_2 uptake by the cultivated hemi epiphytic cactus, *Hylocereusundatus*, Annal of Applied Biology, 144: 1–8.
6. Obeidy AAE, 2006. Mass propagation of pitaya (dragon fruit). Fruits, 61: 313-319.
7. Russell EW and ED Russell, 1986: Soil Supplied Conditions and Plant Growth. William C and Sons Ltd, London.
8. UNDP 1988. Land Resource Appraisal of Bangladesh for Agricultural Development, Agro-ecological Regions of Bangladesh, FAO, Rome, Italy.
9. Vaillant F, Perez A, Davila I, Dornier M, and M Reynes, 2005: Colorant and antioxidant properties of red-purple pitahaya (*Hylocereus*sp.), Fruits, 60: 3–12.
10. Vinas M, Brenes MF, Azofeifa A and VM Jiménez, 2012: In vitro propagation of purple pitahaya (*Hylocereuscostaricensis*[F.A.C. Weber] Britton & Rose) cv. Cebra. In Vitro Cellularand Developmental Biology, 48: 469–477.
11. Zainoldin KH and AS Baba, 2012: The effect of *Hylocerouspolyrhizus* and *Hylocerousundatus* on Physicochemical, Proteolysis and Antioxidant Activity in Yogurt. International Journal of Biological and Life Science, 8: 93-98.

11

NITROGEN REQUIREMENT AND CRITICAL N CONTENT OF STEVIA GROWN IN TWO CONTRASTING SOILS OF BANGLADESH

Md. Maniruzzaman, Md. Akhter Hossain Chowdhury*, KM Mohiuddin and [1]Tanzin Chowdhury

Department of Agricultural Chemistry and [1]Department of Agronomy, Faculty of Agriculture, Bangladesh Agricultural University, Mymensingh-2202, Bangladesh

***Corresponding author:** Md. Akhter Hossain Chowdhury; E-mail: akhterbau11@gmail.com

ARTICLE INFO

Key words
Stevia,
N requirement,
Leaf biomass Yield,
Critical N

ABSTRACT

Nitrogen is recognized as one of the most limiting nutrient for crop growth in Bangladesh and can be supplemented with inorganic fertilizers like urea. The experiment was conducted in the net house of the Department of Agricultural Chemistry, Bangladesh Agricultural University during March to July 2012. The objective was to examine the effects of different levels of N on the growth, leaf biomass yield, N content and to estimate minimum N requirement and critical N content of stevia. The treatments included six N rates (0, 100, 150, 200, 250 and 300 kg ha^{-1}). Plant sampling was done at 15, 30, 45 and 60 days after planting (DAP) to measure plant height, number of branches and leaves, fresh and dry weight of leaves, leaf area and N concentration. The results revealed that all the characters were significantly affected by different N rates. The highest values of all parameters except plant height and N concentration were obtained from 250 kg N ha^{-1} and the lowest values from N control. Nitrogen application at all levels increased leaf dry yield at harvest by 99 to 505% in acid soil and 69 to 438% in non-calcareous soil, respectively over control. The growth of most parameters was rapid at the later stages (30 to 60 DAP). Leaf N content proportionately increased with the increasing rates of N. The highest N concentration was obtained from its highest application (300 kg N ha^{-1}). The minimum amount of N for maximum leaf biomass production in the plants grown in acid and non-calcareous soils was estimated to be ca 273 and 257 kg ha^{-1}, respectively. The critical N concentration to achieve 80% of the maximum production of stevia leaf was also estimated to be ca 1.43 and 1.50% in the leaves of stevia plants grown in acid and non-calcareous soils, respectively.

INTRODUCTION

Stevia, as a new crop to be domesticated in Bangladesh, a package regarding the cultivation aspects, need to be standardized under different agro climatic conditions of Bangladesh. The stevia cultivation has an immense scope for intensive agriculture and precision farming and fits well for high return agriculture (Barathi, 2003). Few sporadic trials on the growth and leaf yield of stevia have been conducted both at pot and field conditions (Khanom, 2007; Nasrin, 2008 and Hasan, 2008). Recently Khan (2014) conducted few experiments on different agronomic aspects like date of planting, pruning, stem cutting etc on the growth and yield of stevia. It is expected that a higher and balanced nutrient supply will result in higher foliage yield. Unfortunately no detailed study has yet been conducted on inorganic fertilizer requirement for large scale cultivation of stevia in Bangladesh. Incorporation of stevia into agricultural production systems depends upon details information regarding the plant, its agronomic potential and nutritional requirement (Ramesh *et al.*, 2007).

Nitrogen (N) is recognized as one of the most limiting nutrients for crop growth. All vital processes in plant are associated with protein, of which N is an essential constituent. It also involved in chlorophyll synthesis, and influences stomatal conductance and photosynthetic efficiency. Consequently, getting more crop production, N application is essential in the form of chemical fertilizer (Ali *et al.*, 2000). Nitrogen is applied to the crop for higher vegetative growth, productivity and quality (Gwal et al., 1999 and Iqbal et al., 2012). Balanced use of N is a key point for higher land profitability and healthy environment. Optimum N requirement need to be screened out for achieving maximum leaf biomass yield of stevia in Bangladesh. Critical values are quite useful and are frequently referred to when interpreting a plant analysis result. The critical N concentration -that is the minimum leaf N concentration required to reach maximal accumulated leaf biomass should be estimated. To the best of our knowledge, the critical N content of stevia is yet to be estimated elsewhere including Bangladesh. Keeping in view the significant role of N in crop production systems, the present research project was designed to study the effects of different levels of N on the growth, leaf yield, N content and its uptake and to estimate minimum N requirement and critical leaf N concentration of stevia under the agro climatic conditions of Bangladesh Agricultural University, Mymensingh-2202.

MATERIALS AND METHODS

A pot experiment was conducted at the net house of the Department of Agricultural Chemistry, Bangladesh Agricultural University, Mymensingh during March to July, 2012 to examine the effects of different levels of N on the growth, leaf yield, N content and its uptake, determine N requirement and critical leaf N concentration of stevia. Two soils viz. acid and non-calcareous of contrasting physical and chemical properties were used (Zaman et al, 2015). Approximately 40 kg soils from each location (*Madhupur* for acid soil and *BAU farm* for non-calcareous soil) were collected from 0 -15cm depth of selected fellow land for the experiment. The samples were made free from plant residues and other extraneous materials, air dried, ground and sieved through a 2mm sieve. 500g sieved soil from each source was preserved in a polythene bag and the physical and chemical properties were determined following standard procedure (Page *et al.* 1982).

Eight kg processed soil was taken in each earthen pot of 23 cm in height with 30 cm diameter at top and 18 cm at bottom leaving 3 cm from the top. *In vitro* produced 45 day old stevia seedlings (*Stevia rebaudiana* Bertoni) were collected from *brac* biotechnology laboratory, Joydebpur, Gazipur and used for the experiment. One stevia seedling was planted in each pot during 1st week of March, 2012. P, K, S, Zn and B were applied as basal doses @ 100, 200, 30, 3 and 1 kg ha^{-1} from TSP, MoP, gypsum, zinc sulphate and boric acid, respectively (Zaman, 2015). Six levels of N viz. 0, 100, 150, 200, 250, and 300 kg ha^{-1} were applied from prilled urea, 1/3rd during pot preparation, 1/3rd at 15 days after planting (DAP) and 1/3rd at 30 DAP. The experiment was laid out in completely randomized design with three replications. Total number of pots was 36 (6 treatment X 2 soil X 3 replication). Intercultural operations like irrigation, soil loosening, weeding, insect pest control, removal of flowers etc. were done as and when necessary. Data were collected at 15, 30, 45 and 60 DAP. The crop was destructively harvested at 60 DAP. After harvesting the crop, leaf samples were separated, cleaned, dried at 60°C for 72 hours, weighed, ground and stored. Plant height, branches plant^{-1}, leaves plant^{-1}, leaf area plant^{-1}, fresh and dry leaf weight of stevia leaves were studied. N content was determined by micro Kjeldahl method (Page *et al.*, 1982). Uptake was calculated from N content and leaf dry yield. N requirement and critical N concentration was also estimated (Chowdhury, 2000). The results obtained were subjected to statistical analysis using standard method of analysis (Steel *et al.*, 1997). The differences among the treatment means were compared by using Duncan Multiple Range test (Gomez and Gomez, 1984).

RESULTS

Effects of different levels of N on various parameters of stevia are described under the following heads

Effect of N on plant height

The data on plant height as affected by N rates in acid and non-calcareous soil is given in Fig 1. Different N rates had significant effect on plant height of stevia. The application of N influenced plant height variably from 15 days after planting (DAP) to 60 DAP. There was a general trend of increase in plant height with increase in N fertilizer at 15, 30, 45 and 60 DAP with the control treatment registering the least.

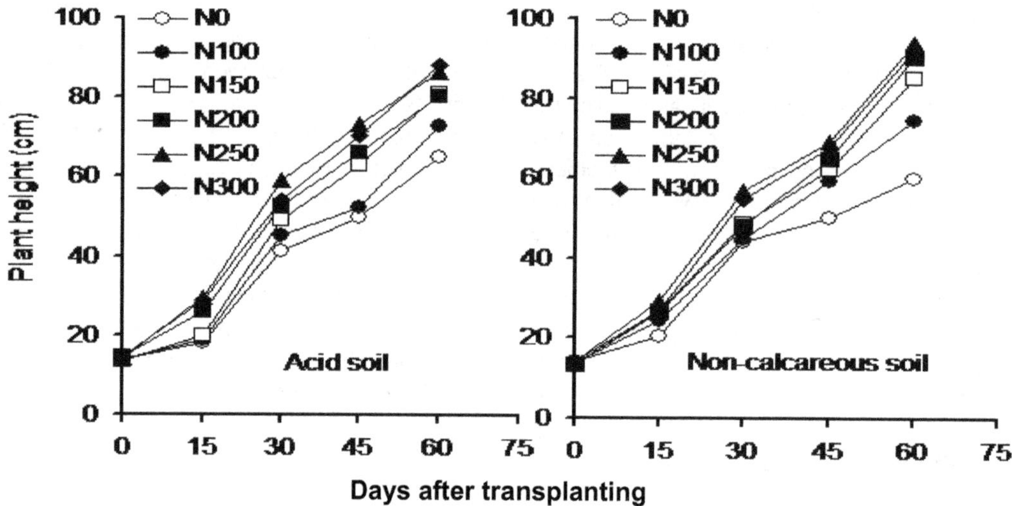

Figure 1. Effects of different levels of N on the plant height of stevia at various DAP

In the first stage of growth, differences between N levels were not significant. After this stage, increased plant height and difference between treatments could be observed. An increase in plant height was observed from planting stage to harvesting in both soils irrespective of treatments. N application at all levels increased plant height by 8 to 23cm in acid soil and 15 to 34cm in non-calcareous soil, respectively. It was observed that the pots subjected to N_{250} and N_{300} were significantly higher than control at 60 DAP although identical to those subjected to N_{200} and N_{250} at the same time. It was also noted that N_{100} and N_{150} were not significantly different at harvest. The highest level of N (N_{300}) produced the tallest plant (88.3cm) in acid soil where as N_{250} produced tallest plant (94cm) grown in non-calcareous soil and the shortest plants formed in the control (without N) in both soils. Height increase was 36% higher in acid soil and 57% higher in non-calcareous soil over control.

Effect of N on branch number

The data on branch number are presented in Figure 2. Branch number plant^{-1} responded significantly due to the application of different levels of N. The result revealed that branches plant^{-1} progressively increased with increasing levels of N application up to 250 kg ha^{-1} in both soils and then declined with further addition. The application of N influenced the number of branches plant^{-1} variably from 15 to 60 DAP irrespective of soils and treatments. Rapid increase in branch number was observed between 15 and 45 DAP and then remained constant or very slowly increased in both soils. The highest number of branches plant^{-1} (7.7 in acid soil and 9.3 in non-calcareous soil) at 60 DAP was counted from the plant receiving 250 kg N ha^{-1} which was identical with N_{200} and N_{300} but significantly different from N_{100} and N_{150}. The lowest branch number was counted from control. N application at all levels increased branch number by 89-156% in acid soil and 60 to 180% in non-calcareous soil, respectively at 60 DAP.

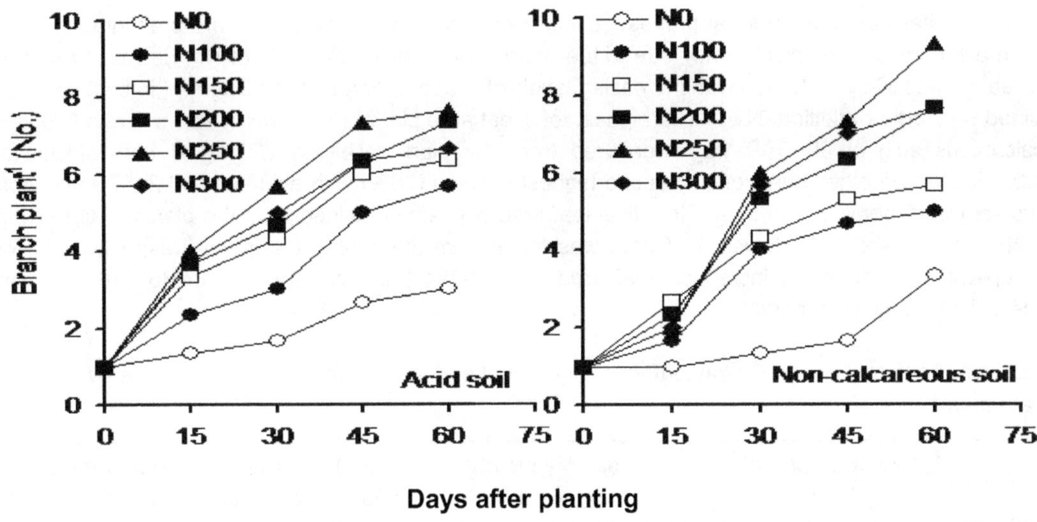

Figure 2. Effects of different levels of N on the branch number of stevia at various DAP

Effect of N on leaf number

The data pertaining to the number of leaves plant^{-1} as influenced by different levels of N in both acid and non-calcareous soils at various DAP have been presented in Fig. 3. Application of N at different doses significantly influenced the number of leaves of stevia plants at all growth stages except 0 DAP irrespective of soils and treatments used.

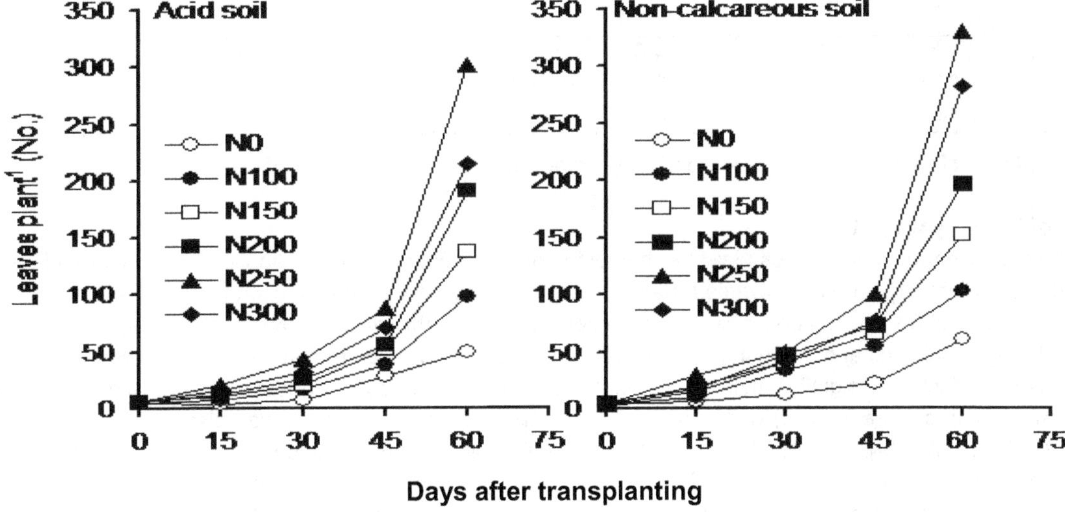

Figure 3. Effects of different levels of N on the leaf number of stevia at various DAP

Leaf number was increased with the increased levels of N up to 250 kg ha^{-1} and then declined by 40% in acid soil and 17% in non-calcareous soil with further addition (N_{300}). Leaf number increase was very slow at the early growth stages (0-30 DAP) while it was rapid between 30 and 60 DAP irrespective of N levels except control. N application at all levels increased the number of leaves by 49 to 251 in acid soil and 42 to 269 in non-calcareous soil, respectively. Maximum number of leaves was recorded with N_{250} which was significantly higher than all other levels of N in both soils. Plants fertilized with N_{200} and N_{300} produced identical number of leaves in acid soil. The minimum number of leaves plant^{-1} was harvested from the plants fertilized with no N irrespective of soils and growth period.

Effect of N on leaf area

The data on total leaf area plant^{-1} at harvest as influenced by different levels of N are presented in Table 1. Leaf area plant^{-1} responded significantly due to the application of different levels of N. The result revealed that leaf area progressively increased with increasing levels of N application up to 250 kg ha^{-1} in both soils and then declined with further addition (N_{300}). The highest total leaf area plant^{-1} (2512cm^2 in acid soil and 2805cm^2 in non-calcareous soil) at 60 DAP was measured from the plant receiving 250 kg N ha^{-1} which was significantly higher than other levels of N. Second highest values (1769cm^2 in acid soil and 2157cm^2 in non-calcareous soil) were obtained from N_{300}. Identical leaf area was also obtained from the plants fertilized with N_{200} and N_{150} in both soils. The lowest leaf area was found from the control treatment irrespective of soils used. N application at all levels increased leaf area by 111-794% in acid soil and 84 to 546% in non-calcareous soil, respectively at harvest.

Table 1. Effects of different levels of N on leaf area, dry weight and yield increase of stevia leaves over control at harvest

N level	Leaf area plant^{-1} (cm^2)		Leaf dry weight (g plant^{-1})		Yield increase over control (%)	Yield increase over control (%)
	Acid soil	Non-calcareous soil	Acid soil	Non-calcareous soil	Acid soil	Non-calcareous soil
N_0	281e	434e	1.52e	1.84f	-	-
N_{100}	592de	799de	3.02d	3.11e	99	69
N_{150}	1003cd	1186cd	4.22c	4.66d	178	153
N_{200}	1401bc	1570c	5.80b	5.92c	282	222
N_{250}	2512a	2805a	9.20a	9.90a	505	438
N_{300}	1769b	2157b	6.57b	8.57b	332	366
CV (%)	4.5	5	4.07	4.13	-	-
$LSD_{0.05}$	240	320	0.61	0.64	-	-
SE±	186	204	0.62	0.70	-	-

CV = Coefficient of variance, LSD = Least significant difference, SE± = Standard error of means

Effect of N on dry weight

The dry weight of stevia leaves plant^{-1} at harvest varied significantly due the application of different levels of N fertilizer (Table 1). Results revealed that dry weight progressively increased with increasing levels of N application up to 250 kg ha^{-1} in both soils and then declined with further addition (N_{300}). The highest dry weight plant^{-1} (9.20g in acid soil and 9.90g in non-calcareous soil) at harvest was measured from the plant receiving 250 kg N ha^{-1} which was significantly higher than other levels of N. Second highest values (6.57g in acid soil and 8.57g in non-calcareous soil) were obtained from N_{300}. Identical dry weight was also obtained from the plants fertilized with N_{200} and N_{300} in acid soil. The lowest values were obtained from the control treatment (1.52g in acid soil and 1.84g in non-calcareous soil. N application at all levels increased leaf dry yield at harvest by 99 to 505% in acid soil and 69 to 438% in non-calcareous soil, respectively over control.

Effect of N on fresh weight

The fresh weight of stevia leaves plant^{-1} at harvest varied significantly due the application of different levels of N fertilizer (Figure 4). Results revealed that fresh weight progressively increased with increasing levels of N application up to 250 kg ha^{-1} in both soils and then declined with further addition (N_{300}).

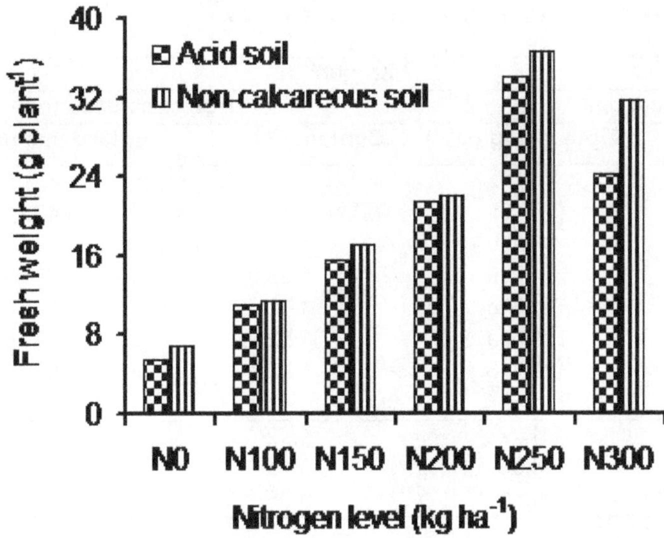

Figure 4. Effects of different levels of N on the fresh weight of stevia leaves at harvest

The highest fresh weight plant^{-1} (34.10g in acid soil and 36.73g in non-calcareous soil) at harvest was measured from the plant receiving 250 kg N ha^{-1} which was significantly higher than other levels of N. Second highest values (24.19g in acid soil and 31.70g in non-calcareous soil) were obtained from N_{300}. Identical fresh weight was also obtained from the plants fertilized with N_{200} and N_{300} in acid soil. The lowest values were obtained from the control treatment (5.61g in acid soil and 6.84g in non-calcareous soil irrespective of soils used. N application at all levels increased fresh weight at harvest by 5.46 to 28.49g plant^{-1} in acid soil and 4.68g to 30.25g plant^{-1} in non-calcareous soil, respectively.

N content and uptake

There was a significant effect of different levels of N on it's content and uptake by stevia leaf (Table 2). N content of the leaf was increased with the increased levels of N irrespective of soils used. The highest N content (1.51% in acid soil and 1.62% in non-calcareous soil) was obtained when N was applied @ 300 kg ha^{-1} in both soils which was statistically identical with the N contents of the leaves of stevia plant fertilized with N_{200} and N_{250} but significantly different from other treatments. The lowest N content was obtained from the plants receiving no N fertilizer in both soils.

The effects of different levels of N on its uptake were significant (Table 2). The trend was similar like N contents of stevia leaves. The N uptake varied from 16.48 to 131.56 mg pot^{-1} in acid soil and 20.42 to 150.48 mg pot^{-1} in non calcareous soil. N uptake as expected increased as N levels increased up to 250 kg ha^{-1} and then decreased with further addition (N_{300}). Like N content, the lowest uptake was observed in the control treatment of both soils.

Table 2. Effects of different levels of N on its content and uptake by stevia leaf at harvest

N level	Nitrogen			
	Acid soil		Non-calcareous soil	
	Content (%)	Uptake (mg pot^{-1})	Content (%)	Uptake (mg pot^{-1})
N_0	1.08c	16.48f	1.11c	20.42d
N_{100}	1.10c	33.22e	1.15c	35.77d
N_{150}	1.23bc	51.92d	1.28bc	59.65c
N_{200}	1.29abc	74.82c	1.38abc	81.70b
N_{250}	1.43ab	131.56a	1.52ab	150.48a
N_{300}	1.51a	99.21b	1.62a	138.83a
CV (%)	1.20	4.72	1.32	4.93
$LSD_{0.05}$	0.13	8.05	0.15	9.73
SE±	0.05	9.61	0.05	12.01

CV = Coefficient of variance, LSD = Least significant difference, SE± = Standard error of means

Critical leaf N concentration of stevia leaf

Critical values are quite useful and are frequently referred to when interpreting a plant analysis result. To determine critical N concentration in stevia leaf we followed the "Critical nutrition concentration" concept advanced by Ulrich (1952) for plant. Critical values as used by Ulrich and Hills (1973) are determined from the relationship of nutrient concentration and relative yield at the time of sampling.

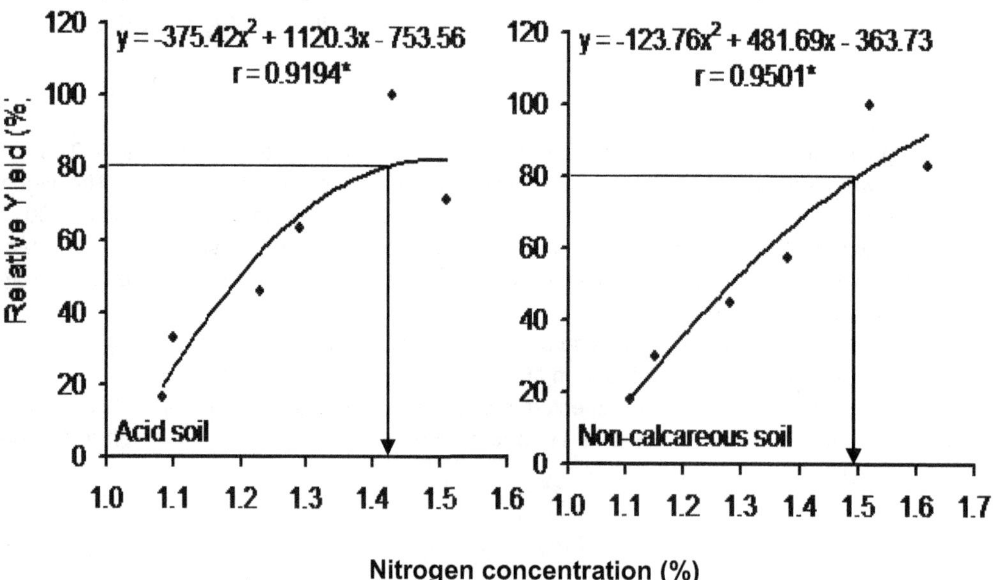

Figure 5. Correlation between leaf N concentration and relative leaf biomass yield of stevia grown in acid and non-calcareous soils. Values are the means of all treatments. *Correlated significantly at $P<0.01$

The critical N concentration in stevia leaf was estimated from the relative amount of leaf biomass to achieve 80% of the maximum production of stevia leaf (Kouno et al., 1999). For both the soil, relative leaf biomass yield was plotted on the ordinate (Y axis) against the respective N concentration of stevia leaf on the abscissa (X axis) in fig. 5. The N concentration corresponding to the arbitrary point at 80% to achieve the maximum leaf biomass production was estimated by the fitted curve to be ca 1.43 and 1.50% in the leaves of stevia plants grown in acid and non-calcareous soils, respectively.

Nitrogen requirement of stevia

A crop's requirement for a specific nutrient is commonly defined as "the minimum content of that nutrient associated with the maximum yield" or "the minimum rate of intake of the nutrient associated with the maximum growth rate" (Loneragan,, 1968). Critical N levels vary greatly depending upon fertilizer applied and whether or not the investigation was performed in the field or green house, choice of crop etc. To determine the requirement of N in soil to obtain 80% of maximum leaf biomass yield, the applied N was plotted on the X axis against the relative leaf biomass yield on the Y axis. From the fitted curve, the corresponding estimated minimum amount of N for leaf biomass production in the plant grown in acid soil and non-calcareous soil to be ca 273 and 257 kg ha^{-1}, respectively (Figure 6)

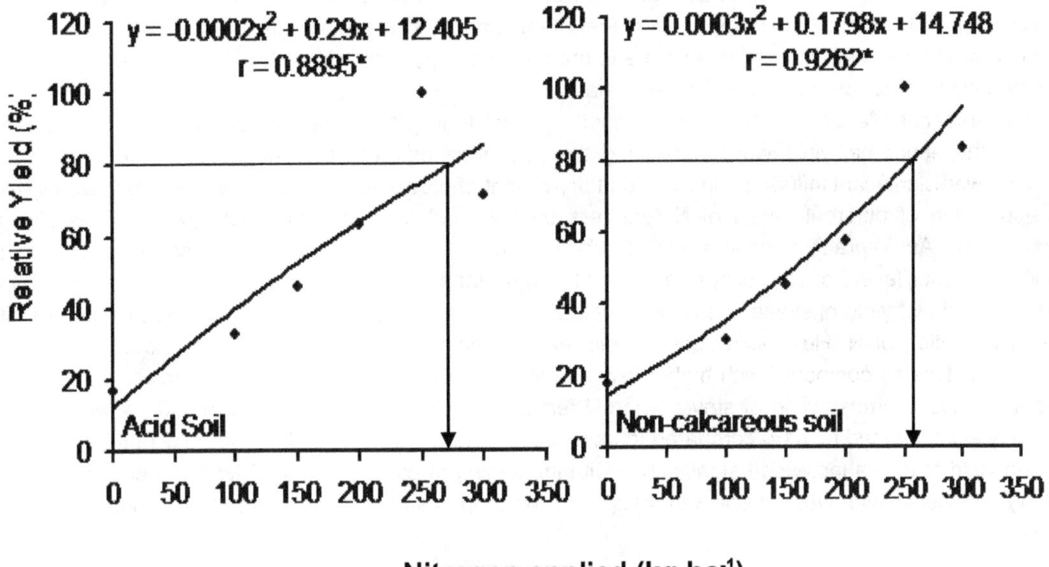

Figure 6. Correlation between applied N and relative leaf biomass yield of stevia grown in acid and non-calcareous soils. Values are the means of all treatments. *Correlated significantly at P<0.01.

DISCUSSION

Leaf yield and yield attributes

The results show that the studied parameters of the stevia plant gradually increased with the progress of growth up to 60 DAP irrespective of treatments. The changes in yield and yield contributing characters varied with treatments. Plant height is an indicator of vegetative growth which is directly influenced by plant production management strategies. The increase in N levels increased the plant height, number of branches and leaves plant^{-1} progressively up to N_{250} at 60 days of growth. Absolute control without N recorded significantly lowest plant height, least number of branches and leaves plant^{-1}. Plant height at harvest was significantly influenced by higher levels of N which in turn was responsible for higher number of branches plant^{-1} and number of leaves plant^{-1} resulting into higher leaf yield. The results are in accordance with the findings of Chalapathi et al. (1999) who also reported increased plant height and number of branches plant^{-1} with higher nutrient levels in sandy loam soils at Bangalore.

Crop performance to a great extent is governed by the number of branches plant^{-1}. It is, therefore, imperative that if the number of branches plant^{-1} is higher, the numbers of leaves are expected to be higher; ultimately the leaf yield will be higher. This finding is also similar with the results of Islam *et al.* (2013) who reported that number of branch plant^{-1} was increased due to application of different doses of inorganic fertilizers. These growth parameters might have possibly contributed positively to the higher leaf yield with higher N application.

Green leaves are the site of photosynthetic activity taking place in the plants. The number of leaves plant^{-1} would also substantiate the fact that increased number of leaves plant^{-1} would contribute to the final yield of the plant particularly the crops like stevia in which only leaves are used for commercial product. Kawatani *et al* (1980) at Japan had also reported the increased number of branches and leaves plant^{-1} of stevia with higher N nutrition. Increased number of leaves plant^{-1} with increased levels of N fertilizers was also reported by Buana and Goenadi (1985) in Brazil. Nasrin (2008) obtained higher values of yield and yield parameters of stevia with N @ 250 kg ha^{-1}. This finding is also similar with the results of Islam *et al.* (2013) who reported that number of leaves plant^{-1} in tomato was increased due to application of different doses of inorganic fertilizers.

Leaf area is an important growth indices determining the capacity of plant to trap solar energy for photosynthesis and has marked influence on the growth and yield of plant. The leaf area was significantly influenced by varied levels of inorganic nutrients. Like yield attributes, leaf area also followed increasing trends with the progress of plant growth with maximum value at 60 DAP irrespective of treatments. Highest values were obtained from N_{250}. Higher leaf area of stevia with higher N levels could be attributed to more number of branches and leaves plant^{-1} due to higher plant height. Significantly lower leaf area, fresh and dry leaf yield was obtained with the absolute control as against all other levels of N due to the lowest number of branches and leaves plant^{-1}. Khanom (2007) reported highest leaf area of stevia plant grown in non-calcareous soil applying chemical fertilizers. From another experiment conducted by Nasrin (2008) using different levels of N, she obtained highest leaf area applying 250 kg N ha^{-1}.

Dry matter accumulation by the crop is another important growth parameter to be considered for determining the economic yield while assessing the effects of different treatments. Different levels of N fertilizers showed significant influence on fresh and dry weight of stevia leaves. The biomass yield was highest due to application of different levels of N fertilizers both in acid and non-calcareous soil. This is also in confirmation with Angkapradipta *et al.* (1986a), where increased biomass production was achieved due to application of higher levels of N. Murayama *et al.* (1980) in Japan experimentally proved that no fertilization resulted in lowest leaf yield of stevia. Increased dry leaf yield was also reported by Shock (1982) in Japan with moderate application of N. He also reported lower leaf dry yield with absolute control without any fertilizer which was 62% less as compared with higher levels of N. Research conducted at Egypt showed a significant increase in dry leaf biomass yield of stevia when N fertilizer was increased from 100 to 300 kg ha^{-1} wherein the dry leaf yield increased by 64% compared to lower dose (Allam *et al*, 2001). In conformity of these findings growth and yield of stevia increased significantly with increasing rates of N up to 60 kg ha^{-1} per crop with the highest dry leaf yield which was on par with 40 kg ha^{-1} per crop in sandy loam soils at Bangalore (Chalapathi *et al*, 1999).

N content and uptake

The nutrient content of a plant varies not only among its various plant parts but changes with age and stage of development. There are also varietal differences which will affect the nutrient content found in various plant parts. A plant analysis interpretation is based on a comparison of the nutrient concentration found in a particular plant part taken at a specific time with known desired value or ranges in concentration. N contents and its uptake by stevia leaf varied significantly in both soils with their additions. The increase in concentration was proportional with the rate of application but the nutrient uptake did not follow the same trend. The highest nutrient contents were obtained from highest N addition (N_{300}) but the highest nutrient uptake was obtained from N_{250}.

Higher nutrient uptake may be related to higher biomass yield. This may be due to the highest dry leaf yield harvested from that treatment. Because nutrient uptake was calculated from their concentrations and corresponding dry leaf yield. In contrast, the lowest content and uptake of nutrients was obtained from control treatments. These results are in conformity with those of Angkapradipta *et al.* (1986). They reported that the stevia plant N content increased due to increased concentration in plant which could be attributed to higher availability and uptake. Nasrin (2008) also reported higher N content and uptake with higher levels of N fertilizer. Shivraj *et al.* (1997) concluded that increased nutrient uptake by stevia resulted from increased application of N fertilizers.

Critical leaf N concentration and N requirement of stevia

The critical level of N in many plants is around 3 percent. For several crops, when the N level in leaves drops below 2.75 percent, N deficiency symptoms appear and yield and quality decline. The primary exceptions are for the very young plants when the critical level may be 4 percent or more, and for leguminous plants, such as soybeans, peanuts, alfalfa, etc., where the critical N percentage is 3 to 4.25 percent. For some tree fruits and ornamentals, N levels may be as low as 2 percent before deficiency occurs. Deficiencies as well as excesses can be a problem. Nitrogen leaf levels in some varieties of pecans exceeding 3.50 percent may result in early defoliation. Nitrogen leaf levels greater than 4.50 to 5 percent retard fruit set in greenhouse tomato. High N levels (>3.50 percent) in forage crops such as fescue is thought to be related to the incidence of grass tetany. Small changes in N content for some crops can result in large effects on yield, plant growth, and the quality of forage and fruit. Therefore, it is important that the N level be maintained within the prescribed limits of the sufficiency range by the proper use of N fertilizer (Bryson and Mills, 2015).

CONCLUSION

The results indicated that all the parameters examined in this study were significantly affected by different doses of N. Across N application rates, the highest values of most parameters except plant height and N content were obtained from 250 kg ha^{-1} and the lowest values from control. N application at all levels increased leaf dry yield at harvest by 99 to 505% in acid soil and 69 to 438% in non-calcareous soil, respectively over control. The increase of most parameters was fast at the later stages (30 to 60 DAP) of plant growth. Leaf N content proportionately increased with the increase of N level. The highest N content was obtained from its highest application. The minimum amount of N for maximum leaf biomass production in the plants grown in acid and non-calcareous soils was estimated to be ca 273 and 257 kg ha^{-1}, respectively. The critical N concentration in stevia leaf was estimated from the relative amount of leaf biomass to achieve 80% of the maximum production of stevia leaf to be ca 1.43 and 1.50% in the plants grown in acid and non-calcareous soils, respectively.

ACKNOWLEDGEMENT

The research work was supported by Bangladesh Agricultural Research Council (BARC), Farmgate Dhaka.

REFERENCES

1. Allam AI, Nassar A and Besheti SY, 2001. Nitrogen fertilizer requirements of *Stevia rebaudiana* under Egyptian conditions. Egyptian Journal of Agricultural Research, 79: 1005-1018.
2. Ali A, Choudhry MA, Malik MA, Ahmad R and Saifullah, 2000. Effect of various doses of nitrogen on the growth and yield of two wheat cultivar. Pakistan Journal of Biological Science, 3: 1004-1005.
3. Angkapradipta P, Tuti-Watsite and Faturachim P, 1986a. The N, P and K fertilizer requirements of *Stevia rebaudiana* Bert. on latosolic soil Menara perkebaunan, Horticulture Abstracts, 56: 9217.
4. Barathi N, 2003. Stevia- The calorie free natural sweetener. *Natural Product Radiance* 2 pp. 120-122.
5. Bryson and Mills. 2015. Plant Analysis Handbook IV, 4th edn., Micro Macro Publishing, 183 Paradise Blvd. Suite 108 Athens, GA 30607
6. Buana L and Goenadi DH, 1985. A study on the correlation between growth and yield of stevia. Menara perkebunan, 53: 68-71.
7. Chalapathi MV, Thimmegowda S, Deva Kumar N, Gangadhar Eswar Rao G, Chandraprakash J. 1999. Influence Of fertilizer levels on growth, yield and nutrient uptake of ratoon crop of Stevia. Crop Research, 21: 947 – 949.
8. Chowdhury MAH, 2000. Dynamics of microbial biomass sulphur in soil and its role in sulphur availability to plants. PhD Thesis, Laboratory of Plant Environmental Science, Graduate School of Biosphere Sciences, Hiroshima University, Japan.
9. Gomez KA and Gomez AA, 1984. Statistical Procedure for Agricultural Research. 2nd edn. International Rice Research Institute. Los Banos, Philippines. pp. 207-215.
10. Gwal HB, RJ Tiwari, RC Jain and FS Prajapati, 1999. Effect of different levels of fertilizer on growth, yield and quality of late sown wheat. RACHIS Newsletter, 18: 42-44.
11. Hasan HM 2008: Agronomic management practice for the improvement of growth and yield of stevia (*Stevia rebaudiana* Bert.). MS Thesis, Department of Agronomy, Bangladesh Agricultural University, Mymensingh.
12. Iqbal J, K Hayat, S Hussain, 2012. Effect of seeding rates and nitrogen levels on yield and yield components of wheat (*Triticum aestivum* L.). Pakistan Journal of Nutrition, 11: 531-536.

13. Islam MR, Chowdhury MAH, Saha BK and Hasan MM 2013: Integrated nutrient management on soil fertility growth & yield of tomato. Journal of Bangladesh Agricultural University, 11: 33-40.
14. Kawatani T, Kaneki Y, Tanabe T and Takahashi T 1980. On cultivation of Kaa-He-E (*Stevia rebaudiana* Bert). VI. Response of stevia to potassium fertilization rates and to the three major elements of fertilizer. Japanese Journal of Tropical Agriculture, 24: 105-112.
15. Khan MAR, 2014. Production technology of stevia (*stevia rebaudiana*) by stem cutting. PhD Thesis, Department of Agronomy, Bangladesh Agricultural University, Mymensingh-2202.
16. Khanom S, 2007. Growth, leaf yield and nutrient uptake by stevia as influenced by organic and chemical fertilizers grown on various types of soil. MS Thesis, Department of Agricultural Chemistry, Bangladesh Agricultural University, Mymensingh-2202.
17. Kouno, K., Lukito, HP and Ando T, 1999. Minimum available N requirement for microbial biomass P formation in a regosol. Soil Biology and Biochemistry, 31: 797-802.
18. Loneragan JF, 1968. Nutrient requirements of plants. Nature, 220: 1307-1308.
19. Murayama SM, Kayan OR, Miyazato K, Nose A 1980. Studies on the cultivation of *Stevia rebaudiana* Bert. II. Effects of fertilizer rates, planting density and seedling clones on growth and yield. Science Bulletin of the college of Agriculture, University of the Ryukyus, Okinawa, 27: 1-8.
20. Nasrin D, 2008. Effect of nitrogen on the growth, yield and nutrient uptake by stevia. MS Thesis, Department of Agricultural Chemistry, Bangladesh Agricultural University, Mymensingh.
21. Page AL, Miller RH and Keeney DR (eds). 1982: Method of Soil Analysis, Part-2 Chemical and Microbiological Properties, 2nd edn., American Society of Agronomy, Inc. Madison, Wisconsin, USA.
22. Ramesh K, Singh V and Ahuja PS 2007: Production potential of *Stevia rebaudiana* (Bert.) Bertoni under intercropping systems. Archives of Agronomy and Soil Science, 53: 443-58.
23. Schock CC 1982: Experimental cultivation of Rebaudia's stevia in California. University California, Davis. Agronomy Progress Report, 122.
24. Shivaraj B, Chalapathi MV and Parma VRR 1997: Nutrient uptake and yield of stevia (*Stevia rebaudiana*) as influenced by methods of planting and fertilizer levels. Crop Research, 14: 205-208.
25. Steel RGD, Torrie JH and Dickey D, 1997. Principles and Procedures of Statistics: A biometrical approach. 3rd Ed. McGraw-Hill Book Co., New York, USA.
26. Ulrich A, 1952. Physiological basis for assessing the nutritional requirements of plants. Annual Review of Plant Physiology, 3: 207-228.
27. Ulrich A and Hills FJ, 1973. Plant analysis as an aid in fertilizing sugar crops: Part i. Sugar beets. In: L.M. Walsh and J.D. Beaton (ed) Soil testing and plant analysis, pp. 271-288. Madison, Wisconsin, USA.
28. Zaman MM, 2015. Nutrient requirement leaf yield and stevioside content of stevia (*Stevia rebaudiana* Bertoni) in some soil types of Bangladesh. PhD Thesis, Department of Agricultural Chemistry, Bangladesh Agricultural University, Mymensingh.
29. Zaman MM, Chowdhury MAH and Chowdhury T, 2015. Growth parameters and leaf biomass yield of stevia (*Stevia rebaudiana*, Bertoni) as influenced by different soil types of Bangladesh. Journal of the Bangladesh Agricultural University, 13: 33-40.

12

EFFECT OF PLANT GROWTH PROMOTING RHIZOBACTERIA (PGPR) IN SEED GERMINATION AND ROOT-SHOOT DEVELOPMENT OF CHICKPEA (*Cicer arietinum* L.) UNDER DIFFERENT SALINITY CONDITION

Mohammad Mosharraf Hossain[1*], Keshob Chandra Das[2], Sabina Yesmin[3] and Syfullah Shahriar[4]

[1]Department of Soil Science, Sher-e-Bangla Agricultural University, Dhaka-1207, Bangladesh; [2]SSO, Molecular Biotechnology Division, National Institute of Biotechnology, Savar-1349, Dhaka, Bangladesh; [3]Mushroom Development Officer, Mushroom Development Institute, Savar-1340, Dhaka, Bangladesh; [4]Department of Soil Science, Sher-e-Bangla Agricultural University, Dhaka-1207, Bangladesh

***Corresponding author:** Md. Mosharraf Hossain; E-mail: soilsaudhaka@yahoo.com

ARTICLE INFO

Key words

Chickpea,
Indole acetic acid,
NaCl,
PGPR

ABSTRACT

Plant growth promoting rhizobacteria (PGPR) are beneficial bacteria that colonize plant roots and enhance plant growth by a wide variety of mechanisms. Ten isolates of bacteria designated as SS01, SS02, SS03, SS04, SS05, SS06, SS07, SS08, SS09 and SS10 were successfully isolated and morphologically and biochemically characterized. Subsequently to investigate the effect of PGPR isolates on the growth of chickpea, a pot culture experiment was conducted in 2013 at National Institute Biotechnology, Bangladesh net house. Prior to seeds grown in plastic pots, seeds were treated with PGPR isolates and seedlings were harvested after 21 days of inoculation. All the isolates were gram negative in reaction, catalase positive, produced indole acetic acid (IAA) as well as performed phosphate solubilization, able to degrade cellulose and have the adaptability in wide range of temperature and showed positive growth pattern in medium. Most of isolates resulted in a significant increasing of shoot length, root length and dry matter production of shoot and root of chickpea seedlings. Application of PGPR isolates significantly improves the percentage of seed germination under saline conditions. The present study, therefore suggested that the use of PGPR isolates SS04, SS10 and SS08 as inoculants biofertilizers might be beneficial for chickpea cultivation in saline condition.

INTRODUCTION

Salinity, an abiotic stress is a severe problem for temperate and tropical agriculture system, which is increasing day by day and affecting 20% of global agriculture (Mayak et al., 2004). The harmful effects of presence of salts in soil result in increased level of ethylene in root, ionic imbalance and hyper-osmotic condition in plants (Niu et al., 1995; Mayak et al., 2004). Many efforts have been made to minimize the severe effects of salt stress on the crop growth and productivity. Biological approaches such as inoculation of seeds/plants with plant growth promoting bacteria (PGPR) and application of growth regulators to induce resistance against stress have already been attempt (Hayat et al., 2010). The most suitable solution or method/approach in this regard is to use the salt tolerant bacterial isolates that may induce salt tolerance thus being useful in facilitating plant growth and yield under salt stress (Bacilio et al., 2004). Rhizosphere is the hotspots of microorganisms and plant growth promoting rhizobacteria (PGPR) are a heterogeneous group of bacteria that can be found in the rhizosphere, at root surfaces and in association with roots, which can improve the extent or quality of plant growth directly and/or indirectly (Gray and Smith, 2005). In last few decades, a large array of bacteria including the species of *Pseudomonas, Burkholderia, Agrobacterium, Erwinia, Xanthomonas, Azospirillum, Bacillus, Enterobacter, Rhizobium, Alcaligenes, Arthrobacter, Acetobacter*, *Acinetobacter, Achromobacter, Aerobacter, Artrobacter, Azotobacter, Clostridium, Klebsiellla, Micrococcus, Rhodobacter, Rhodospirrilum, Flavobacterium* and *Serratia* have been reported to enhance plant growth (Kloepper *et al.*, 1989; Kim and Kim, 2008; Joshi and Bhatt, 2011). PGPR have been demonstrated to increase growth and productivity of many commercial crops including rice, wheat, maize, barley, pea, groundnut, fababean, cucumber, tomato, sorghum, cotton, black pepper, and banana (Ashrafuzzaman *et al.*, 2009; Khalid *et al.*, 2004; Maleki *et al.*, 2010; Mehnaz et al., 2001). The mechanism by which PGPR promote plant growth are not fully understood, but are thought to includes the ability to produce phytohormones, asymbiotic N_2 fixation against phytopathogenic microorganisms by production of siderophores, the synthesis of antibiotics, enzymes and fungicidal compounds (Ahmed *et al.*, 2006). Significant increases in growth and yield attributing agronomical important crops in response to inoculation with PGPR have been reported (Biswas et al., 2000). Another major benefit of PGPR is to produce antibacterial compounds that are effective against certain plant pathogen and pests. Under salt stress, PGPR have positive effect in plants on such parameters as germination rate, tolerance of drought, yield and plant growth (Kokelis-Burelle et al., 2006). Plant growth promoting rhizobacteria (PGPR) induced plants salt stress tolerance has been well studied and is considered to be the cost-effective solution to the problem. PGPR isolated from saline soils improve the plant growth at high salt (Mayak *et al.*, 2004; Barassi *et al.*, 2006). These PGPR tolerance wide range of salt stress and enable plants to withstand salinity by hydraulic conductance, osmotic accumulation, sequestering toxic Na^+ ions, maintaining the higher osmotic conductance and photosynthetic activities (Dodd and Alfocea, 2012). The bacteria obtained from saline environment include, *Flavobacterium*, *Azospirillium*, *Alcaligenes*, *Actinobacterium*, *Pseudomonas* (Rodriguez *et al.*, 1985; Moral *et al.*, 1988; Ilyas *et al.*, 2012), *Sporosarcina*, *Planococcus* (Ventosa *et al.*, 1983), *Bacillus* (Upadhyay *et al.*, 2009), *Thalassobacillus*, *Halomonas*, *Brevibacterium*, *Oceanobacillius*, *Terribacillus*, *Enterobacter*, *Halobacillus*, *Staphylococcus* and *Virgi bacillus* (Roohi *et al.*, 2012) can tolerate in high salinity in soil. Yachana and Subramanian (2013) found that seeds treated with *Pseudomonas pseudoalcaligenes* and *Bacillus pumilus* under saline conditions, as well as non-saline conditions, showed higher germination and survival percentages as compared to non-treated seeds. Kumar *et al.*, (2009) stated that plants inoculated at 1% salinity with the mixture of both PGPRs showed a marginal decrease in germination, survival and plant height, while dry weight showed a marginal increase.

Chickpea is the most important staple food in several developing countries and chemical fertilizers is the most important input required for chickpea cultivation. In order to make its cultivation sustainable and less dependent on chemical fertilizers, it is important to know how to use PGPR that can biologically fix nitrogen, solubilize phosphorus and induce some substances like indole acetic acid (IAA) that can be contribute to the improvement of chickpea growth. Thus the aim of this study was to determine the effect of PGPR strains that are compatible with chickpea. We also investigated the influence of PGPR with salinity on seed germination.

MATERIALS AND METHODS

Soil samples were collected from chickpea field of Tangail sadar upazila of Tangail district. After collection of soil samples, it was stored in plastic container, properly leveled and carried to the laboratory for further use. Ten grams of rhizosphere soil were taken into a 250 ml of conical flask and 90 ml of sterile distilled water was added to it. After serial dilution upto 10, an aliquot of this suspension was spread on the plates of nutrient (NA) agar medium (NH_4Cl 5.0 g; K_2HPO_4 3.0 g; Na_2SO_4 2.0 g; KH_2PO_4 1.0 g; NH_4NO_3 1.0 g; $MgSO_4$, $7H_2O$ 0.1 g; glucose 2.0 g; distilled water 1 litre and pH 7.0±0.2) in the molecular microbiology laboratory of National Institute of Biotechnology (NIB), Savar, Bangladesh. After 3 days of incubation at 28°C, bacterial colonies were streaked to other NA agar plates and incubated at 28°C for 3 more days. Typical single bacterial colonies were observed over the streak. Well isolated single colonies were picked up and different characteristics of colonies such as shape, size, elevation, surface, margin, color, odor and pigmentation etc. were observed and recorded. All the morphological and biochemical characters of the isolates were determined based on Bergey's Manual of Systematic Bacteriology (Garrity *et al.*, 2001). A loopful of bacterial culture from each isolates was diluted into a test tube containing 1 ml sterile distilled water and was vortexed for 2/3 minutes. A loopful suspension was then taken on a glass slide and smeared. The slide was air dried and fixed by heating on a Bunsen flame. The slide was flooded with crystal violet solution for 3 min. The slide was washed gently in flow of tap water and air dried. The slide was observed under microscope and recorded the shape. Motility of bacteria was observed by hanging drop method. A drop of suspension was taken on a cover slip. The cover slip was hanged on a hollow slide with vaseline. The slide was then observed under microscope to test the motility of bacteria. The culture of 10 isolates were streaked on NA agar plates and incubated at 10, 20, 28, 37 and 45°C and also in the NA plate with 0, 3.0, 6.0 and 12.0 mM NaCl solutions in the medium. The bacterial isolates were designated as SS01, SS02, SS03, SS04, SS05, SS06, SS07, SS08, SS09 and SS10. A single colony of bacterial culture was grown on nutrient broth medium. A loopful of the respective culture was transferred to the 100 ml of conical flask then incubated for 7 days on a rotary shaker in 80 rpm. The IAA production and phosphate solubilization were then examined according to the method given by Bric *et al.*, (1991) and Pikovskaya, (1984), respectively. Phosphate solubilizing capacity of the isolates was estimated using Pikovskaya's medium (Pikovskaya, 1948). The cellulose degradation test was performed by using carboxy methyl cellulose (CMC) agar plates according to the method of Kasana *et al.*, (2008). Seeds of chickpea were collected from pulse research centre of Bangladesh Agricultural Research Institute (BARI), Gazipur. Prior to germination, the chickpea seeds were surface sterilized in 3% H_2O_2 and then rinsed with distilled water. The seeds were then surface sterilized with 0.024% sodium hypochlorite for 2 minutes and rinsed thoroughly in sterile distilled water. Seeds were inoculated by overnight soaking with suspensions of bacteria (approximately 10^7-10^8 cfu/ml). Seeds soaked in sterilized distilled water were used as the control. The soaked seeds with rhizobacterial isolates emerged with three different NaCl solutions which were derived from sterile distilled water by adding 0 (control), 3, 6 and 12 mM NaCl, respectively. Chickpea seeds were placed over the sterile filter paper into a petri dish and covered with tight fitting lid. Then the petri-plates were kept in an incubator maintaining the moisture and temperature of 28^0C-30^0C. Seed germination assay was laid out in Completely Randomized Design (CRD) with three replications. Germinated seeds were recorded and discarded at 24 h interval over 10 days.

Germination percentage was calculated by the following formula (Li, 2008):

$$\text{Germination percentage (GP)} = \frac{\text{Total number of seeds germinated}}{\text{Total number of seeds taken for germination}} \times 100$$

An amount of 0.3 kg sand was placed into a pot. Ten PGPR inoculated seeds were sown at 4 to 5 cm depth of sand in each plastic pot and laid out in a completely randomized block design (RCBD) with three replications. The chickpea plants were harvested after 21 days of seed sowing through separating of plants from soil. Shoot length (cm $plant^{-1}$) and root length (cm $plant^{-1}$) and dry weight of shoots and roots of each plant were recorded after drying in an oven for 1 day at 70°C. The data was analyzed statistically by MS-STATC statistical program. The significance of differences between mean values was evaluated by DMRT according to the methods of Gomez and Gomez (1984).

RESULTS AND DISCUSSION

Isolation and Characterization of PGPR

Ten bacterial isolates were successfully isolated from the rhizosphere of chickpea. They were designated as SS01, SS02, SS03, SS04, SS05, SS06, SS07, SS08, SS09 and SS10 as shown in Table 1 and the morphological characteristics of PGPR isolates widely varied. The isolates were found to be first growers. All the isolates produced round shape and raised colonies having smooth shiny surface with smooth margin (Table 1). They differed in color but all were odorless and no pigmentation was observed in the colonies of NA agar plates (Table 1). Diameters of the colonies of isolates varied from 0.2 to 2.0 mm (Table 1). PGPR colonize plant roots and exert beneficial effects on plant growth and development by a wide variety of mechanisms.

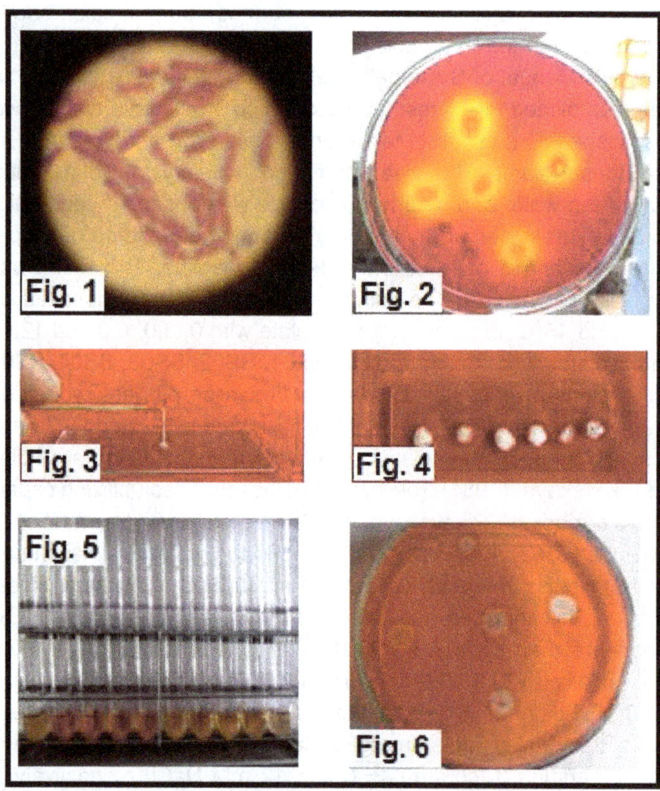

Figure 1. Gram staining of PGPR isolates; **Figure 2.** Phosphate solubilization plate assay test of PGPR isolates; **Figure 3.** KOH test of PGPR isolates; **Figure 4.** Catalase test of PGPR isolates; **Figure 5.** Indole acetic acid (IAA) production by PGPR isolates; **Figure 6.** Cellulose degradation plate assay test of PGPR isolates

Table 1. Morphological characteristics of 3-day old colony of PGPR isolates

Isolates	Shape	Size (mm)	Elevation	Surface	Margin	Colour
SS01	Round	0.9-1.1	Raised	Smooth shiny	Smooth	Off whitish
SS02	Round	0.9-1.1	Raised	Smooth shiny	Smooth	Pinkish
SS03	Round	1.0-1.5	Raised	Smooth shiny	Smooth	Brownish
SS04	Round	1.0-1.5	Raised	Smooth shiny	Smooth	Yolk brown
SS05	Round	1.9-2.0	Raised	Smooth shiny	Smooth	Yellowish
SS06	Round	1.0-1.5	Raised	Smooth shiny	Smooth	Yolk yellowish
SS07	Round	1.5-2.0	Raised	Smooth shiny	Smooth	Whitish
SS08	Round	0.2-0.5	Raised	Smooth shiny	Smooth	Yellowish
SS09	Round	0.9-1.1	Raised	Smooth shiny	Smooth	Whitish
SS10	Round	0.5-1.0	Raised	Smooth shiny	Smooth	Off whitish

Table 2. Cell shape, motility, odor, pigmentation, gram staining, catalase test, KOH test and cellulose degradation test of isolated PGPR of chickpea rhizosphere

Isolates	Cell shape	Motility	Odour	Pigment	Gram staining	KOH test	Catalase test	Cellulose degradation
SS01	Rod	Motile	Odourless	None	-	-	+	+
SS02	Rod	Motile	Odourless	None	-	-	+	+
SS03	Rod	Motile	Odourless	None	-	-	+	+
SS04	Rod	Motile	Odourless	None	-	-	+	+
SS05	Rod	Motile	Odourless	None	-	-	+	+
SS06	Rod	Motile	Odourless	None	-	-	+	+
SS07	Rod	Motile	Odourless	None	-	-	+	+
SS08	Rod	Motile	Odourless	None	-	-	+	+
SS09	Rod	Motile	Odourless	None	-	-	+	+
SS10	Rod	Motile	Odourless	None	-	-	+	+

Table 3. Growth of PGPR isolates at different temperature (10^0C-45^0C) conditions

Isolates	Temperature				
	10^0C	20^0C	28^0C	37^0C	45^0C
SS01	+	++	++	+	-
SS02	+	++	++	+	-
SS03	+	++	++	++	+
SS04	++	++	++	++	++
SS05	+	++	++	+	-
SS06	+	++	++	+	-
SS07	+	++	++	+	-
SS08	+	++	++	+	-
SS09	+	++	++	+	-
SS10	+	++	++	++	-

(- = No growth, + = weak growth and ++ = good growth)

Table 4. Growth of PGPR isolates at different NaCl (0-12.0 mM) concentrations

Isolates	NaCl concentration (mM)			
	0 mM	3.0 mM	6.0 mM	12.0 mM
SS01	++	++	++	+
SS02	++	++	++	+
SS03	++	++	++	++
SS04	++	++	++	++
SS05	++	++	++	+
SS06	++	++	++	++
SS07	++	++	++	+
SS08	++	++	++	+
SS09	++	++	++	++
SS10	++	++	++	++

(+ = weak growth and ++ = good growth)

Microscopic Observation of PGPR Isolates

Microscopic observations were performed to examine some characteristics of PGPR isolates such as shape, gram reaction and motility (Table 2). On the one hand, all the isolates were found in rod shaped, motile and gram negative (Figure 1) in reaction. On the other hand, all the isolates were found to be positive in response to catalase test (Figure 4) and in cellulose degradation test (Figure 6). Catalase activity in the bacterial isolates may potentially be very advantageous. Bacterial isolates with catalase activity positive are highly resistant to environmental, mechanical and chemical stress (Glick *et al.*, 1998). In case of KOH test, all the isolates were tested as negative (Figure 5) and also the isolates were positive in phosphate solubilization NA plate assay test (Figure 2). It was also found to be that growth of isolates on NA agar plates varied in temperature and NaCl concentration (Table 3 and 4). The growth of all isolates was good in the temperature range of 20 to 28°C. In addition, SS03 and SS04 isolates were found to grow at 45°C and isolates SS03, SS04, SS06, SS09 and SS10 were able to grow in 12.0 mM NaCl solutions.

Production of IAA and Solubilization of Phosphorus

As shown in Figure 7, all the isolates produced indole acetic acid (IAA) in various ranges. On the contrary, SS01 was found to be a weak IAA producer in comparison to the minimum IAA producer isolates viz. SS02, SS03, SS04, SS05 and SS09. SS06 and SS07 produced the medium quantity IAA producer. On the other hand, only SS10 produced the highest IAA among the isolates. In phosphate solubilization, SS01 solubilized the lowest amount of phosphate and isolates SS04, SS06, SS07 and SS09 were found to be the medium phosphate solubilizers. Moreover, it was noticed that all the isolates solubilize phosphate in the nutrient agar medium plate assay (Fig. 2). Furthermore, SS05 solubilized the highest amount of phosphate followed by SS02, SS03 and SS08 (Fig.7). It has been reported that IAA production by PGPR can vary among different species and it is also influenced by culture condition, growth stage and substrate ability (Mirza *et al.*, 2001). PGPR have been shown to solubilize precipitated phosphates and enhance phosphate availability to chickpea that represent a possible mechanism of plant growth promotion under field condition (Verma *et al.*, 2001).

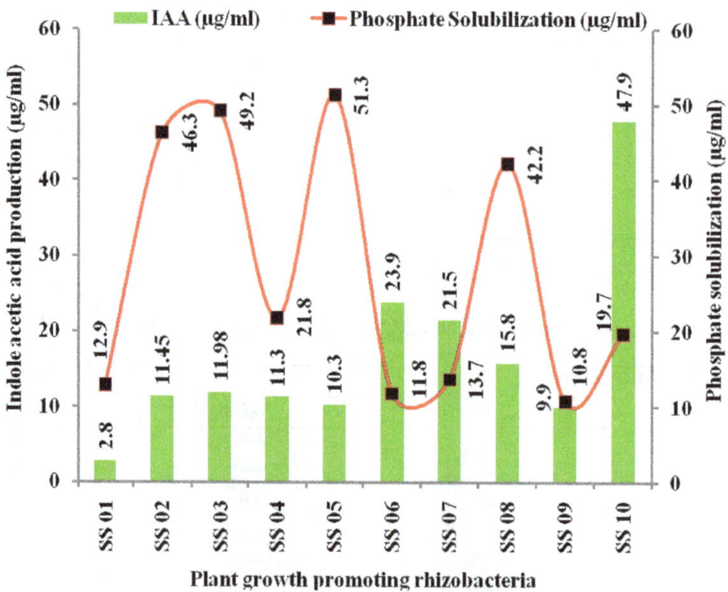

Figure 7. Indole acetic acid (IAA) production and phosphate solubilization by the isolates

Seed Germination

The first observation of this study was that increasing NaCl concentration decreased the germination percentage in chickpea seeds (Fig. 10). When NaCl treatments compared to each other it was seen that the effect of PGPR on germination percentage varies with bacterial isolates. The effect of PGPR on germination rate of seeds under saline conditions was statistically significant ($p<0.05$). Shannon and Grieve, (1999) reported that salinity showed the germination rate and at low concentration the only was on germination rate and not total percentage of seeds.

The results of our study clearly showed that PGPR improved germination percentage and rate according to the control in spite of the use of high concentrations of NaCl. Nelson, (2004) noted that PGPR were able to exert a beneficial effect upon plant growth such as increasing the germination rate.

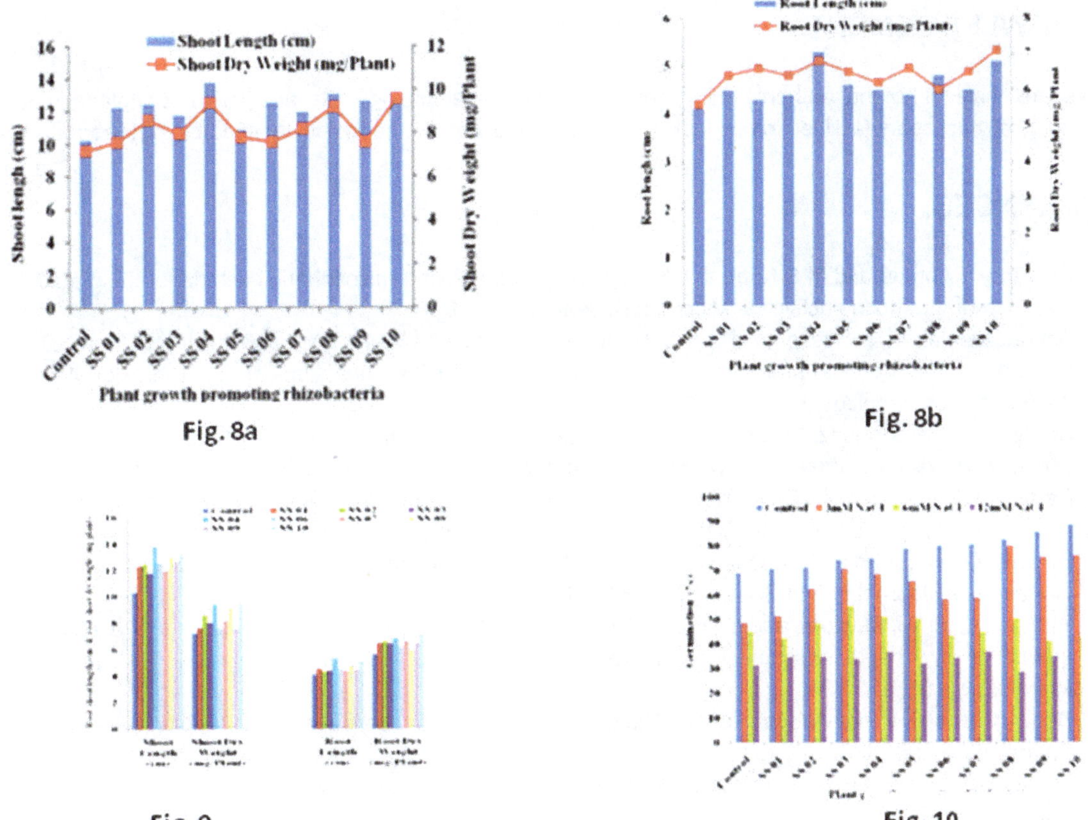

Figure 8a. Effect of PGPR on shoot length (cm) and shoot dry weight (mg/plant) of chickpea; **Figure 8b.** Effect of PGPR on root length (cm) and root dry weight (mg/plant) of chickpea; **Figure 9.** Effect of PGPR on shoot length (cm), shoot dry weight (mg/plant), root length (cm) and root dry weight (mg/plant) of chickpea; **Figure 10.** The effect of PGPR on seed germination of chickpea under salinity condition

Length and Dry Weight of Shoot and Root

The PGPR isolates significantly affected the length of shoots and roots of chickpea seedlings. Results reveal that the shoot length increased in PGPR treated plants over un-inoculated control (Figure 8a, 8b and 9). The highest shoot length 13.80 cm was recorded in SS04 isolate which was statistically similar to isolates SS08 (13.10 cm plant^{-1}) and SS10 (13.20 cm plant^{-1}). A significant increase in shoot dry matter of chickpea seedling was observed in response to PGPR isolates. The highest shoot dry matter was recorded in isolate SS10 (9.60 mg plant^{-1}) followed by SS04 (9.40 mg plant^{-1}) and SS08 (9.20 mg plant^{-1}). Root length ranged from 4.10 to 5.30 cm plant^{-1}. The isolate SS04 produced the highest root length (5.30 cm plant^{-1}), in comparison to other isolates SS05 and SS08 also showed superior root length, respectively (Figure 8b and 9). A significant variation in root dry weight was observed in response to different PGPR isolates. In this study, the effectiveness of PGPR isolates on shoot length, root length and dry weight of shoot and root were investigated. Most of the isolates significantly increased shoot length, root length and dry matter production of shoot and root of seedlings (Figure 8a, 8b and 9). Bacteria produce IAA to promote root growth by stimulating cell division or elongation (Patten and Glick, 2002).

Our results suggested that PGPR are able to enhance the production of IAA, solubilization of phosphorus and resistance to pathogen and pests, thereby improving growth of chickpea plant. The use of PGPR as inoculants biofertilizers is an efficient approach to replace chemical fertilizers and pesticides for sustainable chickpea cultivation in Bangladesh and other developing countries. Further investigations, including efficiency test under green house and field conditions needed to clarify the role of PGPR as biofertilizers that exert beneficial effects on plant growth and development.

ACKNOWLEDGEMENT

We are thankful to the authority of National Institute Biotechnology, Savar, Dhaka, Bangladesh for providing us all necessary facilities to conduct the research and work in the molecular microbiology laboratory.

REFERENCES:

1. Ahmad F, I Ahmad and M S Khan, 2006. Screening of free-living rhizospheric bacteria for their multiple plant growth promoting activities. Microbial Research, 36: 1-9.
2. Ashrafuzzaman M, F Hossen, A Razi, M Ismail, MA Hoque Z M Islam, SM Shahidullah, S Meon, 2009. Efficiency of plant growth-promoting rhizobacteria (PGPR) for the enhancement of rice growth. African Journal of Biotechnology, 8: 1247–1252.
3. Bacilio, M., H. Rodriguez, M. Moreno, J. P. Hemandez and Y. Bashan. 2004. Mitigation of salt stress in wheat seedlings by a gfp-tagged *Azospirillum lipoferum*. Biology and Fertilizer of Soils, 40: 188-193.
4. Barassi CA, G Ayrault, CM Creus, RJ Sueldo and MT Sobrero, 2006. Seed inoculation with Azospirillum mitigates NaCl effects on lettuce. Science of Horticulture, 109: 8-14.
5. Biswas JC, LK Ladha and FB Dazzo, 2000 Rhizobia inoculation improves nutrient uptake and growth of low land rice. Journal of Soil Science, 64: 1644-1650.
6. Bric, J. M., R. M. Bustock and S. E. C. Silversone. 1991. Rapid *in situ* assay for indole acetic acid production by bacterial immobilization on a nitrocellulose membrane. Applied Environmental Microbiology, 57: 535-538.
7. Dodd IC and F Perez-Alfocea, 2012. Microbial alleviation of crop salinity. Journal of Experimental Botany, 63: 3415-3428.
8. Garrity GM, Boone DR and Castenholz RW, 2001. Bergey's manual of systematic bacteriology, 2nd ed., Springer-Verlag, New York, NY.
9. Glick BR, DM Penrose and J Li, 1998. A model for the lowering of plant ethylene concentrations by plant growth promoting bacteria. Journal of Theoretical Biology, 190: 63-68
10. Gomez KA and Gomez AA, 1984. Statistical procedure for agricultural research, (2nd Ed.), John willey and sons, Inc. New York. pp. 1-680.
11. Gray EJ and Smith DL, 2005. Intracellular and extracellular PGPR: commonalities and distinctions in the plant–bacterium signaling processes. Soil Biology and Biochemistry, 37: 395-412.
12. Hayat R, S Ali, U Amara, R Khalid and I Ahmed, 2010. Soil beneficial bacteria and their role in plant growth promotion. a review. Annals of Microbiology, DOI 10.1007/s13213-010-0117-1.
13. Ilyas N, A Bano, S Iqbal and NI Raza, 2012. Physiological, biochemical and molecular characterization of *Azospirillum* spp. isolated from maize under water stress. Pakistan Journal of Botany, 44: 71-80
14. Joshi P and AB Bhatt, 2011. Diversity and function of plant growth promoting rhizobacteria associated with wheat rhizosphere in North Himalayan region. International Journal Environmental Science, 1: 1135-1143.
15. Kasana RC, R Salwan, H Dhar, S Dutt and A Gulati, 2008. A rapid and easy method for the detection of microbial cellulases on agar plates using Gram's iodine. Current Microbiology, 57: 503-507.
16. Khalid A, M Arshad and Z.A Zahir, 2004. Screening plant growth promoting rhizobacteria for improving growth and yield of wheat. Journal of Applied Microbiology, 96: 473-480.
17. Kim JT and SD Kim, 2008. Suppression of bacterial wilt with *Bacillus subtilis* SKU48-2 strain. Korean Journal of Microbiology and Biotechnology, 36: 115-120.
18. Kloepper JW, R Lifshitz and RM Zablotwicz, 1989. Free-living bacterial inocula for enhancing crop productivity. Trends of Biotechnology, 7: 39-43.

19. Kokelis-Burelle N, JW Kloepper and MS Reddy, 2006. Plant growth promoting rhizobacteria as transplant amendments and their effects on indigenous rhizosphere microorganisms. Applied Soil Ecology, 31: 91-100.
20. Kumar A, S Sharma and S Mishra, 2009. Effect of alkalinity on growth performance of Jatropha curcas inoculated with PGPR and AM fungi. Journal of Phytology, 1: 177-184
21. Li Y, 2008. Effect of salt stress on seed germination and seedling growth of three salinity plants. Pakistan Journal of Biological Sciences, 11: 1268-1272.
22. Maleki M, S Mostafee, L Mokhaternejad and M Farzaneh, 2010. Characterization of *Pseudomonas fluorescens* strain CV6 isolated from cucumber rhizosphere in Varamin as a potential biococntrol agent. Australian Journal of Crop Science, 4: 676-683.
23. Mayak S, T Tirosh and BR Glick, 2004. Plant growth promoting bacteria confer resistance in tomato plants to salts stress. Plant Physiology and Biochemistry, 42: 565-572.
24. Mehnaz S, MS Mirza, J Haurat, R Bally, P Normand, A Bano and KA Malik, 2001. Isolation and 16S rRNA sequence analysis of the beneficial bacteria from the rhizosphere of rice. Canadian Journal of Microbiology, 47: 110-117.
25. Mirza MS, W Ahmad, F Latif, J Haurat, R Bally, P Normand and KA Malik, 2001. Isolation, partial characterization and effect of plant growth promoting bacteria on micro-propagated sugarcane *in vitro*. Plant and Soil, 237: 47-54.
26. Moral AD, B Prado, E Quesda, T Gacria, R Ferrer and Ramos-Comenzana, 1988. Numerical taxonomy of moderately halophilic Gram negative rods from an inland saltern. Journal of General Microbiology, 134: 733-741.
27. Nelson L M, 2004. Plant growth promoting rhizobacteria (PGPR): Prospects for new inoculants Online. Crop Management. doi: 10.1094/ CM-2004-0301-05-RV.
28. Niu X, RA Bressan PM Hasegawa and JM Pardo, 1995. Ion homeostasis in NaCl stress environments. Plant Physiology, 109: 735-742.
29. Patten CL and BR Glick, 2002. Role of *Pseudomonas pudita* indole acetic acid in development of the host plant root system. Applied Environmental Microbiology, 68: 3795-3801.
30. Pikovaskaya RI, 1948. Mobilization of phosphorus in soil in connection with the vital activities of some microbial species. Microbiology, 17: 362-370.
31. Rodriguez-ValeraF, A Ventosa, G Juez and LF Imhoff, 1985. Variation of environmental features and microbial populations with the salt concentrations in a multi-pond saltern. Microbial Ecology, 11: 107-111.
32. Roohi A, I Ahmed, M Iqbal and M Jamil, 2012. Preliminary isolation and characterization of halo-tolerant and halophilic bacteria from salt mines of Karak, Pakistan. Pakistan Journal of Botany, 44: 365-370.
33. Shannon MC and CM Grieve, 1999. Tolerance of vegetable crops to salinity. Science of Horticulture, 78: 5-38.
34. Upadhyay SK, DP Singh and R Saikia, 2009. Genetic diversity of plant growth promoting rhizobacteria isolated from rhizosphere soils of wheat under saline condition. Current Microbiology, 59: 489-496.
35. Ventosa A, A Ramose and M Kocur, 1983. Moderately halophilic Gram-positive cocci from hyper-saline environment. Systematic Applied Microbiology, 4: 564-570.
36. Verma SC, JK Ladha and A. K. Tripathi, 2001. Evaluation of plant growth promoting and coloization, ability of endophytic diazotrophs from deep water rice. Journal of Biotechnology, 91: 127-141.
37. Yachana, J and RB Subramanian, 2013. Paddy plant inoculated with PGPR show better growth physiology and nutrient content under saline conditions. Chilean Journal of Agricultural Research, 73: 213-219.

13

EXPLORATION OF PHYSICO-CHEMICAL PARAMETERS AND IONIC CONSTITUENTS FROM GROUNDWATER USED IN IRRIGATION OF TANGAIL DISTRICT, BANGLADESH

Elina Aziz, Md. Younus Mia* and Nowara Tamanna Meghla

Department of Environmental Science and Resource Management, Mawlana Bhashani Science and Technology University, Santosh, Tangail-1902, Bangladesh

***Corresponding author:** Md. Younus Mia, E-mail: mdmia1998@gmail.com

ARTICLE INFO

Key words

Physico-chemical parameters, Groundwater, Irrigation, Tangail district

ABSTRACT

The study was conducted for exploration of physico-chemical parameters and ionic constituents of groundwater used in irrigation of four upazilas namely Tangail Sadar, Kalihati, Delduar and Nagarpur upazila of Tangail district during the months of March, April and May of 2015. The physico-chemical parameters (pH, EC and TDS), ionic constituents (Ca^{2+}, Mg^{2+}, Na^{+}, K^{+}, Cl^{-}, CO_3^{2-}, HCO_3^{-}, PO_4^{3-} and SO_4^{2-}) and trace metal (Fe and Mn) were analyzed to assess the quality of irrigation in relation to soil properties and crop growth. The pH of groundwater indicates slightly alkaline in nature. As regards to EC the groundwater was in 'good' class and medium salinity hazards in quality for irrigation and the concentration of TDS indicates water as fresh water. The concentration of Ca^{2+}, Mg^{2+}, Na^{+}, K^{+}, Cl^{-}, CO_3^{2-}, HCO_3^{-}, PO_4^{3-} and SO_4^{2-} of groundwater were recorded within the permissible limit for irrigation and these ions might not create hazardous impact on soil ecosystem for growing crops. The trace amount of Fe and Mn was detected in irrigation water. In the study area, the groundwater was within the recommended limit and would not create problem for irrigation and that have not long term effects on irrigating agricultural crops which could be safely used for irrigation purposes.

INTRODUCTION

Groundwater is an important source of fresh water for irrigation use in many regions of Bangladesh and is a component of the nation's freshwater resources. About 75% of cultivated land is irrigated by groundwater and the remaining 25% by surface water. Groundwater about 70-90% is used for agricultural purposes and the rest for drinking and other water supplies (BBS, 2006). The chemical quality of groundwater is considered as an important criterion for long-term irrigation because of containing the relatively high content of various ions as dissolved chemical constituents as compared to surface water (Ayers and Westcot, 1985). All water contain varying amount of different types of ion (cations and anions). Among them the main soluble constituents are Ca^{2+}, Mg^{2+}, Na^+ and K^+ as cations and Cl^-, SO_4^{2-}, PO_4^{3-}, CO_3^{2-} and HCO_3^- as anions. Out of the soluble constituents, Ca^{2+}, Mg^{2+}, Na^+, Cl^-, SO_4^{2-}, and HCO_3^- are of prime importance in judging the water quality for irrigation (Golterman, 1971). If the polluted groundwater is applied for irrigation, some ions may accumulate in soils as well as crops and deteriorates soil environment ultimately affecting crop production (Ayers and Westcot, 1985). The quantity and quality of different ions in groundwater and surface water system influences its better utilization for irrigation. Nowadays, agrochemicals are applied indiscriminately under intensive agriculture and as a result, lead to ionic contamination of groundwater and surface water (Lal and Stewart, 1994; Schwartz and Zhang, 2012). Recently, the increased attention has been paid in water for monitoring and management of these ions for using safe water.

Tangail region groundwater is mainly applied for irrigation. During the dry season farmers are completely depended on groundwater for irrigation. The cropping sequences like HYV rice, vegetables and rabi crops are found to be cultivated under irrigation. For this purpose, the quality of irrigation water must be evaluated and to identify the concentrations of ionic compounds that are important for plant growth. Keeping the above fact in judgment, the study was conducted to explore physico-chemical parameters and ionic constituents from groundwater and to evaluate the suitability of water with irrigation standard for crop production.

MATERIALS AND METHODS

Study area

The study area was located at the four upazilas namely Tangail Sadar, Kalihati, Delduar and Nagarpur upazila of Tangail district (Figure 1) that occupies an area of 33426, 29560, 18454 and 26270 acres respectively. The total land areas of Tangail district are 835422 acre in which cultivable area 577701 acre and irrigated area 285914 acre (BBS, 2012).

Sample collection

Groundwater samples were collected during the months of March, April and May (2015) from four upazilas of Tangail district. The each of four upazilas were divided into threes sampling points and from each points twelves samples were collected from each of three months (March, April and May) and finally total thirty six groundwater samples were collected from the agricultural sites of the study area. Groundwater samples were collected from 12 shallow tubewells. At the time of sampling, all groundwater samples were free from colour, odour and unpleasant taste. All samples were filtered through filter paper to remove undesirable solids and suspended materials before chemical analysis. The collected groundwater samples were tightly sealed as quickly as possible to avoid air exposure and analyzed immediately.

Sample Analysis

The pH, EC (Electrical conductivity) and TDS (Total Dissolved Solids) of groundwater were measured by pH, EC and TDS meter, respectively (APHA, 2012). The concentrations of Ca^{2+} and Mg^{2+} in water samples were estimated by EDTA tritimetric method (Page *et al.,* 1982) whereas K^+ and Na^+ contents were determined by flame photometric method (Ghosh *et al.*, 1983). The concentrations of Cl^-, CO_3^- and HCO_3^- in water samples were analyzed by titrimetric method (Tandon, 1995). The content of PO_4^{3-} and SO_4^{2-} in water samples were determined by spectrophotometric method (Tandon, 1995) where as the trace metal Fe and Mn were analyzed by atomic absorption spectrophotometric method (APHA, 2012).

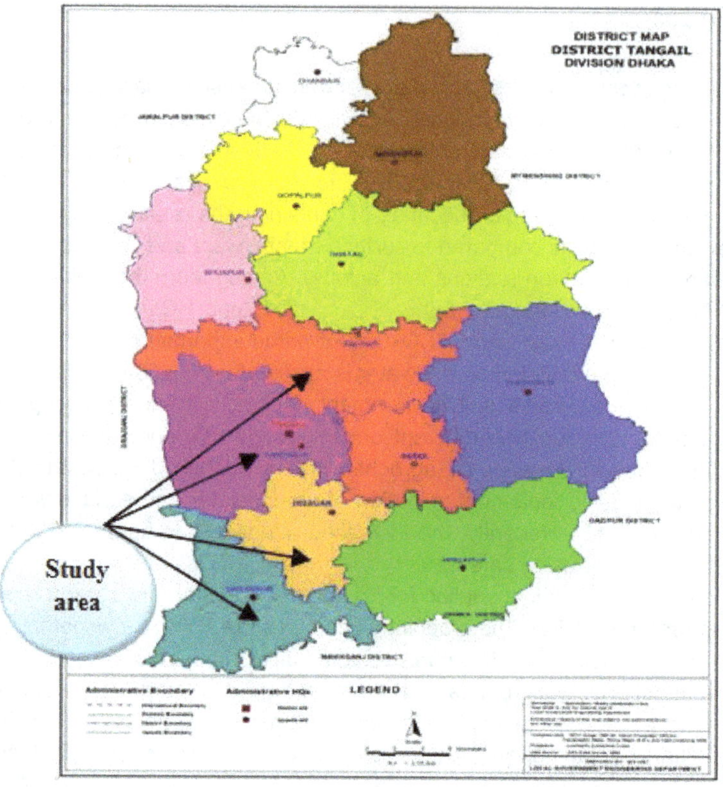

Figure 1. Map showing the study area at the four upazila in Tangail district (Banglapedia, 2008)

RESULTS AND DISCUSSION

The pH of all groundwater samples were ranged from 7.53±0.05 to 8.07±0.06 which indicating slightly alkaline in nature and was suitable for crop production (Table 1) where as the acceptable limit of pH in groundwater for irrigation is from 6.50 to 8.40 (Ayers and Westcot, 1985). The higher value of pH represent that there is high chloride, carbonate, bicarbonate etc. (Michael, 1978).

As regards to electrical conductivity (EC) of all groundwater samples was found within the limit of 373.67±34.64 to 463.33±20.21 µScm^{-1} (Table 1). All groundwater samples were considered as medium salinity hazard (C2, EC= 250-750 µScm^{-1}) and could be safely used for crops growing on soils with moderate level of permeability and leaching as mentioned by Richards (1968).

Table 1. The average value of physico-chemical parameters of groundwater for irrigation

Parameters	Sampling period	Sampling Site				Min. Max
		Tangail Sadar	Kalihati	Delduar	Nagarpur	
	March	7.63±0.04	7.53±0.05	7.81±0.05	7.90±0.04	7.53±0.05
pH	April	7.69±0.03	7.60±0.04	7.89±0.02	7.98±0.02	8.07±0.06
	May	7.74±0.04	7.64±0.03	7.94±0.04	8.07±0.06	
EC	March	406.33±25.11	463.33±20.21	395.00±27.84	456.67±36.53	373.67±34.64
(µScm^{-1})	April	388.67±23.03	446.67±17.56	382.00±30.45	435.33±39.00	463.33±20.21
	May	382.00±25.63	439.00±16.22	373.67±34.64	427.33±27.56	
TDS	March	288.33±7.64	233.33±12.58	233.33±12.58	273.67±12.66	210.00±5.57
(ppm)	April	272.67±14.19	240.33±18.18	218.33±7.64	255.33±12.66	288.33±7.64
	May	261.67±7.64	229.33±13.01	210.00±5.57	242.00±15.36	

The measured total dissolved solids (TDS) varied from 210.00±5.57 to 288.33±7.64 ppm (Table 1). All groundwater samples were considered as fresh water (TDS < 1000 mg L^{-1}) in quality (Freeze and Cherry, 1979) and that could be used for irrigation purposes.

The concentration of Ca^{2+} of all groundwater samples were ranged from 34.01±1.50 to 63.38±3.58 mg L^{-1} (Table 2). Irrigation water containing less than 100 mg L^{-1} of Ca content is "suitable" for raising crop plants, as mentioned by Todd (1980). On the basis of this acceptable limit, Ca^{2+} status of all collected groundwater was not treated as contaminants for irrigation purpose. According to Ayers and Westcot (1985), irrigation water containing below 121.5 mg L^{-1} (20 me L^{-1}) of Mg^{2+} is suitable for irrigating crops. The measured Mg^{2+} content of groundwater samples were varied from 21.52±0.30 to 40.04±0.46 mg L^{-1} (Table 2) that was within the acceptable limit and could safely be used for irrigation.

Table 2. The average values of ionic constituents of groundwater for irrigation

Parameters	Sampling period	Sampling Sites				Min. Max
		Tangail Sadar	Kalihati	Delduar	Nagarpur	
Ca^{2+}	March	40.46±2.42	34.01±1.50	45.34±1.61	53.09±2.45	34.01±1.50
(mg L^{-1})	April	48.64±2.41	42.23±2.30	52.59±4.02	59.97±3.19	63.38±3.58
	May	52.23±3.27	45.78±2.27	55.86±3.85	63.38±3.58	
Mg^{2+}	March	32.05±0.34	28.45±0.58	35.42±0.63	40.04±0.46	21.52±0.30
(mg L^{-1})	April	27.70±0.82	23.48±0.31	31.89±1.34	34.90±0.56	40.04±0.46
	May	24.60±0.92	21.52±0.30	29.36±1.86	32.85±1.00	
Na^{+}	March	19.26±0.31	28.32±1.12	20.94±0.78	22.78±0.94	19.26±0.31
(mg L^{-1})	April	23.72±0.88	23.49±1.06	23.79±1.00	25.07±0.80	26.10±0.78
	May	25.60±0.33	22.89±1.12	24.63±1.49	26.10±0.78	
K^{+}	March	0.69±0.08	0.63±0.06	0.78±0.06	0.76±0.05	0.63±0.06
(mg L^{-1})	April	0.76±0.07	0.69±0.08	0.87±0.08	0.82±0.05	0.92±0.06
	May	0.82±0.05	0.73±0.07	0.92±0.06	0.87±0.04	
Cl^{-}	March	0.65±0.03	0.57±0.07	0.70±0.03	0.77±0.03	0.57±0.07
(me L^{-1})	April	0.71±0.04	0.64±0.04	0.76±0.02	0.84±0.04	0.88±0.03
	May	0.76±0.03	0.68±0.05	0.80±0.02	0.88±0.03	
HCO_3^{-}	March	0.72±0.11	0.64±0.09	0.89±0.10	1.07±0.08	0.64±0.09
(me L^{-1})	April	0.77±0.10	0.70±0.11	0.96±0.13	1.14±0.05	1.17±0.06
	May	0.81±0.11	0.73±0.13	1.00±0.14	1.17±0.06	
PO_4^{3-}	March	0.16±0.02	0.19±0.01	0.20±0.02	0.21±0.02	0.15±0.01
(mg L^{-1})	April	0.17±0.01	0.20±0.02	0.21±0.01	0.22±0.02	0.22±0.02
	May	0.15±0.01	0.18±0.01	0.18±0.02	0.20±0.02	
SO_4^{2-}	March	2.21±0.08	2.25±0.10	2.52±0.11	2.73±0.07	2.14±0.05
(mg L^{-1})	April	2.14±0.05	2.17±0.06	2.43±0.13	2.60±0.09	2.73±0.07
	May	2.18±0.08	2.20±0.09	2.47±0.09	2.65±0.10	

In groundwater samples, the status of Na^{+}, K^{+} and Cl^{-} ions ranged from 19.26±0.31 to 26.10±0.78 mg L^{-1}; 0.63±0.06 to 0.92±0.06 mg L^{-1} and 0.57±0.07 to 0.88±0.03 me L^{-1}, respectively (Table 2). On the basis of Ayers and Westcot (1985), the accepted usual limits of Na^{+}, K^{+} and Cl^{-} ions are 920.00 mg L^{-1}, 2.00 mg L^{-1} and 4.00 me L^{-1} respectively. Considering these limits, all groundwater samples containing these ions had no remarkable impact on soil properties and crop growth when applied to soil as irrigation water.

The concentration of CO_3^{2-} did not get in all the groundwater samples. So the pollution of irrigation water by this ion dose not arises at all. Bohn et al. (1985), as reported that the concentration of CO_3^{2-} should be negligible at pH < 9.0 which might be the reason of finding the samples carbonate free, as because the pH of all groundwater samples varied from 7.53 to 8.07. The concentration of HCO_3^{-} of all groundwater samples were ranged from 0.64±0.09 to 1.17±0.06 me L^{-1} (Table 2) where as the recommended maximum limit of HCO_3^{-} for irrigation water used continuously on soil is 1.50 me L^{-1}, as mentioned by Evangelou (1998). So, the groundwater of study area was suitable for irrigation on the basis of HCO_3^{-} content and has no possibility of soil hazard.

The concentration of PO_4^{3-} ion in all groundwater samples were measured from 0.15±0.01 to 0.22±0.02 mg L^{-1} (Table 2) and was within the recommended limit (2.0 mg L^{-1}) as reported by Ayers and Westcot (1985). The groundwater is not hazardous for long-term irrigation and showing no impact on soil properties and crop growth. In all groundwater samples, the SO_4^{2-} content ranged from 2.14±0.05 to 2.73±0.05 mg L^{-1} (Table 2) and was not problematic when applied to soil as irrigation water, because of persisting below the permissible limit (20.00 mg L^{-1}) for irrigation (Ayers and Westcot, 1985).

In the study area, trace amount of Fe and Mn (<0.01 mg L^{-1}) were detected in all groundwater samples. Ayers and Westcot (1985), reported that the maximum recommended concentration of Fe and Mn in water used for irrigation is 5.0 mg L^{-1} and 0.20 mg L^{-1}, respectively that indicate the groundwater as fresh water and no ionic constituents are considered as hazardous to crop growth for irrigation

CONCLUSION

From the investigation, it can be concluded that groundwater of the study area would not create any problem for irrigation which related to soil properties and crop growth and that have not long term effects on irrigating agricultural crops, that could be safely used for irrigation purpose in Tangail district, hence all of the groundwater samples were within the standard limit. Finally it can be recommended that, the water quality must be tested with time interval for better planning of any irrigation system to sustain the crop production.

REFERENCES

1. APHA (American Public Health Association), 2012. Standard Methods for the Examination of Water and Waste water. 19th ed., Washington DC, USA, pp. 1-30, 40-175.
2. Ayers RS and Westcot DW, 1985. Water Quality for Agriculture. FAO Irrigation and Drainage Paper, 29: 196.
3. Banglapedia, 2008. National Encyclopedia of Bangladesh, Asiatic Society of Bangladesh.
4. BBS, 2006. Statistical Yearbook of Bangladesh, Bangladesh Bureau of Statistics, Statistics Division, Ministry of Planning, Government of the People's Republic of Bangladesh, Dhaka.
5. BBS (Bangladesh Bureau of Statistics), Cultural survey report of Tangail Zila, June, 2012.
6. Bohn HL, McNeal BL and 0' Conner GA, 1985. Soil Chemistry. 2nd ed., John Wiley and Sons, New York, pp. 239-245.
7. Evangelou VP, 1998. Environmental Soil and Water Chemistry: Principles and Applications. John Wiley and Sons, Inc., New York, USA, pp. 478-485.
8. Ghosh AB, Bajaj JC Hasan, R and Singh D, 1983. Soil and Water testing Methods: A Laboratory Manual. Division of Soil Science and Agricultural Chemistry, IARI, New Delhi, India, pp. 1-48.
9. Golterman HL (ed.), 1971. Methods for Chemical Analysis of Fresh waters. Blackwell Scientific Publications, Oxford, pp. 41-42.
10. Lal R and Stewart BA (ed.), 1994. Soil Processes and Water Quality. Lewis Publishers, Boca Raton, Florida, USA, pp. 2-3.
11. Michael AM, 1978. Irrigation: Theory and Practice. Vikas Pub. House Pvt. Ltd. pp. 448-452.
12. Page AL, Miller RH and Keeney DR (eds.) 1982. Methods of Soil Analysis. Part-2. 2nd edn., American Society of Agronomy, Wisconsin, USA, pp. 98-765.
13. Richards LA (ed). 1968. Diagnosis and Improvement of Saline and Alkali Soils. Agricultural Hand Book 60, USDA and IBH. Publishing Co. Ltd. New Delhi, India, pp. 98-99.
14. Schwartz FW and Zhang H, 2012. Fundamentals of Ground Water. Wiley India Pvt. Ltd. New Delhi, India, pp. 374-377.
15. Tandon HLS (ed.), 1993. Methods of Analysis of Soils, Plants, Waters and Fertilizers. Fertilizer Development and Consultation Organization, New Delhi, India, pp. 90-91.
16. Todd DK and Mays LW, 2004. Ground water Hydrology. 3rd ed., John Wiley and Sons Inc., New York, USA.

14

COMBINED USE OF DIETARY PROBIOTIC AND ACIDIFIER FOR THE PRODUCTION OF ANTIBIOTIC FREE BROILER

Abdullah-Al-Masud, Md. Shawkat Ali and Muslah Uddin Ahammad*

Department of Poultry Science, Faculty of Animal Husbandry, Bangladesh Agricultural University, Mymensingh-2202, Bangladesh

***Corresponding author:** Muslah Uddin Ahammad; E-mail: muslah.ps@bau.edu.bd

ARTICLE INFO | **ABSTRACT**

Key words

Probiotic,
Acidifier,
Antibiotic growth,
promoter,
Broiler

The effect of feeding probiotic (Bio-Top; *Bacillus subtilis* and *Bacillus licheniformis*), acidifier (Sal-Stop), antibiotic growth promoter (AGP) or probiotic plus acidifier was investigated in commercial broiler. A total of four hundred Cobb 500 day-old straight run chicks were randomly distributed to 5 different dietary groups having 4 replications each. The number of birds in each replication was 20. The five dietary groups were as control (basal diet; BD), BD containing AGP at a level of 20g/100kg, BD containing probiotic at a level of 200g/100kg, BD containing acidifier at a level of 200g/100kg; and BD containing an equal amount of probiotic plus acidifier (200g/100kg). Broilers that received either probiotic, acidifier or a mixture of probiotic and acidifier (1:1) exhibited higher body weight gain, lower feed conversion ratio (FCR) and higher cost-effectiveness compared with the broilers fed on control diet ($P<0.05$). However, feeding of diet containing both probiotic and acidifier resulted in the highest growth rate and net profit in all dietary regimens. Broilers fed on probiotic and acidifier in a mixture had FCR similar to other treatment groups. This study indicated that the diet containing probiotic-acidifier mixture seems to be more cost-effective in promoting growth performance of broilers, as an alternative to the AGP, as compared to the use of probiotic or acidifier alone in the diet.

INTRODUCTION

Broiler production is one of the most important and promising sector in poultry industry in terms of advantage of quick return that plays a vital role in the economic growth of Bangladesh. It has been proven that the genetic potentiality of the fast growing commercial broilers is achieved in the shortest possible time by the application of modern nutri-biotechnology. However, the optimum growths of broilers are seriously hampered by the invasion of pathogenic microorganisms. In order to cope with the challanges of growth-inhibiting microorganisms, some antibiotics like bacitracin, virginiamycin, flavomycin, avilamycin, tiamulin, colistin sulphate, oxytetracycline, aureomycin, chlortetracycline, neomycin sulphate, erythromycin and enrofloxacin have been used for several decades in broiler feed at a sub-therapeutic level. Antibiotics are double edge weapon. Antibiotics that are used as AGP in broiler feed have been shown to increase meat yield and improve feed efficiency with substantial reduction in pathogenic bacteria in the host gut (Gaskins et al., 2002). They are also widely used in veterinary field for reducing the incidence of diseases. However, indiscriminate use of antibiotics in broiler production leads to the development of antibiotic resistant pathogenic bacteria, thereby causing resistance to medicines, persistence of infections and treatment failure. Several studies provided evidence that inappropriate and excessive use of antibiotics has led to the accumulation of their residues in edible broiler carcass which poses a major threat and potential risk to public health (Donoghue, 2003; El-Kahky and Allam, 2005; Nisha, 2008; Shareef et al., 2009; Jallailudeen, 2015). In Bangladesh, a recent study has shown that high levels of residues of major antibiotics like tetracycline, ciprofloxacin, enrofloxacin and amoxicillin were found mostly in liver, kidney, thigh meat and breast meat of broilers (Sattar et al., 2014). The European Union has reported that about 25,000 patients died each year from infections caused by drug-resistant bacteria, which is equivalent to €1.5 billion of medical healthcare costs (Ziggers, 2011). For this reason, most of the poultry meat consumer groups are avoiding meat from birds fed on diets containing antibiotics. In consequence, the European Union has banned the use of antibiotics in animal production since 2006 and other developed countries have limited the antibiotic use in poultry production. However, the ban of AGP demands the search for more suitable and safer alternatives to antibiotics that would promote growth, feed utilization and gut health without having any residual effect on poultry products. Recently, many feed additives referred to as natural growth promoters or non-antibiotic growth promoters have been evaluated which include probiotics, prebiotics, synbiotics, acidifiers, phytobiotics, etc. (Ricke, 2003; Hruby and Cowieson, 2006; Kocher, 2006; Alavi et al., 2012).

The term probiotic derived from Greek word "pro bios" which means "in favor of life" (Coppola and Turnes, 2004). According to the definition by FAO/WHO, probiotics are live microorganisms which when administered in adequate amounts confer a health benefit on the host (Fuller et al., 1989). Studies have shown that probiotics act as substitute for AGP (Tomasik and Tomasik, 2003), which have been used on poultry to increase weight gain (Kapil et al., 2015). Probiotics added to poultry feed help in the production of vitamin B complex and digestive enzymes, and stimulation of intestinal immunity, increasing protection against toxins produced by pathogenic microorganisms (Alexopoulos et al., 2004). In broiler, probiotic species such as *Lactobacillus*, *Streptococcus*, *Bacillus*, *Bifidobacterium*, *Enterococcus*, *Aspergillus*, *Candida* and *Saccharomyces* are widely used to prevent poultry pathogens and diseases and to improve growth performance of broilers (Zulkifli et al., 2000; Kalavathy et al., 2003; Kabir et al., 2004; Timmerman et. al, 2006; Awad et al., 2009; Lee et al., 2010; Kapil et al., 2015). However, combination of *Bacillus subtilis and Bacillus licheniformis* are widely used as health-boosting probiotic strains, which have been shown to aid in nutrient digestion and absorption and in creating the favourable conditions for beneficial bacteria (Kapil et al., 2015).

On the other hand, recent studies in poultry nutrition have shown that acidifiers (feed additives containing low-molecular-weight organic acids e.g. acetic, citric, lactic, formic, sorbic, ascorbic, propionic, fumeric and malic acid) act as antimicrobial and intestinal pH regulator which when used in poultry diets cause maximum utilization of feed and reduction in harmful bacterial population in the hosts' gut, improvement of performance and promotion of health status of poultry fed diets devoid of AGP (Skinner et al., 1991 and Kil et al., 2011). The organic acids (non-dissociated or non-ionised and more lipophilic form) can penetrate the bacteria cell wall and disrupt the normal physiology of certain types of bacteria (Kapil et al., 2015). It has been reported that acidifiers reduce Coliforms and Clostridia loads in the gut and increase the number of beneficial bacteria such as Lactobacilli in the ileum (Akyurek et al., 2011).

Alhough there is an ample evidence for beneficial bacteria competing and excluding potential pathogens in the intestinal tract of chickens, there is evidence that probiotic strains fail to compete with harmful bacteria when the gut pH increases toward a more alkaline range due to inclusion of high protein and minerals in the diet or in a situation when harmful bacterial population is already high in the gut before probiotic administration (Edens et al., 1997; Edens, 2003; Kapil et al., 2015). Therefore, a favourable microenvironment (to stabilize the low pH) is required which might be provided with the supplementation of acidifier in broiler diet containing probiotic to exhibit full potentiality of probiotic bacteria in the gut. However, limited evidence is available on the benefit of combining probiotics with acidifier in broiler diet in comparison to their benefit alone. In order for the elucidation of the effect of probiotic containing beneficial bacterial spore and acidifier on broiler performance and profitability of broiler rearing, the present study compared the effectiveness of individual supplements and synergistic combination of probiotic (Bio-Top) and acidifier (Sal-Stop) as alternative to AGP in broiler diet.

MATERIALS AND METHODS

Experimental house

The experiment was conducted in a gable type open sided house. The room area was 500sq.ft. The room was partitioned into 20 pens of equal size using wire net. Area of each pen was 20sq. ft. (10 feet x 2 feet).

Experimental birds

A total of 400 day-old straight run Cobb 500 commercial broiler chicks were purchased from the Mono Hatcheries Ltd., Mowna, Gazipur, Dhaka, to carry out this research work. The experiment was conducted at the Bangladesh Agricultural University Poultry Farm for a period of 35 days from 06 August 2015 to 09 September 2015.

Table 1. Ingredient composition of broiler starter diet

Ingredient	T_1	T_2	T_3	T_4	T_5
Maize	54.99	54.97	54.79	54.79	54.79
Soya bean meal	31.0	31.0	31.0	31.0	31.0
Protein concentrate (Propak)	7.0	7.0	7.0	7.0	7.0
Di calcium phosphate	1.35	1.35	1.35	1.35	1.35
Limestone	0.80	0.80	0.80	0.80	0.80
Soybean oil	4.0	4.0	4.0	4.0	4.0
L-Lysine	0.10	0.10	0.10	0.10	0.10
DL-Methionine	0.12	0.12	0.12	0.12	0.12
Vitamin-mineral-amino acid premix	0.25	0.25	0.25	0.25	0.25
Choline chloride	0.03	0.03	0.03	0.03	0.03
Common salt	0.36	0.36	0.36	0.36	0.36
Antibiotic	-	0.02	-	-	-
Probiotic	-	-	0.20	-	0.10
Acidifier	-	-	-	0.20	0.10
Total	**100**	**100**	**100**	**100**	**100**

Where, T1 = Control (Basal diet; BD), T2 = BD containing antibiotic growth promoter (200g/ton), T3 = BD containing probiotic (200g/100kg), T4 = BD containing acidifier (200g/100kg), and T5 = BD containing equal amount of probiotic and acidifier (1:1; 200g/100kg).

Formulation of diet

Broiler starter and grower diets were formulated and manufactured in Poultry Farm feed unit. Starter diet was provided from day old to 21 days and grower diet was provided from 22 days to 35 days. Composition of the ingredients used in starter and grower diet is shown in Table 1 and Table 2, respectively. The nutrient requirements of the experimental broilers were satisfied as per standard recommended by the Cobb 500 strain producing company. Nutrient composition of the formulated starter and grower diet is shown in Table 4.

Table 2. Ingredient composition of broiler grower diet

Ingredient	T_1	T_2	T_3	T_4	T_5
Maize	60.00	59.98	59.80	59.80	59.80
Soya bean meal	24.09	24.09	24.09	24.09	24.09
Protein concentrate (Propak)	8.0	8.0	8.0	8.0	8.0
Di calcium phosphate	1.35	1.35	1.35	1.35	1.35
Limestone	0.70	0.70	0.70	0.70	0.70
Soybean oil	5.0	5.0	5.0	5.0	5.0
L-Lysine	0.10	0.10	0.10	0.10	0.10
DL-Methionine	0.12	0.12	0.12	0.12	0.12
Vitamin-mineral-amino acid premix	0.25	0.25	0.25	0.25	0.25
Choline chloride	0.03	0.03	0.03	0.03	0.03
Common salt	0.36	0.36	0.36	0.36	0.36
Antibiotic	-	0.02	-	-	-
Probiotic	-	-	0.20	-	0.10
Acidifier	-	-	-	0.20	0.10
Total	**100**	**100**	**100**	**100**	**100**

Where, T1 = Control (Basal diet; BD), T2 = BD containing antibiotic growth promoter (200g/ton), T3 = BD containing probiotic (200g/100kg), T4 = BD containing acidifier (200g/100kg), and T5 = BD containing equal amount of probiotic and acidifier (1:1; 200g/100kg).

Experimental ingredients

Probiotic (Bio-Top; Shinil Biogen Co. Ltd., South Korea) product used in this study is imported in Bangladesh by Pharma and Firm Co. Ltd., Dhaka. Probiotic was included in the basal diet at a rate of 200g/100kg of mixed feed. The acidifier (Sal-Stop; Impextraco, Belgium) product used in this study is imported in Bangladesh by Century Agro Ltd., Dhaka, Bangladesh. Acidifier was included in the basal diet at a rate of 200g/100kg of mixed feed. It is a synergistic combination containing salts of propionic acid, acetic acid, formic acid, sorbic acid, lactic acid, phosphoric acid and their free acids, completed with emulsifiers and natural extracts. It contains free fatty acids on a silica carrier. Renamycin (Renata Ltd., Dhaka, Bangladesh) was used in this study as an AGP. Each gram of Renamycin soluble powder contains oxytetracycline USP 200mg. The inclusion rate of the product was 200g/ton of mixed feed.

Table 3. Composition of probiotic (Bio-Top)

Name of ingredient	Amount
BioPlus 2B	25g
a) *Bacillus licheniformis*	$4x10^{10}$ CFU/g
b) *Bacillus subtilis* CH201	$4x10^{10}$ CFU/g
Zinc Oxide	20g

Table 4. Nutrient composition of starter and grower diet

Nutrients	Starter diet	Grower diet
Analyzed		
DM (%)	89.68	87.09
CP (%)	23.66	20.59
CF (%)	3.81	2.88
EE (%)	5.71	4.81
Calculated		
ME (kcal/kg)	3050	3150
Lysine (%)	1.24	1.06
Methionine (%)	0.50	0.63
Metionine + Cystein (%)	1.00	0.92
Calcium (%)	1.21	1.05
Available Phosphorus (%)	0.45	0.42

Management of the experimental birds

The experiment was conducted during August to September 2015. The environmental temperature was a little bit lower than the recommended brooding temperature. So, additional heat was provided to chicks. The chicks were brooded in respective pens using one 100-watt electric bulb in each pen. The chicks were provided with a temperature of 35°C at first week of age, decreasing gradually at a rate of 3°C per week and continued up to 4 weeks of age. The room temperature and humidity was measured by an automatic thermo-hygrometer. Fresh and dried rice husk was used as litter material and spreaded over the floor at a depth of 3 cm. After first two weeks, upper part of the litter mixed with droppings was removed and replaced with new litter. At the end of 3 weeks, old litter was totally replaced by new litter. After 14 days, litter was stirred in every alternative day to dry up quickly and to remove harmful gases. Starter diet was provided for the first 21 days and then grower diet was provided to the broiler up to 35 days of age. In all cases, feeds were offered *ad libitum* to all broilers. Feed was supplied four times daily; once in the morning, noon, afternoon and again at night in such a way that feeder was not kept empty. Fresh and clean water was made available at all times. The broilers were exposed to a continuous lighting period of 23 hours and a dark period of 1 hour in each 24 hours. The dark period provision was made to keep the broilers familiar with darkness in the failure of electricity, if happens. The vaccination schedule that was followed during the experimental period is given in Table 5.

Table 5. Vaccination schedule for the experimental broilers

Age of broilers (day)	Name of Vaccine	Trade Name*	Dose	Route of vaccination
4	IB+ND	MA5+Clone30	One drop	Ocular
10	IBD	D-78	One drop	Ocular
21	IBD	D-78	One drop	Ocular

IB, Infectious Bronchitis; ND, Newcastle Disease; IBD, Infectious Bursal Disease
*Intervet International, B.V. BOXMEER, The Netherlands.

Processing of broilers

At the end of feeding trial, one male and one female broilers having near to pen average weight were taken from each pen for recording meat yield parameters. Broilers were slaughtered and allowed to bleed for 2 minutes and immersed in hot water (semi-scalding; 51-55°C) for 120 seconds in order to loose feathers followed by removal of feathers by hand pinning. Then head, shank, viscera, giblet (heart, liver and gizzard) and abdominal fat were removed for determination of meat yield parameters. Dressed broilers were cut into different parts such as breast, thigh, drumstick and wing. Finally, every cut up parts were weighed and recorded separately for male and female broilers of all replications.

Data collection

Birds were weighed at the first day of experiment (initial body weight) and weekly basis for all birds from each replication. Average body weight gain of the broiler in each replication was calculated by deducting initial body weight from the final body weight. The amount of feed consumed by the birds in each replication of each treatment group were calculated for every week by deducting the amount of feed left over from the amount supplied for a particular week. FCR was calculated as the unit of feed consumed per unit of body weight gain. The cost of broiler production for each treatment group was calculated based on the market price of feed ingredients, cost of chicks, cost of probiotic, acidifier etc., at the time of this experiment. For analysis of exact cost involved in broiler production, some other factors such as electricity, litter material, vaccination, medication, labour and even the depreciation cost of structure were also considered. Profit was calculated per broiler and per kg broiler basis by excluding total cost of production from the total sale price of bird.

Statistical analysis

Data on body weight, body weight gain, feed consumption and FCR of broilers were subjected to analysis of variance (ANOVA) in a completely randomized design (CRD) employing SAS (2009) statistical package program.

RESULTS AND DISCUSSION

Live weight of broiler

Day-old chicks (DOC) were distributed randomly in the pens of different dietary treatments. All DOC, irrespective of their dietary regimen, had similar (P>0.05) initial live weights (Table 6). Broilers gained weight with the advancement of age, but the live weights up to the age of two weeks were not significantly different (P>0.05) for treatment groups. However, broilers fed on either BD, AGP, probiotic, acidifier alone or probiotic plus acidifier differed significantly (P<0.05) for 3rd, 4th and 5th week live weight. Broilers received both probiotic and acidifier exhibited highest live weight at all ages, followed by broilers fed on probiotic, acidifier, AGP and control diet.

Table 6. Live weight of broilers fed on different dietary treatments

Age (Day)	Dietary Treatment					Level of Significance
	T_1	T_2	T_3	T_4	T_5	
0	52.2±0.09	52.6±0.09	52.5 ± 0.09	52.7±0.09	52.6±0.09	NS
7	180.9±0.91	182.8±0.91	185.1±0.91	184.2±0.91	186.1±0.91	NS
14	476.3±4.14	482.2±4.14	492.1±4.14	488.1±4.14	500.5±4.14	NS
21	867.9±9.75^{a}	884.6±9.75^{b}	905.6±9.75^{c}	895.8±9.75^{c}	925.8±9.75^{d}	*
28	1236.1±23.45^{a}	1265.6±23.45^{b}	1303.3±23.45^{c}	1296.5±23.45^{c}	1376.2±23.45^{d}	*
35	1653.3±38.37^{a}	1719.6±38.37^{b}	1793.1±38.37^{c}	1766.0±38.37^{d}	1883.9±38.37^{e}	*

Where, T_1 = Control (Basal diet; BD), T_2 = BD containing antibiotic growth promoter (200g/ton), T_3 = BD containing probiotic (200g/100kg), T_4 = BD containing acidifier (200g/100kg), and T_5 = BD containing equal amount of probiotic and acidifier (1:1; 200g/100kg). NS = Non-significant, P>0.05; * Significant difference, P<0.05.

The results obtained in this study are in consistence with the findings of Bai et al. (2013), who compared probiotic fed broilers with antibiotic, and probiotic plus antibiotic fed broilers. They reported that probiotic and its combination with antibiotic resulted in higher live weight of broilers during grower period (21-42 days) compared to control diet. Sabatkova et al. (2008) investigated efficacy test of probiotic (BioPlus 2B; *Bacillus subtilis* and *B. licheniformis*) and Avilamycin for growth performance and slaughter yields of broilers. They reported that the supplementation of probiotic enhanced higher weight gain than did the Availamycin. Several studies have also shown that broilers fed on probiotic (*Bacillus subtilis and B. licheniformis*) gained higher body weight during the grower phase (21-42 days) than the broilers on control diet (Ahmad and Taghi, 2006; Shim et al., 2012 and Salim et al., 2013). Chowdhury et al. (2009) reported that addition of citric acid to broiler diet increased weight gain significantly compare to the control group. However, the present study revealed that the combination of probiotic with acidifier reseulted in faster growth rate of broilers than did the probiotic alone which are in agreement with the findings of Bandy (2001); Lima et al. (2002) and Kalavathy et al. (2003).

Feed intake

Although feed intake by the broilers varies numerically between the groups, there were no significant differences (P>0.05) in feed consumption due to dietary treatments (Table 7). However, it is clear that inclusion of probiotic in the basal feed resulted in higher feed intake by the broilers, irrespective of the age, compared to other test ingredients in the diet. Higher feed consumption by the broilers under probiotic (Bio-Top) regimen in this study is in agreement with the results of earlier studies (Mohan, et al., 1996; Panda et al., 1999 and Zulkifli, et al., 2000), who demonstrated that probiotic stimulated the gut to intake more feed. In the current study, a tendency of increased feed intake on probiotic was followed by acidifier in the diet (Table 7). This phenomenon is consistent with the observation of Nezha et al. (2007); Panda et al. (2008) and Faria et al. (2009), who found improvement in feed intake by feeding organic acid.

Table 7. Feed consumption of broilers fed on different dietary treatments

Age	Dietary Treatment					Level of
(Day)	T_1	T_2	T_3	T_4	T_5	Significance
7	140.0±0	140.0±0	140.0±0	140.0±0	140.0±0	NS
14	455.0±2.10	455.0±2.10	466.3±2.10	459.0±2.10	456.8±2.10	NS
21	1071.3±2.42	1060.8±2.42	1069.5±2.42	1063.8±2.42	1058.8±2.42	NS
28	2015±3.06	2026.3±3.06	2027±3.06	2029±3.06	2033.5±3.06	NS
35	2652.5±3.83	2655.3±3.83	2673.8±3.83	2655.8±3.83	2662.5±3.83	NS

Where, T_1 = Control (Basal diet; BD), T_2 = BD containing antibiotic growth promoter (200g/ton), T_3 = BD containing probiotic (200g/100kg), T_4 = BD containing acidifier (200g/100kg), and T_5 = BD containing equal amount of probiotic and acidifier (1:1; 200g/100kg). NS = Non-significant, P>0.05; * Significant difference, P<0.05.

Feed conversion ratio

It is revealed (Table 8) that intake of AGP, probiotic, acidifier or probiotic plus acidifier did not affect the differences in FCR values (P>0.05) up to the 2nd week of age of broilers. Intake of basal feed resulted in highest FCR (Table 8), followed by AGP, acidifier, probiotic and probiotic plus acidifier intake at all ages of broilers. The differences in FCR among the dietary treatment groups appeared to be significant (P<0.05) following first two weeks of age. The significant effect of probiotic on FCR in broiler is in close agreement with the findings by Sabatkova et al. (2008); Ashayerizadeh et al. (2009); Zhou et al. (2010) and Shim et al. (2012). They found that supplementing basal diet with probiotic containing *Bacillus subtilis* and *B. licheniformis* improved feed conversion efficiency in broilers. Panda et al. (2008) reported that dietary preparation of *Bacillus subtilis* and *B. licheniformis* (at the rate of $6x10^8$ spores/kg of diet) significantly enhanced feed efficiency in White Leghorn Breeders. The better effect of organic acid on FCR was observed by Muzaffer et al. (2003) and Zhang et al. (2005).

Cost-effectiveness of production

The calculated costs of total production obtained in the present study in terms of per kg broiler were very high, high, medium and low for the basal diet, AGP, acidifier, probiotic and probiotic plus acidifier, respectively (Table 9). The return per kg broiler was the highest for probiotic+acidifier group followed by probiotic, acidifier, AGP and control group. It is therefore clear that inclusion of either probiotic+acidifier, probiotic or acidifier in the basal diet was profitable over either AGP or control group. In addition, it is also obvious that combined use of probiotic and acidifier in the basal diet resulted in more profit compared to the inclusion of either probiotic or acidifier alone in the basal diet. Consequently, addition of probiotic and acidifier together to the diet was the most cost-effective, followed by their singly use in feed. The present study clearly indicates that feeding of probiotic, acidifier and their combined use had beneficial effect on profitability of broiler. The inclusion of both probiotic and acidifier provided with highest profit. This result is agreed with the results of Roy et al. (2013), who reported that feeding probiotic to broilers was more profitable than AGP.

Table 8. Feed conversion ratio of broilers fed on different dietary treatments

Age (Day)	Dietary Treatment					Level of Significance
	T_1	T_2	T_3	T_4	T_5	
7	1.09±0.007	1.08±0.007	1.06±0.007	1.07±0.007	1.05±0.007	NS
14	1.07±0.009	1.06±0.009	1.06±0.009	1.06±0.009	1.02±0.009	NS
21	$1.31±0.017^a$	$1.28±0.017^b$	$1.25±0.017^c$	$1.26±0.017^c$	$1.21±0.017^c$	*
28	$1.70±0.027^a$	$1.67±0.027^b$	$1.62±0.027^c$	$1.63±0.027^c$	$1.54±0.027^c$	*
35	$1.65±0.035^a$	$1.62±0.035^b$	$1.53±0.035^c$	$1.55±0.035^d$	$1.45±0.035^e$	*

Where, T_1 = Control (Basal diet; BD), T_2 = BD containing antibiotic growth promoter (200g/ton), T_3 = BD containing probiotic (200g/100kg), T_4 = BD containing acidifier (200g/100kg), and T_5 = BD containing equal amount of probiotic and acidifier (1:1; 200g/100kg). NS = Non-significant, P>0.05; * Significant difference, P<0.05.

Table 9. Cost of production and profit of broilers on different dietary treatment groups

Variables	Dietary Treatment				
	T_1	T_2	T_3	T_4	T_5
Total Feed intake (kg/broiler)	2.653	2.655	2.674	2.654	2.663
Final body weight (kg/broiler)	1.653	1.719	1.793	1.766	1.883
Basal feed price (Tk/kg)	36.10	36.10	36.10	36.10	36.10
Feed cost (Tk./kg broiler)	36.10	36.25	36.47	36.56	36.93
Feed cost (Tk./broiler)	95.76	96.26	97.51	97.10	98.33
Other costs (TK.)	65.00	65.00	65.00	65.00	65.00
Total production cost (Tk./broiler)	160.76	161.26	162.51	162.1	163.33
Total production cost (Tk./kg broiler)	97.25	93.81	90.64	91.79	86.74
Sale price (Tk.140.00/kg broiler)	231.42	240.66	251.02	247.24	263.62
Profit (Tk./broiler)	70.66	79.4	88.51	85.14	100.29
Profit (Tk./kg broiler)	42.75	46.19	49.36	48.21	53.26
Profit (Tk./kg broiler) over the control	-	3.44	6.61	5.46	10.51

Where, T_1 = Control (Basal diet; BD), T_2 = BD containing antibiotic growth promoter (200g/ton), T_3 = BD containing probiotic (200g/100kg), T_4 = BD containing acidifier (200g/100kg), and T_5 = BD containing equal amount of probiotic and acidifier (1:1; 200g/100kg). NS = Non-significant, P>0.05; * Significant difference, P<0.05.

CONCLUSIONS

Neither probiotic nor acidifier used in the current study had any significant effect on live weight, feed intake and FCR of brolers up to two weeks of age. Use of probiotic and acidifier singly or combindly resulted in increased live weight and decreased FCR after 14 days of age. However, combined use of probiotic and acidifier was superior in terms of live weight, FCR and cost-effectiveness over their use in the diet singly. It is therefore concluded that combining the probiotic with the acidifier exhibited an additive benefit in growth performance of broilers and profitability of broiler production.

REFERENCES

1. Ahmad K and G Taghi, 2006. Effect of probiotic on performance and immunocompetence in broiler chicks. Journal of Poultry Science, 43: 296-300.
2. Akyurek H, ML Ozduven, AA Okur, F Koc and HE Samli, 2011. The effects of supplementing an organic acid blend and/or microbial phytase to a corn-soybean based diet fed to broiler chickens. African Journal of Agricultural Research, 6: 642–649.
3. Alavi SAN, A Zakeri, B Kamrani and Y Pourakbari, 2012. Effect of prebiotics, probiotics, acidifier, growth promoter antibiotics and synbiotic on humural immunity of broiler chickens. Global Veterinaria, 8: 612-617.
4. Alexopoulos IL, A Georgoulakis, SK Tzivara, A Kritas, SC Siochu and Kyriakis, 2004. Field evaluation of the efficacy of a probiotic containing *Bacillus licheniformis* and *Bacillus subtilis* spores, on the health status and performance of sows and their litters. Journal of Animal Physiology and Nutrition, 88: 281-292.
5. Ashayerizadeh A, N Dabiri, O Ashayrizadeh, KH Mizadeh, H Roshanfekr and M Mamooee, 2009. Effects of dietary antibiotic, probiotic and prebiotic as growth promoter on growth performance, carcass characteristics and hematological indices of broiler chickens. Pakistan Journal of Biological Sciences, 12: 52-57.
6. Awad WA, K Ghareeb, AS Raheem and J Böhm, 2009. Effects of dietary inclusion of probiotic and synbiotic on growth performance, organ weights, and intestinal histomorphology of broiler chickens. Journal of Poultry Science, 88: 49-56.
7. Bai SP, AM Wu, XM Ding, Y Lei, J Bai and JS Chio, 2013. Effects of probiotic-supplemented diets on growth performance and intestinal immune characteristics of broiler chickens. Journal of Poultry Science, 92: 633-670.
8. Bandy MT and KS Raisam, 2001. Growth performance and carcass characteristics of broiler chicken fed with probiotics. Indian Journal of Poultry Science, 29: 228-231.
9. Chowdhury R, KM Islam, MJ Khan, MR Karimi, MN Haque, M Khatun and GM Pesti, 2009. Effect of citric acid, avilamycin, and their combination on the performance, tibia ash, and immune status of broilers. Poultry Science, 88: 1616-1622.
10. Coppola MM and CG Turns, 2004. Probioticos e resposta immune. Ciencia Rural, 34: 1297-1303.
11. Donoghue DJ, 2003. Antibiotic residues in poultry tissues and eggs: Human health concern. Poultry Science, 82: 618-621.
12. Edens FW, CR Parkhurst, IA Casas and WJ Dobrogosz, 1997. Principles of *ex ovo* competitive exclusion and *in ovo* administration of *Lactobacillus reuteri*. Poultry Science, 76: 179-196.
13. Edens FW, 2003. An alternative for antibiotic use in poultry: Probiotics. Revista Brasileira de Ciência Avícola, 5: 75-79.
14. El-Kahky MAA and TH Allam, 2005. Detection of some antibiotic residues in camel's meat products. Journal of Egyptian Veterinary Medical Association, 65: 203-209.
15. Faria DE, APF Henrique and R Franzolin, 2009. Alternatives to the use of antibiotic growth promoter for broiler chickens. Probiotic Ciencia Animal Brasileira, 10: 18-28.
16. Fuller R, 1989. Probiotics in man and animals. Journal of Applied Bacteriology, 66: 365–378.
17. Gaskins HR, CT Collier and DB Anderson, 2002. Antibiotics as growth promotants: mode of action. Animal Biotechnology, 13: 29-42.
18. Jallailudeen RL, MJ Saleh, AG Yaqub, MB Amina, W Yakaka and M Muhammad, 2015. Antibiotic residues in edible poultry tissues and products in Nigeria: A potential public health hazard. International Journal of Animal and Veterinary Advances, 7: 55-61.

19. Kabir SML, MM Rahman, MB Rahman, M Rahman and SU Ahmed, 2004. The dynamics of probiotics on growth performance and immune response in broilers. International Journal of Poultry Science, 3: 361-364.
20. Kapil J, KS Sharma, S Katoch, VK Sharma and BG Mane, 2015. Probiotics in broiler poultry feeds: A review. Journal of Animal Nutrition and Physiology, 1: 4-16.
21. Kalavathy R, N Abdullah, S Jalaluddin and YW Ho, 2003. Effects of Lactobacillus cultures on growth performance, abdominal fat deposition, serum lipids and weight of organs of broiler chickens. British Poultry Science, 44: 139-144.
22. Kil DY, WB Kwon and BG Kim, 2011. Dietary acidifiers in weanling pig diets: a review. Revista Colombiana de Ciencias Pecuarias, 24: 231-247.
23. Lee K, HS Lillehoj and GR Siragusa, 2010. Direct-fed microbials and their impacts on the intestinal microflora and immune system of chickens. Journal of Poultry Science, 47: 106-114.
24. Lima ACF, FAR Harnnich, M Macari and JM Pizauro Junior, 2002. Evaluation of the performance of the broiler chickens feed with enzymatic probiotic supplemental. Ars Veterinaria, 18: 153-157.
25. Mohan B, R Kadirevel, A Natarajan and M Bhaskaram, 1996. Effect of probiotic supplementation on growth, nitrogen utilization and serum cholesterolin broilers. British Poultry Science, 37: 395-401.
26. Muzaffer D, O Ferda and C Kemal, 2003. Effect of dietary Probiotic, organic acids and antibiotic supplementation to diet on broiler performance and carcass yield. Journal of Nutrition, 2: 89-91.
27. Nezhad YE, M Shivazad, M Nazerad and MMS Babak, 2007. Influence of citric acid and microbial phytase on performance and phytate utilization in broiler chicks fed a corn–soybean meal diet. Journal of the Faculty of Veterinary Medicine, 61: 407-413.
28. Nisha AR, 2008. Antibiotic residues – A global health hazard. Veterinary World, 1: 375-377.
29. Panda AK, SVR Rao, MR Reddy and NK Praharaj, 1999. Effect of dietary inclusion of probiotics on growth, carcass traits and immune response in broilers. Indian Journal of Poultry Science, 34: 334-346.
30. Panda AK, SSR Rao, S Raju and S Sharma, 2008. Effect of probiotic feeding on egg production and quality, yolk cholesterol and humoral immune response of White Leghorn layer breeders. Journal of the Science of Food and Agriculture, 88: 43-47.
31. Ricke SC, 2003. Perspectives on the use of organic acids and short chain fatty acids as antimicrobials. Poultry Science, 82: 632–639.
32. Roy BC and SD Chowdhury, 2013. Effect of dietary probiotic and antibiotic growth promoter either alone or in combination on the growth performance of broilers during summer. In 18th International Poultry Show and Seminar, WPSA-BB, 153-158.
33. Sabatkova J, I Kumprecht and P Zobac, 2008. The probiotic Bio plus 2B as an alternative of antibiotic in diets for broiler chickens. Acta Veterinaria Brno, 77: 569-574.
34. Salim HM, KH Kang, N Akter, DW Kim, JC Na, HB Jong, HC Choi and WK Kim, 2013. Supplementation of direct-fed microbials as an alternative to antibiotic on growth performance, immune response, cecal microbial population, and ileal morphology of broiler chickens. Journal of Poultry Science, 92: 2084-2090.
35. Sattar S, MM Hassan, SKMA Islam, M Alam, MSA Faruk, S Chowdhury and AKM Saifuddin, 2014.

Antibiotic residues in broiler and layer meat in Chittagong district of Bangladesh. Veterinary World, 7: 738-743.

36. Shareef AM, ZT Jamel and KM Yonis, 2009. Detection of antibiotic residues in stored poultry products. Iraqi Journal of Veterinary Sciences, 23: 45-48.
37. Shim YH, SL Ingali, JS Kim, DK Seo, SC Lee and IK Kwon, 2012. A multimicrobe probiotic formulation processed at low and high drying temperatures: effects on growth performance, nutrient retention and caecal microbiology of broilers. British Poultry Science, 53: 482-490.
38. Skinner JJ, AL Izat and PW Waldroup, 1991. Research note: Fumaric acid enhances performanceof broiler chicken. Poultry Science, 70: 1444-1447.
39. Timmerman HM, A Veldman, E Van den Elsen, FM Rombouts and AC Beynen, 2006. Mortality and growth performance of broilers given drinking water supplemented with chicken-specific probiotics. Poultry Science, 85: 1383-1388.
40. Tomasik PJ and P Tomasik, 2003. Probiotics and prebiotics. Cereal Chemistry, 80: 113-117.
41. Zhang VE, S Alvarez, M Medina and M Medici, 2005. Gut mucosa limmuni stimulation by lactic acid bacteria. Biocell, 24: 223-232.
42. Zhou X, Y Wang, RJ Wu and B Zhang, 2010. Effect of dietary probiotic on growth performance, chemical composition and meat quality of Guangxi Yellow Chicken. Journal of Poultry Science, 89: 588-593.
43. Ziggers D, 2011. Animal Feed News. EU 12-point antibiotic action plan released, 18 November, 2011. http://www.allaboutfeed.net/news/eu-12-point-antibiotic-action-plan-released-12443.html
44. Zulkifli I, N Abdullah, NM Azrin and YW Ho, 2000. Growth performance and immune response of two commercial broiler strains fed diets containing Lactobacillus cultures and oxytetracycline under heat stress conditions. Journal of Poultry Science, 41: 593-597.

15

EFFECT OF REPLACEMENT CORN BY BROWN RICE ON PERFORMANCE OF CHICKEN PRODUCTION IN VIETNAM

Mai Thanh Vu[1,3*], Van Thanh Tran[2,3], My Thi Thuy Nguyen[2,3], Van Cao[1], Tuan Ngoc Minh Nguyen[1]

[1]Hung Vuong University of Phu Tho, Vietnam; [2]Thai Nguyen University, Vietnam; [3] Thai Nguyen University of Agriculture and Forestry, Vietnam

***Corresponding author:** Mai Thanh Vu, E-mail: vumai@hvu.edu.vn

ARTICLE INFO

Key words

Brown rice
Corn
Chicken
Ri lai (Ri x LP)
Vietnam

ABSTRACT

Rice is a major staple food in Vietnam in which brown rice has been recognized as a potential feedstuff for poultry but data on nutritional value of this feed are lacking. In this study the using of brown rice as replacement of corn in chicken diet was evaluated. The body weight, feed intake and feed conversion ratio were recorded. In total, 192 day old chicks of country breed (*Ri lai*) were used in this study. Chickens were divided into four groups and reared at same conditions for 12 weeks. The first group as the control group fed on 100% corn, second group fed on 75% corn and 25% brown rice, third group fed on 50% corn and 50% brown rice and the last group fed on 25% corn and 75% brown rice. The average body weight of chickens among all the treatments was 1.7 kg per bird which was not significantly different (P=0.44). The total feed intake of chickens (4-5 kg) was recorded without significant difference (P=0.23), however the feed conversion ratio were significantly different (P<0.05) between treatments. This study considered as the first report that demonstrates the usefulness of brown rice as a potential alternative of corn for chicken diet in rural areas in Vietnam, especially on the prevailing conditions such as during high price spell of corn.

INTRODUCTION

Vietnam is an agricultural country with around 70% of population living in rural area. Small-scale production of poultry plays an important role in household economy across rural areas (Vang, 2003; Burgos et al., 2008). Although farmers could use the commercial feeds ready to use that are sold in many animal-feed stores, majority of farmers in reality prefer to use corn as the additive to feed chickens according to their practical experience (FAO, 2014). Recently, the price of corn in Vietnam fluctuated from 0.35 USD to 0.5 USD per kilogram leads to impacting heavily on the production cost for farmers (USDA, 2015; Jonas et al., 2015, Luckmann et al., 2015). This might be ineffective if the rural farmers only depend on the corn in poultry production.

Rice is well-known as a major staple food in Vietnam contributed approximately 30% of the total world rice production (Chris, 2014). The rice with good quality is used as human food and the low nutrient rice is rejected for use as feedstuffs for ruminants and monogastric animals. Brown rice (BR) is among these rejected parts of rice production and has high potentiality for poultry production. This kind of rice are usually available in the rural areas in Vietnam but still not effectively used. It is definitely worth to use the brown rice in poultry production in Vietnam. Brown rice was suggested to replace corn in poultry farming in Vietnam due to the actual high price of corn (Kinh, 2015). A study of composition and energy value of brown rice by Asyifah (Asyifah et al., 2012) showed that it is a potential feedstuff for poultry production. The nutrients contained in brown rice are suitable for poultry rearing, especially chickens and broilers due to their high energy value and low fiber content as well as balanced amino acids. Moreover, the production costs could be lower because the source of brown rice in rural areas is commonly available and abundant. Additionally, a recent study on ducks in Vietnam reported that using of brown rice for duck rearing gave no differences in growth performance and could be used to replace corn without negative effect to the growth of ducks (Viet et al., 2015). However, there has been no practical experiment done for chicken rearing in the rural areas.

There were many chicken breeds used in rural farming in Vietnam and the hybrid breed called *"Ri lai" (Ri x LP)* is most common used because of the high living rate and meat quality (Vang et al., 1999; Van, 2002; Duc and Long, 2008). Additionally, the price of the chicken Ri per kilogram is normally much higher and has a preferable taste than broilers after 12 weeks raising. Thus, the chicken Ri was the "chicken of choice" to rearing in rural areas.

In Vietnam, the diets mixed corn with the concentrated feed such as the RTD-VinaS99 (S99) provided by companies were known as a common diet for raising chickens in rural areas. However, recently high changing of corn price negatively affected to chicken production and farmers. Furthermore, the aflatoxin produced from low quality corn could also be a major problem which causes high mortality of chickens and significant damage to their growth performance as well. Therefore, this study was aimed to investigate the effect of using different levels of brown rice and corn as the feed on growth performance within the chicken farming context situation in rural areas in Vietnam.

MATERIALS AND METHODS

Location

The experiments were carried out in the area of animal experiments in Hung Vuong University of Phu Tho, Vietnam from November 2015 to March 2016. This location is about 70 kilometers far from Hanoi, Vietnam.

Corn and brown rice source

Corn and brown rice were provided by local animal feed stores in Phu Tho, Vietnam. The feed RTD-VinaS99 were chosen as a fixed feed in 4 diets. The rate of corn and brown rice in diets was based on the demand of consumption of chickens through periods of growth.

Experimental design

Total of 192 newly hatched chickens of local breed *(Ri x LP)* were purchased from a commercial hatchery. The birds were randomly divided into four groups (one control and three experimental groups) of 48 birds in each. Each group was replicated three times with 16 birds per replicate. The experimental diets were formulated as described in Table 1. The concentrated feed was RTD-VinaS99 used as a fixed feed. The value of nutrition composition of the feeds RTD-VinaS99, corn and brown rice was shown in Table 2.

Table 1. Experimental designs and diets

Experiment	1 - 21 day			22 – 35 day			> 35 day		
	BR (%)	Corn (%)	S99 (%)	BR (%)	Corn (%)	S99 (%)	BR (%)	Corn (%)	S99 (%)
Group 1[a]	0	65	35	0	68	32	0	73	27
Group 2[b]	16.25	48.75	35	17	51	32	18.25	54.75	27
Group 3[c]	32.5	32.5	35	34	34	32	36.5	36.5	27
Group 4[d]	48.75	16.25	35	51	17	32	54.75	18.25	27

[a]control group with only corn and S99, [b]replacing 25% corn by BR, [c]replacing 50% corn by BR and [d]replacing 75% by BR.

Table 2. Nutrition composition of the feeds

Ingredients	Unit	RTD-Vina S99	Corn	Brown rice
Energy of metabolism	kcal/kg	2700	332.9	327.10
Crude protein	%	44	7.93	8.0
Calcium	%	2.5 - 4.5	0.09	0.06
Phosphorus	%	2.0 – 3.5	0.15	0.24
Fiber	%	6.0	3.05	0.60
Lysine	%	2.7	2.5	ND
Methionine + Cysteine	%	2.6	1.56	ND

(Source: provided by the company RTD and National Institution of Animal Science)
ND: not determined

Housing

Birds were housed on rice husk shavings (10 cm height) in 4 similar rooms of the same facility with room climate. Each room was 3.2 x 1.4 m with 12 pens of approximately 2 m^2 each; the studied chicken density was 0.10 m^2/bird. Electronic light was provided for 24 h as a heat source for the first 10 days of age and 10 h of normal light per day during experimental period. The birds were vaccinated against infectious bursal disease (d 7 and d 25) and Newcastle disease (d 2 and d 28). The procedures of caring were following as given by the Viet Nam National Animal Science Institute.

Performance and feed conversion rate

The body weight of chickens was weighed weekly. Feed consumption was summarized to calculate the feed conversion ratio (FCR) and the average daily weight gain (ADG) at the same times.

Statistical Analysis

The data collected from experiments for body weight, feed intake, ADG, and FCR were evaluated and analyzed statistically to show the differences between experimental groups. Statistical analysis was performed as described by Thien (2009) using the software SPSS version 20.0 by the one-way ANOVA test. Differences among treatments were considered to be statistically significant if P-values was less than 0.05. All data were expressed as means.

RESULTS

The results concerning total average body weight gain, the total average feed intake and the feed conversion ratio in different experimental groups were summarized in Table 3.

Table 3. Chicken performance of using corn and brown rice in the diet

Parameter	Group 1 (control group)		Group 2 (75% corn + 25% BR)		Group 3 (50% corn + 50% BR)		Group 4 (25% corn + 75% BR)		P-value
	$\overline{X} \pm \overline{mx}$	Cv (%)	$\overline{X} \pm \overline{mx}$	Cv (%)	$\overline{X} \pm \overline{mx}$	Cv (%)	$\overline{X} \pm \overline{mx}$	Cv (%)	
Hatch (g)	36.23 ± 0.34	0.93	35.92 ± 0.31	0.86	35.56 ± 0.31	0.87	35.83 ± 0.30	0.84	0.51
12 wks (g)	1715.10[a] ± 38.47	2.24	1657.29[ab] ± 37.04	2.22	1641.15[ab] ± 42.48	2.58	1631.77 [ab] ± 30.37	1.86	0.44
TWG (g)	1678.87[a] ± 38.48	2.29	1621.37[ab] ± 37.01	2.28	1605.59[ab] ± 42.39	2.64	1595.94 [ab] ± 39.36	2.46	0.44
TFI (g)	4338.96[a] ± 65.33	1.51	4299.16[ab] ± 57.16	1.32	4365.62[ab] ± 73.19	1.68	4521.25 [ab] ± 59.11	1.31	0.23
FCR	2.58[a] ± 0.03	1.16	2.65[ab] ± 0.03	1.13	2.71[ab] ± 0.06	2.21	2.83[b] ± 0.03	1.06	0.03

$\overline{X} \pm \overline{mx}$: Means ± standard errors mean.

[a,b] Means within each row with the same superscript letter are not significant (P>0.05); CV: Coefficient of Variation; Wks: Weeks; TWG: total weight gain; TFI: total feed intake

Body weight gain

The initial weight of day old chicks was around 35 grams. After 12 weeks, the live weight gain was approximately 1.7 kilogram per bird. The final average weight gain was highest in the group 1 (1715.10 g, P<0.05) and lowest in the group 4 (1631.77 g, P>0.05). There were no significant differences between the body weight gain of chickens using corn and brown rice at different levels (P=0.44).

Feed intake and feed conversion

The feed intake was highest in group 4 (4521.25 g, P>0.05) and lowest in group 2 (4299.16 g, P>0.05). However, there were no significant differences of feed intake between different groups (P = 0.23). On the other hand, the feed conservation ratio was significantly different (P = 0.03) among experimental groups. Differences were observed between group 1 (FCR = 2.58) and group 4 (FCR = 2.83) due to P < 0.05 (P = 0.02), but there were no differences (P > 0.05) between group 1 and group 2 (FCR = 2.65) and group 3 (FCR = 2.71). Additional, all rearing conditions for each experimental group were identical and maintained during the period of the experiment. The mortality rate was 4.2% (2 birds) in the control group and 2.1% (1 bird) in each experimental group.

DISCUSSION

Vietnam is considered as a rich country with rice yields over the world (Bernhard Liese et al., 2014). However, high potential of using agricultural by-products from rice after processing has still been limited in usage in Vietnam. Some parts of rice including polishings (Ali MA et al., 1995, Rahman et al., 2005), rice bran (Wang et al., 1997) and rice grain that is undesirable for human consumption (Alias et al., 2008) demonstrated as feedstuffs for ruminants and monogastric animals. Brown rice was found to have a good potential as feed ingredients in the poultry industry (Asyifah et al., 2012). Analyses of nutrients of brown rice have shown that it has a high energy value (327.10 kcal/kg), low fiber content (0.60%) and is balanced in amino acids (Asyifah et al., 2012). This study considered as the first trial to examine the using of corn and brown rice for chicken feeding.

The live body weights of chickens in 4 treated groups were approximately 1.7 kg after 12 weeks feeding and no differences were found between using corn and brown rice. This strongly suggest that farmers could consider using brown rice as an alternative feedstuff to corn. Moreover, our findings on the growth performance of chickens using the studied diets agree with the biological growth performance of local chicken lines *"Ri lai" (Ri x LP)* in Vietnam.

The total feed intake of chickens was around 4.5 kg and no significant differences were found between groups (P = 0.23). However, the FCR was significantly different between the group 1 and group 4 and difference within the group 1 on their growth performance and feed intake. This may be attributed to the high difference in the sex of chickens within the group 1. There were 38 (79.16%) male chickens while only 10 (20.84%) female chickens within group 1, but there were equal number of male (20-25) and female (23-28) chickens between group 2, group 3 and group 4. This is in good agreement with the study on the ability of the growth performance of male chickens which was observed by Dayhim et al., 1992. Based on our findings, consequently, the diets (no. 1, 2 or 3) can be selected based on each condition in rural areas. Similar to our findings, a recent study of using brown rice with replacement of proportion of corn in diets for super meat ducks by Viet et al. (2015) showed no significant differences in growth performance of ducks fed with various levels of brown rice. Therefore, the growth performance has shown that brown rice can be used in chicken and duck production as a positive solution replacing corn due to rapid changing of prices of corn. It is also a good choice for farmers as well as the areas with redundancy of rice in Vietnam such as Mekong river delta areas.

To the best of our knowledge, there is no published information available on using brown rice with corn in chicken diets to compare the data of the present study about the feed intake in chicken production in Vietnam. However, studies from other countries showed that there were many alternative feeds for chickens rearing instead of corn only. The soybean, for instance, is also an alternative feedstuff to replace corn (Lee and Garlich, 1992). However, the replacement depends on the available feeds in each area and regions.

CONCLUSION

This study showed for the first time that body weight gain, feed intake and feed efficiency did not significantly differ when corn was substituted by mixtures of corn and brown rice in different ratios. This provides a basis for selection of either corn or brown rice according to the prevailing conditions in many rural areas in Vietnam. Moreover, it is considered as a new evidence for optional selection of corn and brown rice in poultry nutrition in Vietnam with the situation of high price of corn. Effective using of brown rice as a substitute of corn is a good solution for chicken production in Vietnam.

COMPETING INTEREST

The authors declare that having no competing interests.

ACKNOWLEDGEMENT

The authors would like to thank Dr. Helmut Hotzel and Dr. Hosny El-Adawy who helped us in improving the manuscript.

REFERENCES

1. Ali MA and S Leeson, 1995. The nutritive value of some indigenous Asian poultry feed ingredients. Anim. Feed Science. Technololy, 55: 237–257.
2. Alias I and T Ariffin, 2008. Potential of feed rice as an energy source for poultry production. Pages 31–38 in Proceedings of Workshop on Animal Feedstuffs in Malaysia: Exploring Alternative Strategies, Putrajaya, Malaysia.
3. Asyifah MN, S Abd-Aziz, LY Phang and MN Azlian, 2012. Brown rice as a potential feedstuff for poultry. Poultry Science, 21: 103-110.
4. Bernhard Liese, Somporn Isvilanonda, Khiem Nguyen Tri, Luan Nguyen Ngoc, Piyatat Pananurak, Romnea Pech, Tin Maung Shwe, Khamsavang Sombounkhanh, Tanja Möllmann, Yelto Zimmer, 2014. Economics of Southeast Asian Rice Production. A report of Agri Benchmark. Working paper No. 1.
5. Burgos S, Hinrichs J, Otte J, Pfeiffer D. and Roland-Holst D, 2008. Poultry, HPAI and livelihoods in Viet Nam - A Review. HPAI Research Brief. Mekong Team Working Paper No. 2.
6. Chris Lyddon, 2014. Focus on Vietnam. World Grain Information.
7. Duc NV, Long T, 2008. Poultry production systems in Vietnam. Working Paper Report. FAO report.
8. FAO, 2014. Poultry and nutrition and feed. Available at: http://www.fao.org/ag/againfo/themes/en/poultry/AP_nutrition.html.
9. Deyhim F, RE Moreng and EW Kienholz, 1992. The effect of testosterone propionate on growth of broiler chickens. Poultry Science 71: 1921-1926.
10. Kinh VL, 2015. Rice should to be used in animal feeding (Vietnamese). Available at: http://www.qdfeed.com/vi/news/Tin-trong-nuoc/Nen-su-dung-lua-gao-trong-TACN-313/.
11. Lee H and Garlich JD, 1992. Effect of overcooked soybean meal on chicken performance and amino acid availability. Poultry Science71: 499-508.
12. Luckmann J, Ihle R, Kleinwechter U and Grethe H, 2015. Do Vietnamese upland farmers benefit from high world market prices for maize? Agricultural Economics, 46: 1–11. doi: 10.1111/agec.12194.
13. Rahman MM, MBR Mollah, FB Islam, and MAR Howlider, 2005. Effect of enzyme supplementation to parboiled rice polish based diet on broiler performance. Livest. Res. Rural Development, 4:17–25.
14. Thien NV, 2009. Statistics using in scientific experiments of animal production (Vietnamese). Text book. Printed in the Agricultural Publisher.
15. USDA Foreign Agricultural Service, Grain Report, 2015. Vietnam Grain and Feed Annual. Available at: http://gain.fas.usda.gov. Accessed on 05/05/2015.
16. Van TT, 2002. A study on effect of breeds and protocols to meat production of Kabir, Sasso, Luong Phuong raised in the way of semi-intensive keeping in Thai Nguyen, Vietnam (Vietnamese version). The Report of Science and Research, Thai Nguyen. pp. 65 – 68.
17. Vang ND, Xuan TC, Tien PD, Nga LT, Hung NM, 1999. Production of the chicken Ri (Vietnamese version). Journal of Animal Science, Vietnamese Association of Animal Husbandry. pp. 99-100.
18. Vang ND, 2003. The Vietnam national country report on animal genetic resources. Available at: hftp://ftp.fao.org. Accessed on 15/01/2010.
19. Viet TQ, Huyen Van Le, Huyen Thi Ninh, Anh Ngoc Nguyen, 2015. A study on the effect of rice and brown-rice to replacing in the diet of CV super meat for growth performance and feed conservation ratio (Vietnamese version). Journal of Agriculture and Rural Development 8th 2015, 79 - 85.
20. Wang GJ, RR Marquardt, W Guenter, Z Zhang, and IZ Han, 1997. Effects of enzyme supplementation and irradiation of rice bran on the performance of growing Leghorn and broiler chickens. Anim. Feed Sci. Technol. 66: 47–61.

16

GENETIC DIVERSITY OF AROMATIC RICE IN BANGLADESH

Md. Mizan Ul Islam*, Parth Sarothi Saha, Tusher Chakrobarty, Nibir Kumar Saha, Md. Sirajul Islam and MA Salam

Agriculture and Food Security Program, BRAC Agricultural Research and Development Centre, Gazipur-1701, Bangladesh

***Corresponding author:** Md. Mizan Ul Islam; E-mail: mizan.islam@brac.net

ARTICLE INFO

Key words

Diversity

Aroma rice

Bangladesh

ABSTRACT

The nature and magnitude of diversity in 53 aromatic rice genotypes was evaluated in rain fed condition at BRAC Agricultural Research and Development Centre, Gazipur in 2012. The Euclidian method of divergence analysis indicated the presence of appreciable amount of genetic diversity in the material. These aromatic rice genotypes were grouped into six clusters where cluster I was the largest. Inter-cluster distances were larger than the intra-cluster distance revealed that there situated considerable genetic diversity among the genotypes. Based on positive value of vector I and II days to 50% flowering, seed length, and grain yield per hill had maximum contribution towards genetic divergence. Maximum yield contributing traits were accumulated in cluster V and as a result higher grain yield (42.0 g/hill) was obtained in this cluster. The genotypes of cluster V can be used in hybridization program to produce higher yielding breeding materials with all other clusters. The genotypes Cluster IV and cluster V were found most suitable for the respective characters and can be used as potential donor for future breeding programs.

INTRODUCTION

Bangladesh is well known for its native wealth of rice genetic resources and among these, large number of aromatic varieties cultivated in different agro-climatic regions of country (Nayak *et al.* 2004; Pandey *et al.* 2011). Rice landraces of Bangladesh possess tremendous genetic variation. Evaluation of this large genetic diversity is very important for their rational use in diverse needs. Genetic diversity is pre requisite for any crop improvement program, as it helps in the development of superior recombinants by providing necessary gene sources (Naik *et al.*, 2006). Success of hybridization and subsequent selection of desirable segregants depends largely on the selection of parents with high genetic variability for different characters. The more diverse the parents, within overall limits of fitness, greater are the chances of obtaining higher amount of heterotic expression in F_1s and broad spectrum of variability in segregating generations. Statistical analysis quantifies the genetic distance among the selected genotype and reflects the relative contribution of specific traits towards the total divergence. The use of Mahalanobis D^2 statistic for estimating genetic divergence has been emphasized by Shukla *et al.* (2006) and Sarawgi and Rita Binse (2007). The crosses between parents with suitable genetic divergence are generally the most responsive for yielding the most promising segregants. The present study was, therefore, undertaken to assess the extent of genetic diversity in 53 aromatic rice genotypes which will help to select prospective parents to develop transgressive segregants.

MATERIALS AND METHODS

The present investigation was under taken during rain-fed season (*Aman*) in 2012 at the BRAC Agricultural Research and Development Centre (BARDC), Gazipur, Bangladedsh. Geographically BARDC is located at 23.975°N latitude and 90.399°E longitude and 14 meters above the mean sea level. The genotypes were grown in 5.4 m x 0.6 m plots following randomized complete block design with three replications. The spacing was 20 cm between rows and 15 cm between plants. The data were recorded for yield and yield attributing characters and quality characters of aromatic rice used in this study. Observations were recorded on five randomly selected plants in each replication from the centre row. Different productive and quality characters viz. days to Days to 50% flowering, plant height (cm), tillers per hill, effective tillers per hill, panicle length (cm), filled grains per panicle, unfilled grains per panicle, total grain, sterility percentage, spikelet density, 1000 seed wt (gm), seed length (mm), seed breadth (mm), kernel length (mm), kernel breadth (mm) and grain yield per hill were recorded. The data were analyzed following Mahalanobis's (1936) generalized distance (D^2) extended by Rao (1952). Intra and inter-luster distances were calculated by the methods of Singh and Chaudhary (1985). All statistical analyses were carried out using GenStat 5.5, STAR, version 2.0.1 2014 and PB Tools, version 1.4. 2014 computer software.

RESULT AND DISCUSSIONS

Genetic Distance

All the inter-cluster distances were larger than the intra-cluster distance, indicating presence of wider diversity among genotypes of different groups (Table 1). The germplasm are traditional but they showed high variability between them which was revealed from the results of intra and inter-cluster distances. Basher *et al.* (2007) reported that inter-cluster distances were larger than intra-cluster distances in a multivariate analysis in rice. Intra-cluster distances were low for all clusters indicated homogeneous nature of the genotypes within the clusters. The results were supported by the findings of Iftekharuddaula *et al.* (2002) in rice. Intra cluster distance is highest in cluster II.

Regarding inter-cluster distance, cluster V showed maximum genetic distance (12.004) from cluster I followed by 11.711 of same cluster V from cluster IV (104). It is obvious that in all the cases cluster V produced the highest inter-cluster distances with other clusters suggesting wide diversity of the genotypes within cluster V with the genotypes of other clusters and the genotypes in these clusters could be used as parents in hybridization program for getting transgressive segregants (Saini and Kaiker, 1987). Lowest inter cluster D^2

values was recorded between cluster I and III (3.349). Highly divergent genotypes would produce a broad spectrum of variability in the subsequent generation enabling further selection and improvement, which would facilitate successful breeding of rice.

Table 1. Intra and inter cluster average distances of 53 aromatic rice genotypes

Cluster	I	II	III	IV	V	VI
I	**1.177**	6.625	3.349	4.01	12.004	8.245
II		**0.97**	5.245	8.339	10.624	7.231
III			**1.631**	4.295	10.22	6.605
IV				**1.4**	11.711	7.493
V					**5.604**	9.189
VI						**3.143**

Clustering of genotypes

The genotypes fall into six clusters (table 02). The distribution pattern indicated that cluster I, the largest cluster, comprised sixteen genotypes followed by cluster III (11 genotypes), cluster IV (10 genotypes), cluster II (9 genotypes), cluster VI (5 genotypes) and cluster V (2 genotypes). The clustering pattern revealed that the genotypes clustered together indicated there was no association between eco-geographical distribution of genotypes and genetic divergence.

Table 2. Clustering pattern of 53 aromatic rice genotypes

Cluster no.	No. of genotypes	Name of the Genotypes
I	1, 6, 11, 12, 14, 19, 20, 21, 23, 24, 25, 34, 40, 41, 43, 44	Badshabhog, Binnaful, Chinikanai-1, Chinikanai-2 Chinigura-PL2, Kalijira (Thin type), Kalijira (Round type), Kalijira (White type), Kalijira-PL3, Kalijira-PL6, Kalijira-PL9, Parbatjira, Tilkapur, Tulshimala, Tulshimala-PL3, Uknimadhu
II	2, 15, 16, 17, 26, 33, 37, 38, 50	Baoi-jhak, Damander-mukh, Gopalbhog, Gandho-kasturi, Kalijira-Japani, Modhumala, Rajbhog, Sabrot, Thai-3PL8-3
III	3, 4, 10, 22, 27, 28, 30, 31, 39, 42, 45	Basmati (Gazipur), Basmati PL90, BRRI dhan 38, Kalijira (Late), Kaloshailla, Kataribhog PL1, Jata Kataribhog, Philippine Kataribhog,Sakkorkhora, Tulshimala-PL1, Shandha
IV	5, 7, 8, 9, 13, 18, 29, 32, 35, 36	Begun bichi, BR5 (Dulabhog), BRRI dhan34, BRRI dhan 37, Chinigura, Kalijira (Normal), Kataribhog PL2, Kaminisharu, Premful, Radhunipagal
V	46, 48	Basmati 370, Jesmine
VI	47, 49, 51, 52, 53	Basmati 386, BRRI dhan50, CNI 9012, GSR 1234, GSR 1242

Cluster Mean

The highest cluster means for grain yield, 1000 seed wt and panicle length were obtained from cluster V (table 03). Highest total grain number, sterility percentage, spikelet density found in cluster IV, whereas the lowest grain yield and number of filled grain per panicle found in cluster III. Mean performance of different clusters for the characters studied revealed that dwarf stature, lower tillers per hill were in cluster VI whereas highest 1000 seed weight, panicle length, tillers number per hill and high yielding genotype were in V. Maximum desirable characters were accumulated in cluster V and as a result higher grain yield (42.0 g/hill) was obtained in this cluster. It was interesting that considering cluster distances, the cluster V showed comparative higher estimates of inter cluster-values with all other clusters.

Table 3. Cluster means for 53 aromatic rice genotypes

Traits	I	II	III	IV	V	VI
Days to 50% flowering	78	81	79	79	91	90
Plant height (cm)	114	109	114	113	121	87
Tillers/hill	11	9	11	9	13	7
Effective tillers/hill	9	8	10	9	12	6
Panicle length (cm)	24.0	25.1	24.8	25.1	27.5	24.4
Filled grains /panicle	169	112	109	171	157	134
Unfilled grains /panicle	22	12	22	42	28	32
Total grain	191	124	131	214	185	166
% Sterility	11.5	9.8	16.7	20.1	15.2	18.5
Spikelet density	8.0	5.0	5.4	8.6	6.9	6.9
1000 seed wt (gm)	10.6	20.4	15.0	10.9	23.9	19.9
Seed length (mm)	6.0	7.5	7.6	6.6	10.6	10.1
Seed breadth (mm)	2.3	2.8	2.4	2.2	2.5	2.3
Kernel length (mm)	4.3	5.4	5.3	4.6	7.5	7.0
Kernel breadth (mm)	2.0	2.4	2.0	1.9	2.0	1.8
Grain yield/Hill	16.7	16.9	16.3	16.6	42.0	17.2

Table 4. Contributions of the characters towards divergence in 53 aromatic rice genotypes

Traits	Genotypes	Replication	Error	SE	CV	Mean	Min	Max	LSD	Vector 1	Vector 2	h^2_b
Days to 50% flowering	79.51	15.92	0.40	0.36	0.8	81	71	98	1.0	0.211	0.026	98.5
Plant height (cm)	479.47	19.42	17.23	2.4	3.8	111	74	135	6.7	-0.010	0.003	89.9
Tillers/hill	13.40	4.58	2.42	0.9	15.6	10	6	14	2.5	-0.010	-0.147	60.1
Effective tillers/hill.	11.68	4.47	1.84	0.78	15.2	9	5	13	2.2	-0.724	-0.147	64.1
Panicle length (cm)	10.40	11.09	2.55	0.92	6.5	24.7	21	29	2.6	-0.027	0.142	50.7
Filled grains /panicle	4861.18	91.49	181.54	7.78	9.4	143	71	278	21.8	0.148	-0.110	89.6
Unfilled grains /panicle	499.99	306.84	100.39	5.78	39.7	25	5	61	16.2	0.241	-0.075	57.0
Total grain	6599.48	693.31	222.40	8.61	8.8	169	91	327	24.2	-0.155	0.106	90.5
% Sterility	112.63	52.25	27.05	3	35.4	14.7	5.1	32.7	8.4	-0.239	0.126	51.3
Spikelet density	10.43	0.11	0.59	0.45	11.2	6.9	3.6	11.4	1.3	-1.316	-0.012	84.7
1000 seed wt (gm)	74.86	0.07	0.07	0.16	1.9	14.6	8.9	27.9	0.4	-0.070	-0.251	99.7
Seed length (mm)	6.96	0.54	0.12	0.2	4.7	7.3	5.6	11.2	0.6	0.873	0.689	95.2
Seed breadth (mm)	0.24	0.03	0.01	0.07	5.1	2.4	2	3.5	0.2	-0.487	-0.438	83.3
Kernel length (mm)	3.60	0.09	0.06	0.15	4.9	5.1	3.6	7.6	0.4	-0.838	0.144	94.9
Kernel breadth (mm)	0.16	0.00	0.01	0.05	4.5	2.0	1.5	2.9	0.2	1.661	-3.679	86.4
Grain yield/Hill	113.54	21.04	6.52	1.47	14.6	17.4	9.4	50.1	4.1	0.274	0.181	84.6

Canonical Vector Analysis

The canonical vector analysis revealed that the vectors (vector I and II) for Days to 50% flowering, seed length, and grain yield per hill were positive (Table 4). Such result indicated that these three characters contributed maximum towards divergence. The greater divergence in the present materials due to these three characters will offer a good scope for improvement of yield through rational selection of parents for producing heterotic rice hybrids.

Correlation

50% flowering time had significant positive correlation with sterility percentage, 1000 seed wt (gm), seed length (mm), kernel length (mm), and grain yield per hill while negatively associated with plant height (cm). Plant height showed some sorts of correlation for tillers per hill and effective tillers per hill, while significant negative correlation for seed length, kernel length (table 05). Nofouzil et al. (2008) reported significant and positive correlation between grain yield and plant height. Significant and positive correlation among number of productive tillers per plant and grain yield was also noticed by Ali *et al.* (2008) in wheat. This result is also in consistent with the findings of Dogan (2009) and Gashaw et al. (2007). They reported significant positive correlation between grain yield and 1000-grain weight.

The correlations of tillers/hill were significantly positive with effective tillers per hill and GYP on the other hand effective tillers per hill and panicle length have significantly positive relation with GYP. There is no negative correlation for tillers/hill, effective tillers per hill and panicle length, where total grain showed only negative correlation with 1000 seed wt, seed length, Seed breadth, Kernel length and Kernel breadth. TSW has positively correlated with all characters. Although sterility percentage and kernel breadth showed no significant correlation with any characters but seed length and seed breadth showed significant positive correlation.

Table 5. Phenotypic correlation coefficients estimation between yield and yield component characters

	50F	PH	TILL	ETILL	PL	TG	%S	TSW	SL	SB	KL	KB
PH	-0.335*											
TILL	-0.074	0.286*										
ETILL	-0.142	0.321*	0.893**									
PL	0.127	0.224	-0.002	0.041								
TG	0.063	0.113	-0.148	-0.157	0.164							
%S	0.301*	-0.078	0.002	-0.019	-0.094	0.101						
TSW	0.564**	-0.25	-0.184	-0.202	0.11	-0.481**	0.04					
SL	0.623**	-0.430**	-0.107	-0.095	0.095	-0.328*	0.167	0.772**				
SB	0.218	0.058	-0.007	-0.055	0.163	-0.424**	-0.077	0.626**	0.167			
KL	0.602**	-0.387**	-0.111	-0.119	0.084	-0.356**	0.111	0.803**	0.977**	0.195		
KB	0.093	0.145	-0.045	-0.1	0.037	-0.315*	-0.084	0.470**	-0.075	0.898**	-0.039	
GYP	0.461**	0.091	0.338*	0.384**	0.295*	0.217	-0.067	0.434**	0.444**	0.198	0.437**	0.099

50F=Days to 50% flowering, PH=plant height (cm), TILL=tillers per hill, ETILL=effective tillers per hill, PL=panicle length (cm), TG=total grain, %S=sterility percentage, TSW=1000 seed wt (gm), SL=seed length (mm), SB=seed breadth (mm), KL= kernel length (mm), KB=kernel breadth (mm).

CONCLUSION AND RECOMMENDATION

Evaluated 53 aromatic rice genotypes of the present study are good sources of valuable genes. There is a high degree of diversity existed among the genotypes which needs to be utilized in future varietal improvement and/or development programs. Specifically, the genotypes of clusters V and IV may be selected as parents for hybridization programs to develop high yielding rice varieties with desirable other good characters. Future breeding goal should concentrate on selecting genotypes with high grain yield, long panicle length and more tillers per hill.

ACKNOWLEDGEMENT

The authors are grateful to technical assistants for data collection, cultivation and maintenance of rice plant.

REFERENCES

1. Ali Y, Atta BM, Akhter J, Monneveux P and Lateef Z, 2008. Genetic variability, association and diversity studies in wheat (Triticum aestivum L.) germplasm. Pakistan Journal of Botany, 40: 2087-2097.
2. Basher MK, Mia MAK, Nasiruddin M, Elahi NE, Julfiquar A W and Rasul MG, 2007. Genetic divergence based on morphophysiological characters of some maintainer and restorer lines of rice (Oryza sativa L.). In Proc. of the 7th Biennial Conference of Plant Breeding and Genetics Society of Bangladesh held on May 26, 2007 at Bangladesh Rice Research Institute, Gazipur. 72p.
3. Dogan R, 2009. The correlation and path coefficient analysis for yield and some yield components of durum wheat (Triticum turgidum var. durum L.) in west Anatolia conditions. Pakistan Journal of Botany, 41: 1081-1089.
4. Gashaw A, Mohammed H and Singh H, 2007. Selection criterion for improved grain yields in Ethiopian durum wheat genotypes. African Crop Science Journal, 15: 25-31.
5. Iftekharuddaula, KM, Akter, K, Bashar, MK and Islam MR, 2002. Genetic parameters and cluster analysis of analysis of panicle traits in irrigated rice. Bangladesh Journal of Plant Breeding and Genetics, 15: 49-55.
6. Naik D, Sao A, Sarawagi SK and Singh P, 2006. Genetic divergence studies in some indigenous scented rice (Oryza sativa L.). Accessions of Central India. Asian Journal of Plant Sciences, 5:197-200.
7. Nayak AR, Chaudhury D and Reddy JN, 2004. Genetic divergence in scented rice. Oryza, 41:79-82.
8. Nofouzil F, Rashidi V and Tarinejad AR, 2008. Path analysis of grain yield with its components in durum wheat under drought stress, International meeting on soil fertility land management and agroclimatology, Turkey. 681-686.
9. Pandey A, Bisht IS and Bhat KV, 2012. Population structure of rice (*Oryza sativa*) landraces from high altitude area of Indian Himalayas. Annals of Applied Biology, 160: 16-24.
10. PBTools, version 1.4. 2014. Biometrics and Breeding Informatics, PBGB Division, International Rice Research Institute, Los Baños, Laguna.
11. Rao CR, 1952. Advanced Statistical method in biometries research, John Wiley and Sons, New York.
12. Saini HC and Kaicker US, 1987. Genetic diversity in opium poppy. Indian Journal of Genetics, 47: 291-296.
13. Sarawgi AK and Rita B, 2007. Studies on genetic divergence of aromatic rice germplasm for agro morphological and quality characters. Oryza, 44: 74-76.
14. Shukla V, Singh S, Singh H and Pradhan SK, 2006. Multivariate analysis in tropical japonica "New plant type" rice (Oryza sativa L.). Oryza, 43: 203-207.
15. Singh R K and Chaudhury BD, 1985. Biometrical methods in quantitative genetic analysis. Kalayoni publishers. New Delhi.
16. STAR, version 2.0.1 2014. Biometrics and Breeding Informatics, PBGB Division, International Rice Research Institute, Los Baños, Laguna.

17

EFFECTIVENESS OF SELECTED MASS MEDIA IN AGRICULTURAL TECHNOLOGY TRANSFER TO THE FARMERS OF BANGLADESH

Suzan Khan, M Hammadur Rahman and Mohammed Nasir Uddin*

Department of Agricultural Extension Education, Faculty of Agriculture, Bangladesh Agricultural University, Mymensingh-2202, Bangladesh

***Corresponding author:** Mohammed Nasir Uddin; E-mail: nasirbau@gmail.com

ARTICLE INFO

Key words

Effectiveness
Mass media
Technology transfer
Farmers
Bangladesh

ABSTRACT

The role of mass media in dissemination of farm information to the farmers is crucial. Thus, the study was undertaken to determine the effectiveness of selected mass media in agricultural technology transfer to farmers of Bangladesh and identify the influential factors affecting the effectiveness of mass media in technology transfer to farmers. The study was conducted in three villages of Gouripur sub-district under Mymensingh district in Bangladesh. One hundred ten farmers were interviewed using a structured questionnaire for data collection during the period of 15 May to 15 June, 2016. Both descriptive and inferential statistics were used to analyze the collected data. Results show that 62.7 per cent of the farmers perceived that effectiveness of mass media in technology transfer was low, while 31.8 per cent and 5.5 per cent of them perceived the issue as "moderately effective" and "highly effective", respectively. Television was most popular mass media compared to radio, leaflet, poster and farm magazine based on the farmers' responses. Out of eight characteristics, farmers' education, extension contact and use of media had positively significant with the effectiveness of mass media. Multiple regression analysis revealed that 39.3% of the total variation in perceived effectiveness of mass media explained by two variables, namely education and use of media and identified as influential factors affecting the effectiveness of selected mass media.

INTRODUCTION

Mass media in agricultural information dissemination generally, are useful in reaching a wide audience at a very fast rate (Swanson and Rajalathi, 2010). These are useful as source of agricultural information to farmers and as well constitute methods of notifying farmers of new developments and emergencies. They could equally be important in stimulating farmers' interest in new ideas and practices (Ani et al. 1997). To a large extent, mass media serve as a veritable instrument for information dissemination in Agriculture. Food and Agricultural Organization of the United Nations (FAO, 2001) reported that in many developing countries, wide adoption of research results by majority of farmers remains quite limited. Efficient use of mass media can be helpful of minimizing the prevailing information gap in agricultural arena. In Bangladesh, various mass communication media are being used to transmit agricultural information to farmers in line with national policy on agriculture. While mass media can be defined as a system that provides information to many people (Sylvia, 2004). Both print and electronic media are included as mass media while print media include circulars posters, leaflets, bulletin, newspapers and journals etc, and the electronic media consist of Television, Radio, Internet, Video and Telephone (Shuwa et al. 2015; Ifenkwe and Ikpekaogu, 2012). Mass media are the essential elements needed for effective transfer of technologies that are designed to boost up agricultural production. While mass media particularly radio, television cell phone etc, have been playing significant role to provide a fast information and knowledge about the agriculture in global arena. Better distribution of mass media can increase agricultural attractiveness and providing more information to stakeholder for increasing production (Rao, 2007). Mass media have potentiality to provide timely and reliable information on weather, input prices, and marketing of the products (Chapman and Slaymaker, 2002) to the stakeholders especially the farmers. In Bangladesh, Islam and Gronlund (2010); Asaduzzamanet al. (2010); Cho and Tobias (2011) conducted some preliminary analyses and explored potentials of using mass media in marketing agricultural products. Osman (2014) conducted a study to determine the extent of use of ICT based media by the farmers in receiving agricultural information. Impact of mass media on farmers' agricultural production in Borno State was conducted by the Shuwa et al 2015.

Although a number of studies have already been conducted in Bangladesh on use of different mass media in technology transfer to farmers (Asif, 2016; Sharmin, 2013), study on effectiveness of these media particularly in agricultural area is scarcely available. However, no systematic study is available on the aspects of effectiveness of mass media in receiving technical information as well as technology transfer to the farmers. In order to fulfill the present information gap, the study was undertaken by the researchers. The major objectives of the study were to determine the effectiveness of mass media in agricultural technology transfer to the farmers and to identify the factors influencing the effectiveness of mass media in technology transfer to farmers.

MATERIAL AND METHODS

Study area, population and sampling

The study was conducted at Gauripur upazila (sub-district) under Mymensingh district of Bangladesh. In specific, three randomly selected villages, namely Charsrirampur, Chondopara and Payra in Dohakhola union of Gouripur upazila were the location of the study. The total number of the farm family heads (367) of the three selected villages constituted the population of the study. Thirty percent (30%) of the farm family heads from each of the three selected villages were randomly selected as the sample of the study. Thus, a total of 110 farmers constituted the sample for the study.

Selection and measurement of explanatory and focus variables

The explanatory variables of the study were eight selected characteristics of the farmers. Opinion on effectiveness of the selected mass media was given by the farmers. Therefore, farmers' personal characteristics such as age, education, household size, farm size, annual family income, organizational participation, extension contact and use of media were taken as explanatory variables of the study. Most of the explanatory variables were measured by developing scales based on the raw scores. Effectiveness of selected mass media in technology transfer was the focus variable of the study. While effectiveness is defined as the influence or capacity or validity of something produce the expected results. Farmers' response on effectiveness of selected mass media was measured using a 4 - point rating scale while score of 3, 2, 1 and 0 was assigned to indicate extent of effectiveness as highly effective, moderately effective, low effective and not at all respectively. The same measuring scale was also used by other researches to determine the effectiveness (Majydyan, 1996; Amin et al. 2013). Respondent's obtained scores of all (5) selected mass media (Radio, Television, Poster, Farm Magazine and Leaflets) were added to compute his/her total score and could range from 0 to 15. While score '0' indicate no effective and '15' indicate highly effective. Finally, the respondents were categorized into three categories according to their effectiveness responses (see Table 3).Both descriptive and inferential analyses were used to analyze the collected data. A co-efficient of correlation was used to understand the relationship between explanatory variables and focus variable while multiple regression models was used to quantify the contribution of the explanatory variables to focus variable effectiveness of the selected mass media.

Data collection

The empirical data were collected through face to face interview by using a pre-tested structured questionnaire during the period of 15 May to 15 June, 2016.

RESULTS AND DISCUSSION

Descriptive information of the farmers

The individual characteristics of the respondents, namely age, education, household size, farm size, annual family income, organizational participation, extension contact and use of media were considered as the explanatory variables of the study. Salient features of the characteristics and basic statistical value of respondents have been presented in Table 1 which is self-explanatory.

Table 1. Descriptive statistics and salient features of the farmers

Characteristics	Scoring system	Range		Mean	Standard Deviation
		Possible	Observed		
Age	Years	Unknown	22-75	45.45	12.08
Education	Years of schooling	Unknown	0-18	6.31	4.57
Household size	No. of members	Unknown	2-12	5.16	1.89
Farm size	Hectares	Unknown	0.16-3.95	0.80	0.76
Annual family income	'000' Tk.	Unknown	78.92-616.09	230.77	90.87
Organizational participation	Scale score	Unknown	0-54	10.06	8.74
Extension contact	Scale score	0-36	1-20	8.95	4.23
Use of media	Scale score	0-15	1-10	3.51	2.41

However, it is indicated that average level of education of the sample was just passed primary level (Class six) which might be one of the limitations of the understanding the using of mass media. The average farm size (0.80 ha) of the respondents also seem to be small farmer while families sized of the farmers were 5.16 which is more than the national average of 4.50. The average annual income of the farmers in the study area was BDT 2, 30,770 ($2,884.63 US), which is more than the national average of household income BDT 1, 37,748 ($1721 US) (HIES, 2010). The similar findings found in the study conducted by Haq (2016).

Use of selected mass media by the farmers

The results presented in Table 2 indicated that nearly almost all farmers (97.3 percent) had no use of radio whereas similar trend shows in case of farm magazine, poster and leaflets. But, Television was the popular mass media as a significant numbers of farmers (94.5%) have been using in receiving agricultural information. Television is audio visual media, available, relatively cheaper and easily understandable of the programmes especially agriculture programs that have been broadcasting were may be the reason of popular media. Similar finding was found by the Prathap and Ponnusamy (2006); Age et al. (2012). Moreover, it was also observed during data collection while farmers were watching the television not only at home but also at tea stalls available on the street and or at market. The similar studies were conducted by the Njoku (2016) and Mirani (2013) while radio, television etc are played an important role in technology transfer. Overall extent of selected mass media used by the farmers has been presented in Table 2.

Table 2. Farmers' category according to their use of selected mass media

Mass media	Category (0 to 3 scale score)	Frequency (n=110)	Percent	Mean	Standard Deviation
Radio/community radio	Not at all (0)	107	97.3	0.03	0.16
	Rarely (1)	3	2.7		
	Less frequently (2)	0	0		
	Regularly (3)	0	0		
Television	Not at all (0)	6	5.5	1.86	0.84
	Rarely (1)	29	26.4		
	Less frequently (2)	49	44.5		
	Regularly (3)	26	23.6		
Farm magazine	Not at all (0)	70	63.6	0.66	0.97
	Rarely (1)	13	11.8		
	Less frequently (2)	21	19.1		
	Regularly (3)	6	5.5		
Poster	Not at all (0)	65	59.1	0.46	0.60
	Rarely (1)	39	35.5		
	Less frequently (2)	6	5.5		
	Regularly (3)	0	0		
Leaflet/booklet	Not at all (0)	63	57.3	0.49	0.62
	Rarely (1)	40	36.4		
	Less frequently (2)	7	6.4		
	Regularly (3)	0	0		

Effectiveness of mass media in receiving agricultural information as perceived by the farmers

The observed scores of effectiveness of mass media in technology transfer ranged from 2 to 10 against the possible range of 0 to 15. The average and standard deviation were 5.21and 1.78, respectively. Based on their responses about the effectiveness of mass media scores, the respondents were classified into three categories as shown in Table 3.

Table 3. Effectiveness of selected mass media in receiving agricultural information as perceived by the farmers

Categories of farmers (score)	Number	Percent	Mean	SD
Low effective (up to 5)	69	62.7	5.21	1.78
Moderately effective (6-8)	35	31.8		
Highly effective (above 8)	6	5.5		
Total	110	100		

Data presented in Table 3 revealed that a significant numbers (62.7 per cent) of the farmers perceived mass media as low effective and 31.8 per cent of them perceived mass media as moderately effective. While only 5.5 per cent farmers were found who opined that mass media were highly effective in transferring technical information to them. Farmers with low literacy, low knowledge, low interactions, low communication to extension agent, etc. were may be the responsible factors having low effectiveness of the mass media. Besides, respondents' age may also the another reason having low effective of mass media as most of the respondents' seems to be middle aged while young people generally have the more interest to use the mass media.

Relationships of selected characteristics of the farmers and their opinion on effectiveness of mass media

A total of eight selected characteristics of the farmers were considered for understanding the relationships between those characteristics and their perceived effectiveness of mass media. To test the relationship, Pearson's correlation coefficients were computed as the results have been presented in Table 4.

Table 4. Correlation between farmers' characteristics and effectiveness of mass media

Focus variable	Selected personal socioeconomic characteristics (Table formation?)	Correlation co-efficient (r)values
Effectiveness of mass media in technology transfer to farmers	Age	-.148
	Education	.480**
	Household size	.063
	Farm size	.090
	Annual family income	.143
	Organizational participation	.134
	Extension contact	.311**
	Use of media	.587**

** Significant at 1% level

Out of eight characteristics farmers' Education, extension contact and use of media had significant and positive relationship with their perceived effectiveness of mass media. On the other hand, age, household size, farm size, annual family income and organizational participation of the farmers did not show any significant relationship with their perceived effectiveness of mass media. A linear multiple regression analysis was done with the explanatory variables and effectiveness of selected mass media which is explained in the following section.

Contribution of explanatory variables to farmers' perceived effectiveness of mass media

A linear multiple regression analysis was computed in order to determine the contribution of explanatory variables to their perceived effectiveness of mass media. A general full regression model analysis was run with all explanatory variables. The findings of the regression analysis are presented in Table 5.

The regression analysis indicates that 39.3% of the total variation of the focus variable (farmers' perceived effectiveness of mass media) explained by two variables, namely education and use of media, while the remaining 60.7% remain unexplained. This indicating that education and use of mass media were the vital factors that affecting the effectiveness of mass media in technology transfer to farmers. The model also indicates that education was positively and significantly correlated to farmers' perceived effectiveness of mass media at 5% level. This means that the farmers' having higher education had more likelihood to understand effectiveness of mass media in technology transfer as compared to those farmers having less education. This is because of the fact that education influences the farmers to understand well about the importance and utility of mass media. Use of media and effectiveness of mass media were showed positive and significant relationship at 1% level. The probability of perceived effectiveness of mass media is higher for those farmers who are the higher users of mass media. Uddin (2007); Rehman et al. (2011); Sheybani and Soleimanpour (2015) found the similar result in their respective studies.

Table 5. Regression coefficient of farmers' opinion on effectiveness of mass media with their selected characteristics

Explanatory variable	Unstandardized Coefficients		Standardized Coefficients	t	Sig.
	B	Std. Error	Beta		
(Constant)	3.708	.720		5.150	.000
Age	-.010	.012	-.065	-.787	.433
Education	.077	.040	.199	1.952	.050
Household size	.104	.079	.111	1.329	.187
Farm size	-.115	.214	-.050	-.540	.591
Annual family income	-.001	.002	-.028	-.291	.772
Organizational participation	-.024	.018	-.117	-1.313	.192
Extension contact	.008	.043	.019	.190	.850
Use of media	.376	.092	.510	4.105	.000
n = 110, R^2 = .393 , R = .627 F-value = 8.190**					

** Significant at 1% level

CONCLUSION AND RECOMMENDATIONS

Majority of the farmers considered the mass media as low effective to technology transfer. Thus, necessary interventions such as campaign on mass media, awareness building, incentives for mass media user, training etc. should take initiatives by the government as well as non-government agencies to increases the effectiveness of the mass media. Television was the popular mass media as perceived by the farmers among other selected mass media. Therefore, most of the cases, information related to technology transfer to farmers should be communicated using this media. Education, extension contact and use of media had significant relationships with farmers' responses on effectiveness of mass media. Among the significant variables education and use of media were identified as the influential factors to the effectiveness of mass media as perceived by the farmers and confirmed by the regression model. Therefore, policy makers should consider these variables while taking the policy to make the mass media effective in agricultural development arena.

REFERENCES

1. Age A, C Obinne and T Demenon, 2012. Communication for sustainable rural and agricultural development in Benue State, Nigeria. Sustainable Agriculture Research, 1: 118-129
2. Amin M, Sugiyanto, K Sukesi and Ismadi 2013. The effectiveness of cyber-extension-based information technology to support agricultural activities in *Kabupaten* Donggala, Central Sulawesi Province, Indonesia. International Journal of Asian Social science, 3: 882-889
3. Ani A, UC Undiandeye and DA Anogie, 1997. The Role of Mass Media in Agricultural Information in Nigeria. Educational Forum, 3: 80-85.
4. Asaduzzaman M, 2011. E-governance initiatives in Bangladesh: Some observations". Nepalese Journal of Public Policy and Governance, 29: 42-54.
5. Asif MAS, 2016. Use of mobile phone by the farmers in receiving information on vegetables cultivation, Unpublished MS Thesis, Department of Agricultural Extension Education, Bangladesh Agricultural University, Mymensingh-2202.
6. Chapman R and T Slaymaker 2002. ICTs and rural development: Review of the literature, Current interventions and opportunities for action. In: working paper 192, Overseas Development Institute, London.

7. Cho KM and D Tobias, 2011. Assessing the requirements for electronically linking farmers with markets. Research pilot project report prepared for USAID. http://www.meas-extension.org/meas-offers/pilot-projects
8. FAO 2001. Reports of Food and Agricultural Organization of the United Nations. International Journal of Agriculture and Biology, 3: 222.
9. Gronlund A, 2010. Using ICT to combat corruption: tools, methods and results. In C. Strand (Ed.), increasing transparency and fighting corruption through ICT: empowering people and communities, pp. 7–26.
10. Haq AZM, 2016. Farmers' education and farmers' wealth in Bangladesh. Turkish Journal of Agriculture: Food Science and Technology, 3: 787-792.
11. HIES, 2010. Household income and expenditure survey. Bangladesh bureau of statistic. Ministry of planning, Government of People's Republic of Bangladesh. Dhaka. Bangladesh.
12. Ifenkwe GE and Ikpekaogu F, 2012. Noise mitigation for effective agricultural extension print message delivery and utilization. Journal of Agricultural Extension and Rural Development, 4: 51-56.
13. Majydyan R, 1996. Perception of the effectiveness of selected communication media used by the BAUEC farmers (unpublished MS thesis). Department of Agricultural Extension Education, Bangladesh Agricultural University, Mymensingh-2202.
14. Mirani Z, 2013. Perception of farmers and extension and research personnel regarding use and effectiveness of sources of agricultural information in Sindh Province of Pakistan. The Journal of Community Informatics, 9: 1-8.
15. Njoku JIK, 2016. Effectiveness of radio-agricultural farmer programme in technology transfer among rural farmers in Imo State, Nigeria. Net Journal of Agricultural Science, 4: 22-28.
16. Osman SM, 2014. Farmers' use of ICT based media in receiving agricultural information (MS thesis). Department of Agricultural Extension Education, Bangladesh Agricultural University, Mymensingh-2202.
17. Prathap PD and KA Ponnusamy, 2006. Effectiveness of four mass media channels on the knowledge gain of rural women. Journal of International Agricultural and Extension Education, 13: 73-81.
18. Rao N, 2007. A framework for implementing information and communication technologies in agricultural development in India. Journal of Technological Forecasting and Social Change, 74: 491-518.
19. Rehman F, S Muhammad, I Ashraf and Hassan S, 2011. Factors affecting the effectiveness of print media in the dissemination of agricultural information. Sarhad Journal of Agriculture, 27: 119-124.
20. Sylvia B (2004). Voices for change. Rural women and communication. Communication for development group extension, Education and communication service. FAO, Rome.
21. Shuwa MI, L Shettima, BG Makinta, A Kyari, 2015. Impact of mass media on farmers agricultural production, case study of Borno State, Agricultural Development Programme, Academic Journal of Scientific Research, 3: 008-014.
22. Sharmin F, 2013. Use of communication media by the fish farmers in commercial fish culture, Master Thesis (unpublished), Department of Agricultural Extension Education, Bangladesh Agricultural University, Mymensingh-2202.
23. Swanson B and R Rajalahti, 2010. Strengthening agricultural extension and advisory systems: procedures for assessing, transforming, and evaluating extension systems. Agriculture and rural development discussion Paper 44. ARD, Washington, DC, USA: World Bank.
24. Sheybani EJH and MR Soleimanpour, 2015. Factors affecting extension-education media effectiveness in agriculture information transmission to farmers in Tehran Province. Journal of Scientific Research and Development, 2: 211-215.
25. Uddin KF, 2007. Use of mass media by the farmers in receiving agricultural information, MS thesis, department of agricultural extension & information system, Sher-E-Bangla Agricultural University, Dhaka.

18

PHOSPHORUS USE EFFICIENCY AND CRITICAL P CONTENT OF STEVIA GROWN IN ACID AND NON-CALCAREOUS SOILS OF BANGLADESH

Md. Maniruzzaman, [1]Tanzin Chowdhury, [2]Md. Arifur Rahman and [3*]Md. Akhter Hossain Chowdhury

Soil Resources Development Institute, Farmgate, Dhaka, Bangladesh; [1]Department of Agricultural Chemistry, Sher-e-Bangla Agricultural University, Sher-e-Bangla Nagar, Dhaka-1207, Bangladesh and [2 & 3]Department of Agricultural Chemistry, Faculty of Agriculture, Bangladesh Agricultural University, Mymensingh-2202, Bangladesh

***Corresponding author:** Md. Akhter Hossain Chowdhury; E-mail: akhterbau11@gmail.com

ARTICLE INFO

Key words

Stevia
Leaf biomass yield
PUE
P requirement
Critical P content

ABSTRACT

Knowledge of phosphorus (P) uptake and its use efficiency by crop plants is essential for adequate management of the plant nutrients to sustain food production with a minimal environmental impact. To study the effects of P on the growth, leaf biomass production, P content and uptake and to estimate P use efficiency (PUE), minimum P requirement and critical leaf P content of stevia, a pot experiment was conducted in the net house of the Department of Agricultural Chemistry, Bangladesh Agricultural University following completely randomized design (CRD) with three replications in acid and non-calcareous soils of Bangladesh. The applied treatments was six *viz.* 0 (P_0), 25 (P_{25}), 50 (P_{50}), 75 (P_{75}), 100 (P_{100}) and 150 (P_{150}) kg P ha^{-1}. Plant samples were collected at 15 days interval to obtain different parameters. Collective results indicated that significantly highest values of different parameters were obtained with P @ 100 kg ha^{-1} and the lowest from P control. Phosphorus application increased leaf dry yield at harvest by 55 to 510% in acid soil and 70 to 488% in non-calcareous soil over control. The rapid growth of the plant was recorded at the later stages (30 to 60 days after planting). Phosphorus content and uptake was directly proportional with the increased levels of P except the treatment P_{150} in both soils. Maximum PUE and fertilizer P use efficiency (FPUE) was observed at P_{100} treatment. Critical P content was estimated to be *ca* 0.19 and 0.30% in the leaves of stevia plants grown in acid and non-calcareous soils, respectively. For maximum leaf biomass production of stevia grown in acid and non-calcareous soils, the minimum requirement of P was also estimated to be *ca* 109 and 104 kg ha^{-1}, respectively. The information of this finding would contribute to optimize the soil P use and improve fertilizer management for stevia cultivation.

INTRODUCTION

Phosphorous (P) is one of the least available, least mobile, mineral nutrient to the plants in many cropping environments, based on its contribution to the biomass as a macronutrient (Goldstein et al., 1988). Many soils have large reserves of total P, often hundred-time more than the P available to the crops (Al-Abbas and Barber, 1964). Phosphorus is needed in metabolic processes such as energy transfer, signal transduction, macro-molecular biosynthesis, photosynthesis, respiration, etc. Relative to nitrogen and potassium, the recovery of P fertilizers by crop plants is usually very low due to soils' high capacity to fix P to soil constituents of little bioavailability (Manske et al., 2001; Lynch, 2007; Balemi and Negisho, 2012).

Stevia (*Stevia rebaudiana* Bertoni.) is a herbaceous perennial small bush contains the secret of stevioside, which make it the sweetest herb in the world (Soejarto et al., 1983). Dry leaves are the economic part in stevia plant. The leaves are having commercial importance due to presence of di-terpene sweet glycosides which are 300-400 times sweeter than sugar without any side effects. Hence, stevia is a potential natural source of no calorie sweeteners, alternative to the synthetic sweetening agents like saccharine, aspartame that are available in the market to the diet conscious consumers and diabetic patients. Stevia crop cultivation made significant impact in the countries like Japan, China, Korea, Mexico, USA, Thailand, Malaysia, Indonesia, Australia, Canada and Russia (Brandel and Rosa, 1992). Studies conducted in India so far could suggest only few management approaches of stevia for improving its productivity. Since the production potential of stevia in India is 2-3 t ha^{-1} of dry leaves as against 1-2 t ha^{-1} in China, it has definite advantage over China (Chalapathi et al., 1997b). Bangladesh being an agro-based country could easily introduce this plant as an industrial crop like sugarcane, sugar beet, tea or coffee and can commercially be cultivated in its relatively high land, char land, home stead area etc. as it grows well in open space having regular sun light.

Plants that are efficient in absorption and utilization of the absorbed nutrients greatly enhance the efficiency of applied fertilizers. Recently, suitable soil for stevia cultivation (Zaman et al., 2015), N and S requirement and critical N and S content of stevia grown in two contrasting soils of Bangladesh has been reported by Zaman et al.,, (2016a and 2016b). It is expected that a higher and balanced nutrient supply will result in higher foliage yield. So there is a need to set up certain protocols for cultivation of stevia in various soil conditions so that farmers can be benefited by selling, and industries also can get healthy leaves throughout the year to isolate the active components and can formulate economical market products.

Nevertheless, to the best of our knowledge, no systematic and detailed study has yet been conducted on P fertilizer requirement for stevia in acid and non-calcareous soils of Bangladesh. Therefore, optimum P requirement need to be determined for achieving maximum leaf biomass yield of stevia in the country. Critical values are quite useful and are frequently referred to when interpreting a plant analysis result. The critical P concentration of stevia is yet to be estimated. Though the level of P in the soil is one of the critical factors determining the growth and yield of the plants, no report has yet been published on the requirement of P and critical P level for the growth and leaf biomass yield of stevia in Bangladesh. Keeping in view the significant role of P in crop production systems, the present piece of research work was undertaken to investigate the effects of different levels of P on the growth, leaf biomass yield, P content and its uptake, minimum P requirement, PUE and critical leaf P concentration of stevia in two contrasting soils under the agro climatic conditions of Bangladesh Agricultural University, Mymensingh.

MATERIALS AND METHODS

Before starting the experiment, initial soil pH, organic carbon, cation exchange capacity (CEC), available P were determined separately. Two soils viz. acid and non-calcareous of contrasting physical and chemical properties were used (Zaman et al., 2015). Approximately 40 kg soils from each location (Madhupur for acid soil and BAU farm for non-calcareous soil) were collected from 0-15cm depth of selected fellow land for the experiment. The samples were made free from plant residues and other extraneous materials, air dried, ground and sieved through a 2mm sieve. 500g sieved soil from each source was preserved in a polythene bag and the physical and chemical properties were determined following standard procedure (Page et al., 1982).

Eight kg processed soil was taken in each earthen pot of 23 cm in height with 30 cm diameter at top and 18 cm at bottom leaving 3 cm from the top. Forty five days old stevia seedlings (*Stevia rebaudiana* Bertoni) were collected from brac biotechnology laboratory, Joydebpur, Gazipur and used for the experiment. One stevia seedling was planted in each pot during 1st week of March, 2012. N, K, S, Zn and B were applied as basal doses @ 250, , 200, 30, 3 and 1 kg ha^{-1} from prilled urea, MoP, gypsum, zinc sulphate and boric acid, respectively (Zaman, 2015). Six levels of P viz. 0, 25, 50, 75, 100, and 150 kg ha^{-1} were applied from TSP. Nitrogen was applied in equal three installments, 1/3rd during pot preparation, 1/3rd at 15 days after planting (DAP) and 1/3rd at 30 DAP. The experiment was laid out in completely randomized design with three replications. Intercultural operations like irrigation, soil loosening, weeding, insect pest control, removal of flowers etc. were done as and when necessary.

Data were collected at 15, 30, 45 and 60 DAP. The crop was destructively harvested at 60 DAP. After harvesting the crop, leaf samples were separated, cleaned, dried for 72 hours, weighed, ground and stored. Plant heights, number of branches and leaves, leaf area, leaf fresh and dry weight were studied. Phosphorus content of stevia leaf was determined colorimetrically using $SnCl_2$ as a reducing agent (Page et al., 1982). Uptake was calculated from P content and leaf dry yield. PUE (grain yield per unit P added) was also calculated (Moll et al., 1982).

Phosphorous uptake, utilization and use efficiency were calculated following the below mentioned formulae:

$$\text{Phosphorus Use Efficiency (PUE)} = \frac{\text{Yield (kg)}}{\text{P in soil (kg)}}$$

$$\text{Phosphorus Uptake Efficiency (PUPE)} = \frac{\text{P in plant (kg)}}{\text{P in soil (kg)}}$$

$$\text{Phosphorus Utilizati on Efficiency (PUTE)} = \frac{\text{Yield (kg)}}{\text{P in plant (kg)}}$$

The mathematical calculation of fertilizer P uptake, utilization and use efficiency was described as:

$$\text{Fertilizer P use Efficiency (FPUE)} = \frac{\text{(Fertilize d pot yield - control pot yield)}}{\text{P applied}}$$

$$\text{Fertilizer P Uptake Efficiency (FPUPE)} = \frac{\text{(P in fertilized plant - P in control plant)}}{\text{P applied}}$$

$$\text{Fertilizer P Utilizati on Efficiency (FPUTE)} = \frac{\text{(Yield of fertilized pot - yield of control pot)}}{\text{(P in fertilized pot - P in control pot)}}$$

Phosphorus requirement and critical P content of stevia was also estimated following Chowdhury, 2000. The results obtained were subjected to statistical analysis using standard method of analysis (Steel et al., 1997). The differences among the treatment means were compared by using Duncan Multiple Range test (Gomez and Gomez, 1984).

RESULTS AND DISCUSSION

Effects of different levels of P on various parameters of stevia are discussed under the following heads-

Plant Height

Different levels of P influenced the plant height of stevia (Figure 1). At harvest, plant height was significantly and rapidly increased with the increased levels of P up to 100 kg ha^{-1} and then slowly decreased with further increase in P levels (150 kg ha^{-1}). P application at all levels increased plant height by 65.0 to 94.0 cm in acid soil and 59.3 to 93.0 cm in non-calcareous soil at 60 DAP. Plant height was significantly increased with the advancement of the growth period irrespective of P levels. However, the tallest plants of 94.0 cm in acid soil and 93.0 cm in non-calcareous soil were obtained from P_{100} which was identical with P_{75} and P_{150} but statistically different from P_{25} and P_{50} and the shortest plant was obtained from the control treatment irrespective of the soils studied.

Height increase was 45% higher in acid soil and 56% higher in non-calcareous soil over control. The increase in P levels increased the plant height progressively up to P_{100} at 60 days of growth. Absolute control (P_0) recorded significantly lowest plant height of stevia.

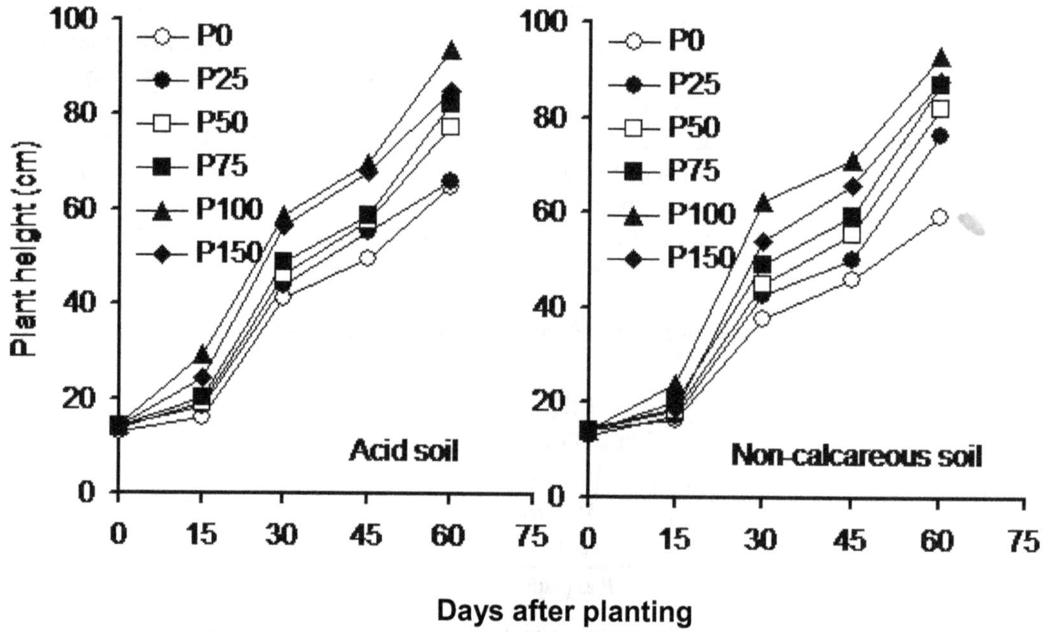

Figure 1. Effects of different levels of P on the plant height of stevia at various DAP

The results are in accordance with the findings of Chalapathi et al., (1999) who also reported increased plant height of stevia with higher nutrient levels in sandy loam soils at Bangalore. Our result is in agreement with the findings of Alam et al., (2003), who reported that the plant height of wheat significantly increased with increasing phosphorus application to the soil. On the other hand, a finding was also found by Rasul (2016), who observed highest plant height of wheat applying P_2O_5 @ 250 kg ha^{-1}.

Branch Number

The number of branch $plant^{-1}$ of stevia was significantly influenced by different levels of P (Figure 2). At harvest, number of branches $plant^{-1}$ was significantly increased with the increased levels of P up to 100 kg ha^{-1} and then slowly decreased with further increase in P levels (150 kg ha^{-1}). Phosphorus application at all levels increased branch number by 75 to 187% in acid soil and 45 to 163% in non-calcareous soil at 60 DAP. The number of branch $plant^{-1}$ was significantly increased with the advancement of the growth period irrespective of P levels and soils used. However, highest number of branches $plant^{-1}$ in both soils was obtained from P_{100} which was statistically identical with all P levels except control and P_{25} and the lowest value was obtained from the control treatment in both soils. Crop performance to a great extent is governed by the number of branches $plant^{-1}$. It is, therefore, imperative that if the number of branches $plant^{-1}$ is higher, the numbers of leaves are expected to be higher; ultimately the leaf yield will be higher. This finding is also similar with the results of Jain and Singh (2003) who reported that number of branches per plant in pea increased with the application of P. Similarly, Atif et al., (2014) found highest number of branches of pea applying P @ 100 kg ha^{-1}. This growth parameter might have possibly contributed positively to the higher leaf yield with higher P application.

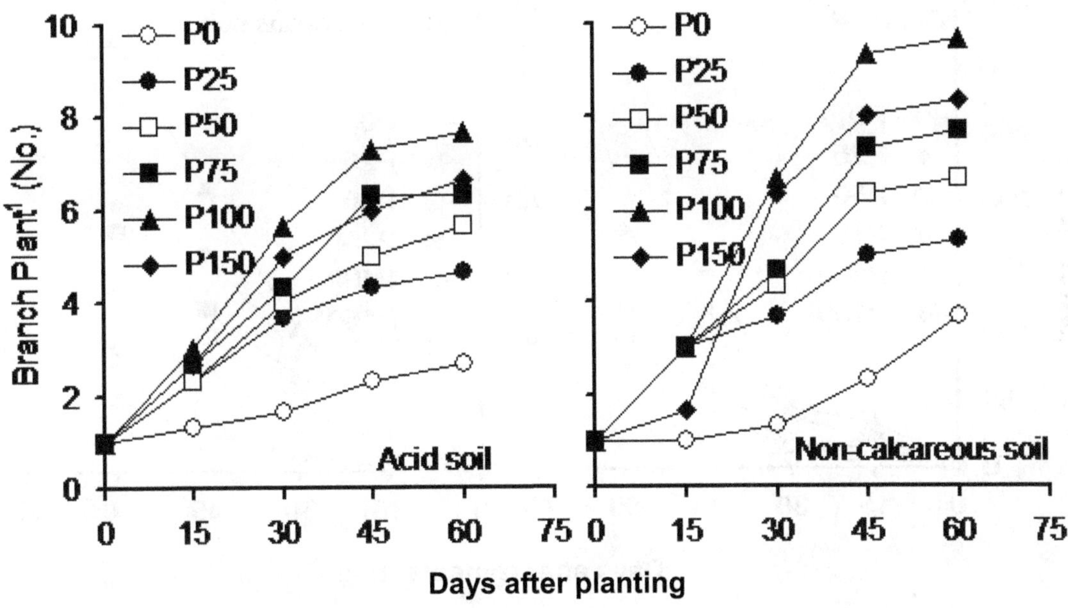

Figure 2. Effects of different levels of P on the branch number of stevia at various DAP

Leaf Number

The number of leaves of stevia in both acid and non-calcareous soils at all growth stages except 0 DAP (Figure 3) is also significantly influenced by different levels of P. Number of leaves plant^{-1} was increased with the increased levels of P up to 100 kg ha^{-1} and then declined by 65% in acid soil and 63% in non-calcareous soil with further addition (P_{150}). Leaf number increase was very slow at the early growth stages (0-30 DAP) while it was rapid between 30 and 60 DAP irrespective of P levels except control. The most dominant increase was observed between 45 and 60 DAP particularly when P was applied @ 100 kg ha^{-1}. P application at all levels increased the number of leaves by 26 to 248 in acid soil and 38 to 270 in non-calcareous soil. Plants fertilized with P_{150} and P_{75} produced identical number of leaves in non-calcareous soil. The minimum number of leaves plant^{-1} was obtained from the plants fertilized with P control irrespective of soils and growth period. Green leaves are the site of photosynthetic activity taking place in the plants. The number of leaves plant^{-1} would also substantiate the fact that increased number of leaves plant^{-1} would contribute to the final yield of the plant particularly the crops like stevia in which only leaves are used as commercial product.

Kawatani et al., (1980) had also reported increased number of branches and leaves plant^{-1} of stevia with higher inorganic nutrition in Japan. This finding is also similar with the results of Islam et al., (2013) who reported that number of leaves plant^{-1} in tomato was increased due to the application of different doses of inorganic fertilizers. Aladakatti et al., (2012) found that plant height of stevia at harvest was significantly influenced by higher levels of nitrogen, phosphorus and potassium which in turn were responsible for higher number of branches plant^{-1} and number of leaves plant^{-1} resulting into higher leaf yield. Increased number of leaves plant^{-1} of stevia with increased levels of N, P and K fertilizers was also reported by Buana and Goenadi (1985) in Brazil.

Leaf Area

The total leaf area plant^{-1} at harvest as influenced by different levels of P is presented in Table 1. Leaf area plant^{-1} responded significantly due to the application of different levels of P. The results revealed that leaf area progressively increased with increasing levels of P application up to P_{100} in both soils and then declined with further addition (P_{150}). The highest total leaf area plant^{-1} (2207cm^2 in acid soil and 2866cm^2 in non-calcareous soil) observed at 60 DAP was measured from the plant receiving 100 kg P ha^{-1} which was significantly higher than other levels of P. Second highest values (1322cm^2 in acid soil and 1752cm^2 in non-calcareous soil) were obtained from P_{150}.

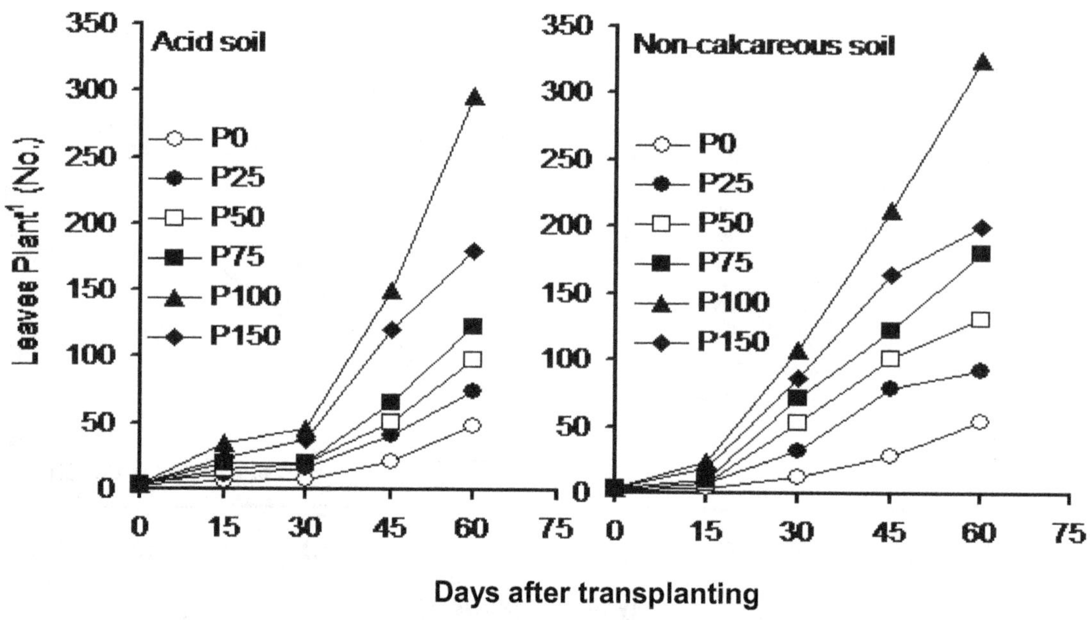

Figure 3. Effects of different levels of P on the leaf number of stevia at various DAP

Identical leaf area was obtained from the plants fertilized with P_{50} and P_{75} in non-calcareous soil. The lowest leaf area was found from the control treatment which was identical with P_{25} irrespective of soils used. P application at all levels increased leaf area by 55 to 510% and 70 to 488% in acid and non-calcareous soils, respectively at harvest. Leaf area is an important growth indices determining the capacity of plant to trap solar energy for photosynthesis and has marked influence on the growth and yield of plant. Like other yield attributes, leaf area also followed increasing trends with the progress of plant growth with maximum value at 60 DAP irrespective of treatments. Higher leaf area of stevia with higher P levels could be attributed to more number of branches and leaves plant^{-1} due to higher plant height. Khanom (2007) reported highest leaf area of stevia plant grown in non-calcareous soil applying chemical fertilizers.

Leaf dry weight

The data pertaining to the dry weight of stevia leaves plant^{-1} at harvest as influenced by different levels of P fertilizer have been presented in Table 1. Results revealed that leaf dry weight significantly and progressively increased with the increased levels of P application up to 100 kg ha^{-1} in both soils and then declined with further addition (P_{150}). The highest leaf dry weight plant^{-1} (9.02g in acid soil and 9.70g in non-calcareous soil) at harvest was measured from the plant receiving 100 kg P ha^{-1} which was significantly higher than other levels of P. Second highest values (5.42g in acid soil and 6.10g in non-calcareous soil) were obtained from P_{150} in both soils. Identical dry weight was also obtained from the plants fertilized with P_{75} and P_{150} in non-calcareous soil. The lowest values were obtained from the control treatment (1.48g in acid soil and 1.65g in non-calcareous soil). Phosphorus application at all levels increased leaf dry yield at harvest by 55 to 510% and 70 to 488% in acid and non-calcareous soils, respectively over control.

Dry matter accumulation by the crop is another important growth parameter to be considered for determining the economic yield while assessing the effects of different treatments. Phosphorus fertilizers showed significant influence on the dry weight of stevia leaves. These results are in conformity with the findings of Pramanik and Singh (2003) who reported that the application of P_2O_5 at 60 kg ha^{-1} significantly increased yield attributes and yield over control in chickpea. The lowest growth, yield and yield attributing characters of grass pea were recorded under control (P_0) treatment. Murayama et al., (1980) in Japan experimentally proved that no fertilization resulted in lowest leaf yield of stevia. Thus, the findings of our study were in agreement with that of Ojeniyi et al., (2007) who reported that application of N, P, K and animal manure increased the dry weight of tomato as compared to control. Sood and Kumar (1994) also reported that green and dry foliage yield increased with increasing levels of N and P, which also confirmed the results obtained in the present study. In case of simultaneous increase biomass yield, it was reported with the application of N, P and K at 60, 30 and 45 kg/ha, respectively, produced higher dry leaf yield with the simultaneous higher nutrient uptake by stevia plant (Chalapathi et al., 1997 and 1999).

Table 1. Effects of different levels of P on leaf area, dry weight and yield increase of stevia leaves over control at harvest

P level	Leaf area plant^{-1} (cm^2)		Leaf dry weight (g plant^{-1})		Yield increase over control (%)	Yield increase over control (%)
	Acid soil	Non-calcareous soil	Acid soil	Non-calcareous soil	Acid soil	Non-calcareous soil
P_0	243e	268e	1.48e	1.65e	-	-
P_{25}	494de	657de	2.30d	2.81d	55	70
P_{50}	708cd	1032d	2.99cd	4.05c	102	145
P_{75}	886c	1532c	3.74c	5.49b	153	233
P_{100}	2207a	2866a	9.02a	9.70a	510	488
P_{150}	1322b	1752b	5.42b	6.10b	266	270
CV (%)	5	5	4.92	4.29	-	-
$LSD_{0.05}$	177	237	0.51	0.56	-	-
SE±	161	209	0.61	0.64	-	-

CV = Coefficient of variance, LSD = Least significant difference, SE± = Standard error of means

Leaf fresh weight

The data pertaining to the fresh weight of stevia leaves plant^{-1} at harvest as influenced by different levels of P fertilizer have been presented in Figure 4. Results revealed that leaf fresh weight significantly and progressively increased with the increased levels of P application up to P_{100} in both soils and then declined with further addition (P_{150}).

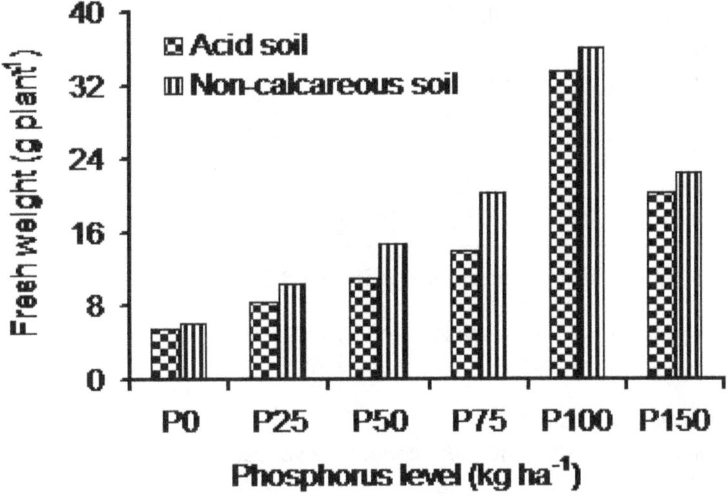

Figure 4. Effects of different levels of P on the fresh weight of stevia leaves at 60 DAP

The highest fresh weight plant^{-1} (33.50g in acid soil and 36.02g in non-calcareous soil) at harvest was measured from the plant receiving 100 kg P ha^{-1} which was significantly higher than other levels of P. Second highest values (20.23g in acid soil and 22.45g in non-calcareous soil) were obtained from P_{150} in both soils. Identical fresh weight was also obtained from the plants fertilized with P_{75} and P_{150} in non-calcareous soil. The lowest values were obtained from the control treatment (5.49g in acid soil and 6.17g in non-calcareous soil.

Phosphorus application at all levels increased fresh weight at harvest by 2.96 to 28.01g plant^{-1} in acid soil and 4.26 to 29.85g plant^{-1} in non-calcareous soil.

Leaf P concentration and uptake

The data on the P concentration and uptake by stevia leaves as influenced by different levels of P have been presented in Table 2. Both the concentration and uptake was significantly influenced by the application of P fertilizers. Phosphorus concentration of the leaf was increased with the increased levels of P irrespective of soils used. The highest concentration (0.21% in acid soil and 0.33% in non-calcareous soil) was obtained when P was applied @ 100 kg ha^{-1} in both soils which was statistically identical with the P contents of the leaves of stevia plant fertilized with P_{75} and P_{150} but significantly different from other treatments. The lowest P content was obtained from the plants receiving no P fertilizer in both soils. Phosphorus uptake was also significantly affected by its additions. The uptake of P did not follow the same trend like P concentration of stevia leaves. P uptake varied from 1.09 to 17.76 mg pot^{-1} in acid soil and 1.71 to 32.12 mg pot^{-1} in non-calcareous soil. The uptake of P as expected increased with increasing P levels up to 100 kg ha^{-1} and then decreased with further additions (P_{150}). The lowest P uptake was observed in the control treatment of both soils. The nutrient content of a plant varies not only among its various plant parts but changes with age and stage of development. Phosphorus contents and its uptake by stevia leaf varied significantly in both soils with their additions. The increase in concentration and nutrient uptake was proportional with the rate of application up to treatment P_{100}. Higher nutrient uptake may be related to higher biomass yield due to the highest dry leaf yield harvested from that treatment. Because nutrient uptake was calculated from their concentrations and corresponding dry leaf yield.

Table 2. Effects of different levels of P on it's content and uptake by stevia leaf at harvest

P level	Phosphorus			
	Acid soil		Non-calcareous soil	
	Content (%)	Uptake (mg pot^{-1})	Content (%)	Uptake (mg pot^{-1})
P_0	0.07b	1.09e	0.10b	1.71e
P_{25}	0.11b	2.45e	0.16b	4.98e
P_{50}	0.14a	4.58d	0.22b	9.36d
P_{75}	0.18a	7.45c	0.28b	15.55c
P_{100}	0.21a	17.76a	0.33b	32.12a
P_{150}	0.19a	10.54b	0.29b	17.88b
CV (%)	3.7	8.3	3.9	7.9
$LSD_{0.05}$	0.06	0.78	0.06	1.12
SE±	0.01	1.42	0.02	2.55

CV = Coefficient of variance, LSD = Least significant difference, SE± = Standard error of means

These results are in agreement with the results found by Sushanta et al., (2014), who found that the total P uptake by wheat increased with increasing P fertilizer application.

Critical P concentration of stevia leaf

We followed the "Critical nutrition concentration" concept advanced by Ulrich (1952) for plant to determine critical P concentration in stevia leaf. Critical values as used by Ulrich and Hills (1973) are determined from the relationship of nutrient concentration and relative yield at the time of sampling. The critical P concentration in stevia leaf was estimated from the relative amount of leaf biomass to achieve 80% of the maximum production of stevia leaf following the procedure of Kouno et al., (1999). For both the soils, relative leaf biomass yield was plotted on the ordinate (Y axis) against the respective P concentration of stevia leaf on the abscissa (X axis) in Figure 5. The Phosphorus concentration corresponding to the arbitrary point at 80% to achieve the maximum leaf biomass production was estimated by the fitted curve to be ca 0.19 and 0.30% in the leaves of stevia plants grown in acid and non-calcareous soils, respectively (Figure 5).

Sharif et al., (1988) reported the critical concentration of P in maize plant as 1.4 mg P/g dry matter at 95% of relative yield. They found the maximum 100% relative yield at P tissue level of 1.8 mg P/g dry matter. Another study of Fageria et al., (2007) reported that the average P content of 0.16% in rice leaves is adequate for rice growth and yield.

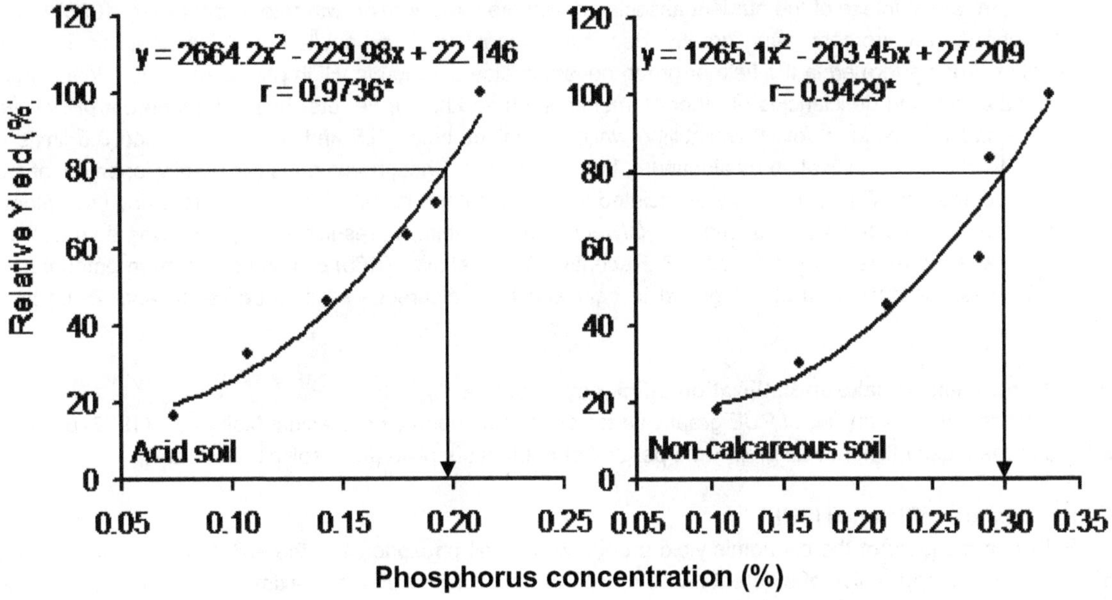

Figure 5. Correlation between leaf P concentration and relative leaf biomass yield of stevia grown in acid and non-calcareous soils. Values are the means of all treatments. **Correlated significantly at P<0.01

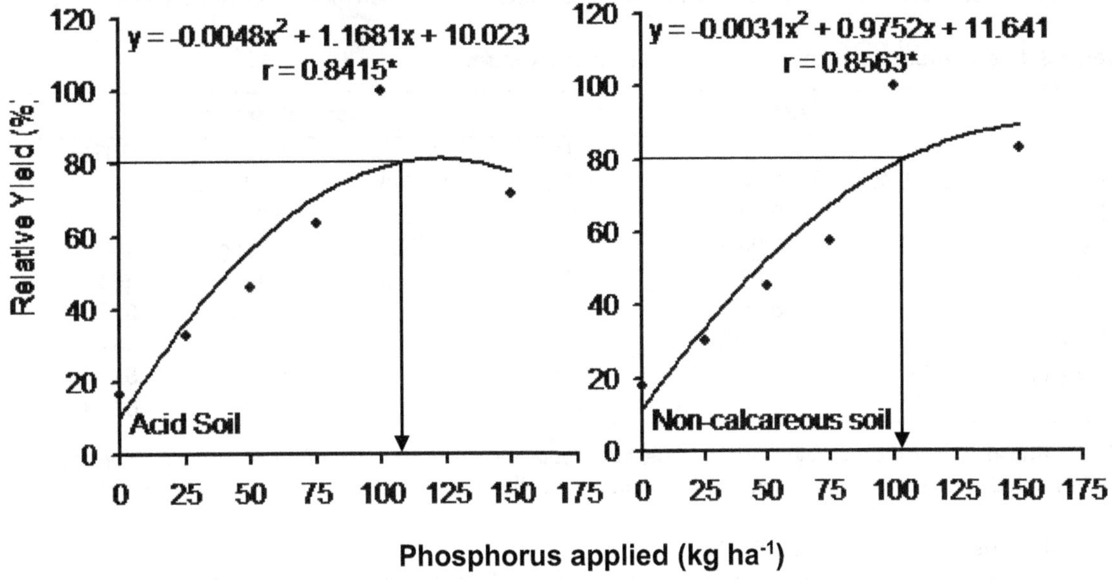

Figure 6. Correlation between applied P and relative leaf biomass yield of stevia grown in acid and non-calcareous soils. Values are the means of all treatments. **Correlated significantly at P<0.01

Phosphorus requirement of stevia plant

To determine the requirement of P in soil to obtain 80% of maximum leaf biomass yield, the applied P was plotted on the X axis against the relative leaf biomass yield on the Y axis. From the fitted curve, the corresponding estimated minimum amount of P for leaf biomass production in the plant grown in acid soil and non-calcareous soil to be ca 109 and 104 kg ha^{-1}, respectively (Figure 6). A crop's requirement for a specific nutrient is commonly defined as "the minimum content of that nutrient associated with the maximum yield" or "the minimum rate of intake of the nutrient associated with the maximum growth rate" (Loneragan, 1968).

Phosphorus requirement vary greatly depending upon fertilizer applied and whether or not the investigation was performed in the field or green house, choice of crop etc. High phosphorus plant levels may cause imbalances and deficiencies of other elements, such as Zn, Cu, Fe, etc. Soils with inherent pH values between 6 and 7.5 are ideal for P availability, while pH values below 5.5 and between 7.5 and 8.5 limits P availability to plants due to fixation by aluminum, iron, or calcium. Phosphorus does not readily leach out of the root zone; potential for P loss is mainly associated with erosion and runoff. Phosphorus deficiency symptoms often occur as young plants are exposed to cool/wet growing conditions, resulting in a phase where vegetative growth exceeds the roots' ability to supply P. Recently, Zaman et al., (2016b) estimated minimum amount of S for leaf biomass production in stevia grown in acid and non-calcareous soils to be ca 40 and 45 kg ha^{-1}, respectively.

Phosphorous use, uptake and utilization efficiency of stevia

Our current understanding of PUE greatly varies depending upon crop species. Moll et al., (1982) using N, defined nutrient use efficiency as grain yield per unit of nutrient supplied (from soil plus fertilizer).

Phosphorus use efficiency (PUE)

PUE is a measure of the economic yield produced per unit phosphorus in the soil. PUE was significantly influenced by the application of different levels of P. The results revealed that maximum PUE of 74.5% in acid soil and 69.6% in non-calcareous soil were observed at P_{100} treatment. Similarly, the minimum PUE of 44.1% in acid soil and 33.9% in non-calcareous soil were obtained with P_{50} treatment and control (P_0), respectively (Table 3). Similarly, at higher P application rates plants used smaller proportion of fertilizer P due to the greater fixation that resulted in low PUE (Sultani et al., 2004; Alam et al., 2005; Rahim et al., 2010). Marschner (1995) reported that P retention in roots is improved sufficient levels, which may increase PUE at low P level.

Table 3. Phosphorus use, uptake and utilization efficiency of stevia leaf

P level	Acid soil			Non-calcareous Soil		
	PUE	PUPE	PUTE	PUE	PUPE	PUTE
P_0	57.35ab	0.041c	1476a	33.94b	0.04c	997a
P_{25}	45.75b	0.048c	976ab	45.05b	0.07bc	631b
P_{50}	44.09b	0.062bc	708b	46.78b	0.10bc	469bc
P_{75}	47.79b	0.085b	566b	48.88ab	0.14b	364c
P_{100}	74.54a	0.156a	480b	69.65a	0.23a	310c
P_{150}	41.90b	0.080b	527b	35.35b	0.10bc	351bc
CV (%)	2.66	5.04	5.22	2.98	5.90	5.06
SE±	3.25	0.01	97.10	3.27	0.02	62.11

PUE = Phosphorus use efficiency, PUPE = Phosphorus uptake efficiency; PUTE = Phosphorus utilization efficiency. Means with dissimilar letter(s) are significantly different from other treatments according to Scheffé-test ($p<0.05$)

Phosphorous uptake efficiency (PUPE)

PUPE is the ratio of kg P in plants and kg P in soil. PUPE increased with the increased levels of P up to 100 kg ha^{-1} and then declined. The highest PUPE (0.156 in acid soil and 0.23 in non-calcareous soil) was observed when the plants were amended with 100 kg P ha^{-1}. The lowest PUPE (0.041 in acid soil and 0.04 in non-calcareous soil) was recorded from the control treatment (Table 3). These results are in line with the previous work of Pongsakul and Gensen (1991).

Phosphorous utilization efficiency (PUTE)

PUTE is the kg of dry matter yield divided by kg of P in plant, measure the efficiency with which P in plant is utilized for producing economic yield. The PUTE rate decreased with the increased level of P application. The PUTE was maximum (1476 in acid soil and 997 in non-calcareous soil) when no P was applied and it was minimum (480 and 310 in acid and non-calcareous soils, respectively) when the soil was amended with P @ 100 kg/ha (Table 3).

Fertilizer P use, uptake and utilization efficiency of Stevia

Fertilizer P use efficiency (FPUE)

The results revealed that maximum FPUE of 31.9% in acid soil and 37.7% in non-calcareous soil was observed at P_{100} treatment. Similarly, the minimum FPUE of 15.2% in acid soil and 23.9% in non-calcareous soil was obtained with P_{25} treatment (Table 3). Thus the finding of our study is in good agreement with the finding of Yamoah et al., (2002) who reported that fertilizer P use efficiency improved significantly, when integrated (organic and inorganic) source of P was used.

Table 4. Fertilizer P use, uptake and utilization efficiency of stevia

P level	Acid soil			Non-calcareous Soil		
	FPUE	FPUPE	FPUTE	FPUE	FPUPE	FPUTE
P_0	-	-	-	-	-	-
P_{25}	15.24b	0.011b	1368a	27.25b	0.027b	1014a
P_{50}	15.96b	0.015b	1107b	24.30b	0.032b	787b
P_{75}	16.51b	0.018b	947b	23.85b	0.039b	642b
P_{100}	31.92a	0.035a	938b	37.71a	0.064a	608b
P_{150}	12.35c	0.013b	951b	14.04c	0.023b	636b
CV (%)	3.91	2.13	5.12	3.24	2.75	4.58
SE±	1.856	58.324	0.002	2.126	52.371	0.004

FPUPE = Fertilizer P uptake efficiency; FPUE = Fertilizer P use efficiency; FPUTE = Fertilizer P utilization efficiency, Means with dissimilar letter(s) are significantly different from other treatments according to Scheffe-test ($p<0.05$)

Fertilizer P uptake efficiency (FPUPE)

FPUPE was significantly influenced by the application of different levels of P. The FPUPE reached its maximum (0.035 in acid soil and 0.064 in non-calcareous soil) value when the soils were amended with 100 kg P ha^{-1} and it was minimum (0.011 in acid soil and 0.023 in non-calcareous soil) value those were fertilized by no P and 150 kg P ha^{-1}, respectively (Table 4).

Fertilizer P utilization efficiency (FPUTE)

FPUTE decreased gradually with increased levels of P application up to treatment P_{100}. The highest FPUTE (1368 in acid soil and 1014 in non-calcareous soil) was recorded in soil fertilized with 25 kg P ha^{-1} and the lowest (938 and 608 in acid and non-calcareous soils, respectively) value was found in soil amended with 100 kg P ha^{-1} (Table 4).

CONCLUSION

The results clearly showed that all the parameters examined in this study were significantly affected by different doses of phosphorus. The highest values of most parameters were obtained from 100 kg P ha^{-1} and the lowest values from control. However the decreasing trends in these parameters with further increase beyond 100 kg P ha^{-1} indicated that higher doses of P could be detrimental to stevia and it should be avoided. Application of P @ 100 kg ha^{-1} gives highest leaf dry yield at harvest (510% in acid soil and 488% in non-calcareous soil). Same responses were found in case of P content and uptake by stevia in both soils. The highest PUE and FPUE were also recorded at P_{100} treatment. The overall results suggest that farmers can be advised to apply P @ 109 kg ha^{-1} in acid soil and 104 kg ha^{-1} in non-calcareous soil to ensure optimum yield and maximum PUE. However, such studies should be conducted with different combination of soils and applied P before widespread recommendations.

ACKNOWLEDGEMENTS

We express our sincere thanks to the concerned authority of Bangladesh Agricultural Research Council (BARC), Farmgate, Dhaka for financial support.

CONFLICT OF INTEREST

The authors declare that there is no conflict of interests regarding the publication of this paper.

REFERENCES

1. Aladakatti YR, YB Palled, MB Chetti, SI Halikatti, SC Alagundagi, PL Patil, VC Patil and AD Janawade, 2012. Effect of nitrogen, phosphorus and potassium levels on growth and yield of stevia (*Stevia rebaudiana* Bertoni). Karnataka Journal of Agricultural Science, 25: 25-29.
2. Alam SM, S Azam, S Ali and M Iqbal, 2003. Wheat yield and P fertilizer efficiency as influenced by rate and integrated use of chemical and organic fertilizers. Pakistan Journal of Soil Science, 22(2): 72-76.
3. Alam SM, SA Shah, and MM Iqbal, 2005. Evaluation of method and time of fertilizer application for yield and optimum P-efficiency in wheat. Songklanakarin Journal of Science and Technology, 27: 457-463.
4. Al-Abbas AH and SA Barber, 1964. A soil test for phosphorous based upon fractionation of soil phosphorous: I Correlation of soil phosphorous fraction with plant available phosphorous. Soil Science Society of America Proceedings, 28: 218-221.
5. Atif MJ, SA Shaukat, ASZ Shah, YA Choudry and SK Shaukat, 2014. Effect of Different Levels of Phosphorus on Growth and Productivity of Pea (*Pisum Sativum* L.) Cultivars Grown as Off-Season under Rawalakot Azad Jammu and Kashmir Conditions. Journal of Recent Advances in Agriculture, 2: 252-257.
6. Balemi T and K Negisho, 2012. Management of soil phosphorus and plant adaptation mechanisms to phosphorus stress for sustainable crop production: a review. Journal of Plant Nutrition Soil Science, 12: 547–561.
7. Buana L and DH Goenadi, 1985. A study on the correlation between growth and yield of stevia. Menara Perkebunan, 53: 68-71.
8. Chalapathi MV, B Shivaraj and VR Parama, 1997a. Nutrient uptake and yield of Stevia (*Stevia rebaudiana* Bertoni) as influenced by methods of planting and fertilizer levels. Crop Research, 14: 205-208.
9. Chalapathi MV, S Timmegowda, VR Prama and TG Prasad, 1997b. Natural non calorie sweetener stevia (*Stevia rebaudiana* Bertoni): A future crop of India. Crop Research, 14: 347-350.
10. Chalapathi MV, S Thimmegowda, DN Kumar, GGE Rao and J Chandraprakash, 1999. Influence of fertilizer levels on growth, yield and nutrient uptake of ratoon crop of stevia. Crop Research, 21: 947-949.
11. Chowdhury MAH, 2000. Dynamics of microbial biomass sulphur in soil and its role in sulphur availability to plants. PhD Thesis, Laboratory of Plant Environmental Science, Graduate School of Biosphere Sciences, Hiroshima University, Japan.

12. Fageria NK and MPB Filho, 2007. Dry matter and grain yield, nutrient uptake, and phosphorus use efficiency of lowland rice as influenced by phosphorus. Communications in Soil Science and Plant Analysis, 38(9-10): 1289-1297.
13. Gomez KA and AA Gomez, 1984. Statistical Procedure for Agricultural Research. 2nd Eds., Los Banos, Philippines: International Rice Research Institute, pp: 207-215.
14. Goldstein AH, DA Baertiein and RG McDaniel, 1988. Phosphate starvation inducible metabolism in *Lycopersicon esculentum*. Plant Physiology, 87(3): 716-720.
15. Islam MR, MAH Chowdhury, BK Saha and MM Hasan, 2013. Integrated nutrient management on soil fertility growth & yield of tomato. Journal of Bangladesh Agricultural University, 11: 33-40.
16. Jain LK and P Singh, 2003. Growth and nutrient uptake of chickpea as influenced by phosphorus and nitrogen. Crop Research, 25: 401-413
17. Kawatani T, Y Kaneki, T Tanabe and T Takahashi, 1980. On cultivation of Kaa-He-E (*Stevia rebaudiana* Bert). VI. Response of stevia to potassium fertilization rates and to the three major elements of fertilizer. Japanese Journal of Tropic Agriculture, 24: 105-112.
18. Khanom S, 2007. Growth, leaf yield and nutrient uptake by stevia as influenced by organic and chemical fertilizers grown on various types of soil. MS Thesis, Department of Agricultural Chemistry, Bangladesh Agricultural University, Mymensingh-2202.
19. Kouno K and S Ogata, 1988. Sulphur supplying capacity of soils and critical S values of forage crops. Soil Science and Plant Nutrition, 34: 327-339.
20. Loneragan JF, 1968. Nutrient requirements of plants. Nature, 220: 1307-1308.
21. Lynch J, 2007. Roots of the second green revolution. Australian Journal of Botany, 55: 493-512.
22. Marschner H, 1995. Mineral nutrition of higher plants, 2nd Eds., Annals of Botany Company, London: Academic Press, pp: 889.
23. Manske G, J Ortiz-Monasterio, MV Ginkel, R González, R Fischer, S Rajaram and P Vlek, 2001. Importance of uptake efficiency versus P utilization for wheat yield in acid and calcareous soils in Mexico. European Journal of Agronomy, 14: 261-274.
24. Murayama SM, OR Kayan, K Miyazato and A Nose, 1980. Studies on the cultivation of *Stevia rebaudiana* Bert. II. Effects of fertilizer rates, planting density and seedling clones on growth and yield. Science Bulletin of the college of Agriculture, University of the Ryukyus, Okinawa, 27: 1-8.
25. Ojeniyi SO, MA Awodun and SA Odedina, 2007. Effect of animal manure amended spent grain and cocoa husk on nutrient status, growth and yield of tomato. Middle East Journal of Scientific Research, 2: 33-36.
26. Page AL, RH Miller and DR Keeney, 1982. Method of Soil Analysis, Part-2 Chemical and Microbiological Properties, 2nd Eds., American Society of Agronomy, Inc. Madison, Wisconsin, USA.
27. Pongsakul P, ES Gensen, 1991. Dinitrogen fixation and soil N uptake by soybean as affected by phosphorus application. Journal of Plant Nutrition, 14: 809-823.
28. Pramanik K and RK Singh, 2003. Effect of levels and mode of phosphorus and biofertilizers on chickpea under dryland conditions. Indian Journal of Agronomy, 48: 294-296.
29. Rahim A, AM Ranjha, Rahamtullah and EA Waraich, 2010. Effect of phosphorus application and irrigation scheduling on wheat yield and phosphorus use efficiency. Soil Environment, 29: 15-22
30. Rasul GAM, 2016. Effect of Phosphorus Fertilizer Application on Some Yield Components of Wheat and Phosphorus Use Efficiency in Calcareous Soil. Journal of Dynamic Agricultural Research, 3: 46-52.
31. Sharif MZ, R Amin, FE Qayum and M Aslam, 1988. Plant tissue concentration and uptake of phosphorus by maize as affected levels of fertilization. Pakistan Journal of Agricultural Research, 9: 335-338.
32. Sushanta S, S Bholanath, M Sidhu, P Sajal and DR Partha, 2014. Grain yield and phosphorus uptake by wheat as influenced by long-term phosphorus fertilization. African Journal of Agricultural Research, 9: 607-612.
33. Soejarto DD, CM Compadre, PJ Medon, SK Kamath and AD Kinghorn, 1983. Potential sweetening agents of plant origin. II. Field search for sweet tasting stevia species. Economic Botany, 37: 71-79.

34. Sood BK, N Kumar, 1994. Effect of nitrogen and phosphorus on forage yield and nutrient uptake of oat-berseem mixture. Crop Research, 8: 239 – 244.
35. Steel RGD, JH Torrie and D Dickey, 1997. Principles and Procedures of Statistics: A biometrical approach, 3rd Eds., McGraw-Hill Book Co., New York, USA.
36. Sultani MI, M Shaukat, IA Mehmood and MF Joyia, 2004. Wheat growth and yield response to various green manure legumes and different P levels in pothowar region. Pakistan Journal of Agricultural Research, 41: 102-108
37. Ulrich A, 1952. Physiological basis for assessing the nutritional requirements of plants. Annual Review. Plant Physiology, 3: 207-228.
38. Ulrich A and FJ Hills, 1973. Plant analysis as an aid in fertilizing sugar crops: Part I. Sugar beets. In: Soil testing and plant analysis, Eds., Walsh, L.M. and J.D. Beaton Madison, Wisconsin, USA, pp: 271-288.
39. Yamoah CF, A Bationo, B Shapiro and S Koala, 2002. Trend and stability analysis of millet yields treated with fertilizer and crop residues in the Sahel. Field Crops Research, 75: 53-62.
40. Zaman MM, 2015. Nutrient requirement leaf yield and stevioside content of stevia (*Stevia rebaudiana* Bertoni) in some soil types of Bangladesh. PhD Thesis, Department of Agricultural Chemistry, Bangladesh Agricultural University, Mymensingh.
41. Zaman MM, MAH Chowdhury and T Chowdhury, 2015. Growth parameters and leaf biomass yield of stevia (Stevia rebaudiana, Bertoni) as influenced by different soil types of Bangladesh. Journal of Bangladesh Agricultural University, 13: 33-40.
42. Zaman MM, MAH Chowdhury, KM Mohiuddin and T Chowdhury, 2016a. Nitrogen requirement and critical N content of stevia grown in two contrasting soils of Bangladesh. Research in Agriculture, Livestock and Fisheries, 3: 87-97.
43. Zaman MM, MAH Chowdhury, T Chowdhury and ABMM Hasan, 2016b. Critical leaf S concentration and S requirement of stevia grown in two different soils of Bangladesh. Fundamental and Applied Agriculture, 1(3): 106-111.

19

SMARTPHONE BASED AUTOMATIC PRICE DETERMINATION OF AGRICULTURAL PRODUCTS

Md Rakib Hassan*

Department of Computer Science and Mathematics, Faculty of Agricultural Engineering and Technology, Bangladesh Agricultural University, Mymensingh 2202, Bangladesh

***Corresponding author:** Md Rakib Hassan; E-mail: rakibkuet@yahoo.com

ARTICLE INFO

Key words

Agricultural products
Market equilibrium price
Sellers
Buyers

ABSTRACT

Smartphones are increasingly becoming an integral part of our life. We carry it every time with us and use it to do various tasks including browsing internet, emails, social websites and others. We also perform online shopping with our smartphones. In this paper, a new automatic price determination algorithm for agricultural products is proposed that can be used by both the sellers and buyers using their smartphones. The buyers will place their demands for their required agricultural products to the sellers of a particular market and the sellers will also place their supplies. The algorithm will then automatically calculate the market equilibrium price using learning rate based iterative distributed price determination algorithm. As a result, both the sellers and buyers can save their time in finding the suitable prices of the agricultural products. The performance results show that the algorithm is stable and reaches the market equilibrium price within a few milliseconds.

INTRODUCTION

In any market, the price of the products vary from seller to seller. The price of the agricultural product also fluctuates based on the supply of the product and the demand of the buyers. The agricultural market is no exception to that. As a result, the buyer has to visit different sellers' profiles to check the price within a market. On another dimension, the price varies significantly from market to market. There are other factors that affect the price of agricultural products, for example, drought, flood, rain, external supplies and production. Therefore, it is very difficult to predict the price of agricultural products without visiting the sellers within a market and also other markets. This problem not only affects the buyers but also the sellers as well. Some independent sellers may be selling the price below the minimum price without knowing the actual price in the market and thus not getting the proper price for their time, energy and monetary investments. So the bottom line is that the price depends on a number of factors. In this paper, an automatic price determination algorithm is proposed that can be used in smartphones which will collect the supplies of the agricultural products from the sellers and the demands of the buyers of their list of agricultural products. Then the algorithm will automatically calculate the market equilibrium price using iterative distributed algorithm. As a result, both the sellers and the buyers will be able to know the actual price of an agricultural product for a particular day. This will benefit both the sellers and buyers by reducing the risk of their losses.

In this paper, market equilibrium is used to determine the price of agricultural products automatically. It is the state when the supply of a product of sellers become equal to the demand from the buyers (Varian, 1992), (Mankiw, 2006). In market equilibrium, the underlying competition among sellers and/or buyers is inherently captured. Therefore, it can demonstrate the general trend of the market involving both the sellers and buyers. In a market, there can exist product shortage or surplus which directly affects the price of the product. If there is shortage, the price will increase whereas the price will decrease for surplus. In real markets, market equilibrium exists in different forms and scales. It can consist of a pair of seller-buyer or a group of multiple sellers and buyers. The area of the sellers and buyers can span from a single small market to the whole world market. For example, house prices may increase in an area if there is a large demand of that area. But if there is no one to live there, the price will fall. Similarly, oil market is a case in the whole world market. Less supply and increased demand of oil increases the price of oil. On the other hand, increase in supply and less demand will reduce the price of oil.

Market equilibrium approaches have been adopted in various fields, such as, microeconomics (Sharpe, 1964), (Black, 1972), automobile industries (Berry, Levinsohn, & Pakes, 1995), share market (Admati, Pfleiderer, & Zechner, 1994), banking (Besanko & Kanatas, 1993), spectrum resource trading (Niyato & Hossain, 2008b), supermarkets (Smith, 2004), house prices and rents (Ayuso & Restoy, 2006), etc. Market equilibrium is also analyzed with other approaches. For example, (Niyato & Hossain, 2008a) compares the market equilibrium algorithm with competitive and cooperative market. Market equilibrium is found to generate the most stable solution among the three models and it also ensures the smallest profit for the sellers which is also beneficial for the buyers. Besides, the market equilibrium has less communication overhead compared to the other models.

In this paper, market equilibrium is used because it represents the general trend of the market and captures the underlying competition of the sellers and/or buyers. Besides, it is more stable than other methods and its communication overhead is also the lowest.

MATERIALS AND METHODS

Different hardware and software tools have been used to model the agricultural market scenario with multiple sellers and buyers. The used hardware device was a laptop with Intel core i7 processor, 8 GB RAM and Windows 8.1 operating system. The software tools include emulated android devices running different android operating systems, Matlab (Attaway, 2013) and Java programming language (Schildt, 2014). The system model is shown in Figure 1.

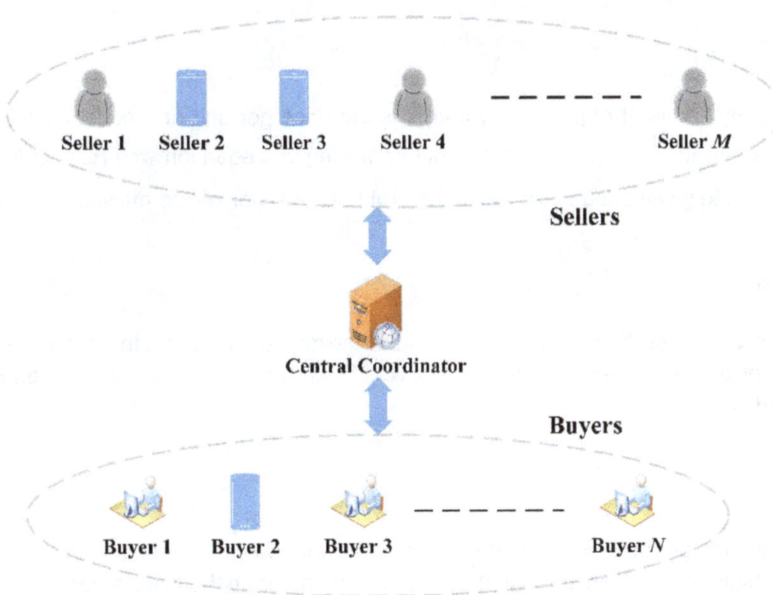

Figure 1. System model for agricultural market scenario

Figure 1 shows that there can be M number of sellers and N number of buyers. Sellers may or may not have smartphones. There is also a central coordinator that collects all the supplies from the sellers and all the demands from the buyers using online and offline forms. The central coordinator is a server where the distributed market equilibrium algorithm calculates the price of the agricultural products. Each of the sellers will have different products to sell in the market. The buyers will have different demands for different products. Market equilibrium price will be calculated for each of the products based on the supplies from the sellers and demands from the buyers.

To model the supply and demand of an agricultural product, a utility function (Pindyck & Rubinfeld, 2009), (Niyato and Hossain, 2010) is used which defines the satisfaction of the buyers. The more the utility for a product, the higher the demand from a buyer. Based on the basic supply and demand theory of microeconomics (Pfitzner, 1993), the utility function for the buyer can be defined as:

$$u(x_i) = log(x_i) - p_i x_i + c \qquad (1)$$

where, x_i is the amount of product the seller is willing to sell; p_i is the offered price per unit of product and c is a constant. Utility or satisfaction of the buyer will initially increase and then saturate for higher amount of the agricultural product and for lower price offered.

Demand function for the buyer can be obtained from the utility function in (1) which also defines the profit for the buyer. By setting $\frac{\delta u}{\delta x_i} = 0$, we can calculate the demand function for the buyer. This demand function represents the required amount of product x_i by the buyer for the price p_i.

$$\mathcal{D}_i = x_i = \frac{1}{p_i} \qquad (2)$$

The profit or utility of the seller can be defined as the income earned from selling the product to the buyer. Thus, the supply function for the seller can be computed as follows:

$$\mathcal{S}_i = \log(X_i - x_i) + p_i x_i \qquad (3)$$

where, X_i is the total amount of product x_i and p_i is the price per unit of product charged by the primary service. The supplied product can be obtained by differentiating this equation with respect to x_i, $\frac{\delta \mathcal{S}_i}{\delta b_i} = 0$. Thus the supply function for the given price p_i for which the profit of the seller will be maximized can be written as:

$$\mathcal{S}_i = x_i = X_i - \frac{1}{p_i} \qquad (4)$$

In market equilibrium, the supply function becomes equal to the demand function. In this state, there is no excess supply from the seller or excess demand from the buyer. Thus, the market equilibrium can be expressed as $\mathcal{D}_i = \mathcal{S}_i$ from (2) and (4):

$$\frac{1}{p_i} = X_i - \frac{1}{p_i} \qquad (5)$$

The seller and the buyer may have limited information about each other's supply and demand, the market equilibrium can be obtained iteratively through the central coordinator. In the initial stage, the buyer will observe the advertised price from the seller for a particular product and submit its demand through the central coordinator. The seller submits the price $p_i[0]$ to the coordinator which then computes the required product amount using the demand function of (2) and the coordinator then send the price to the seller. The central coordinator thus updates the price using the following equation:

$$p_i[t+1] = p_i[t] + \alpha_i(\mathcal{D}_i[t] - \mathcal{S}_i[t]) \qquad (6)$$

where, α_i is the learning rate that weighs the difference between demand and supply. The higher the value of α_i, the more weight is given to the difference than the previous price. A positive value of the difference implies excessive demand from the buyer and negative value indicates excessive supply. Thus the price will be gradually adjusted based on the supply and demand. This iteration will continue until the price difference $|p_i[t+1] - p_i[t]|$ becomes less than a predefined threshold value.

RESULTS AND DISCUSSION

Supply adaptation for different learning rates is shown in Figure 2. When the value of learning rate α is 0.03, the supply changes to a single value after a few iterations. When the value of learning rate α becomes larger (e.g., 0.05 and 0.07), the supply value fluctuates continuously, because smaller values of learning rate ensures that the value reaches equilibrium state gradually.

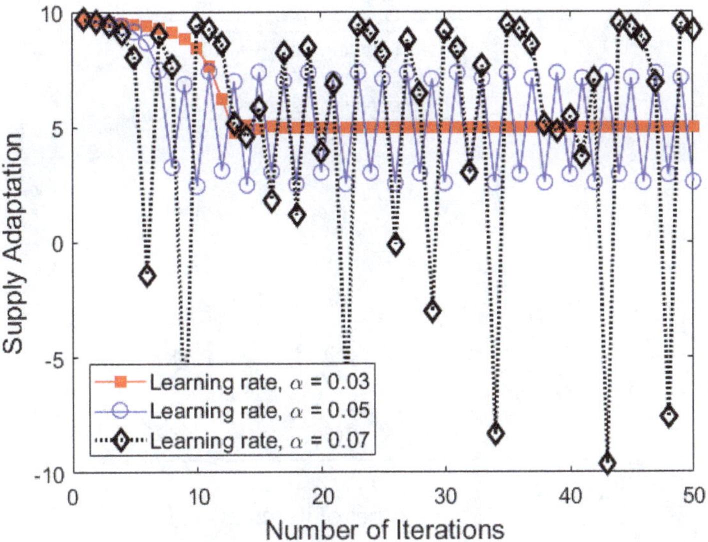

Figure 2. Supply adaptation with different learning rates

Similarly, demand adaptation for different learning rates is shown in Figure 3. The demand's value also reaches a stable value after a few iterations for learning rate 0.03. Figure 4 shows the adaptation of price for different learning rates. All these figures represent the stages of reaching the market equilibrium gradually.

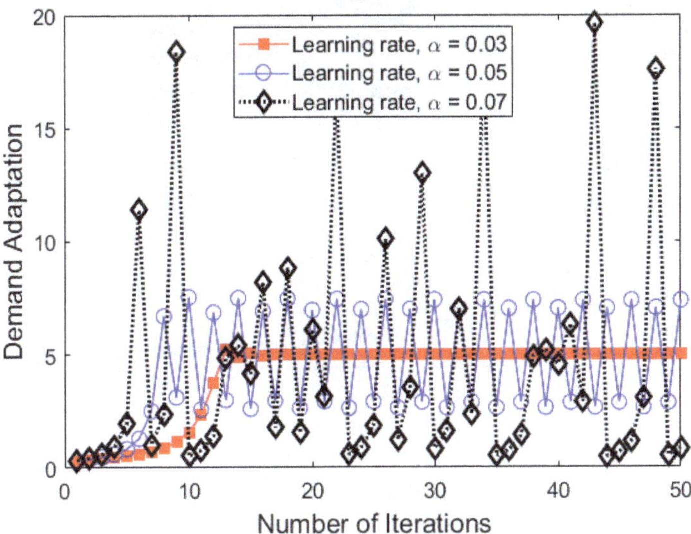

Figure 3. Demand adaptation for different learning rates

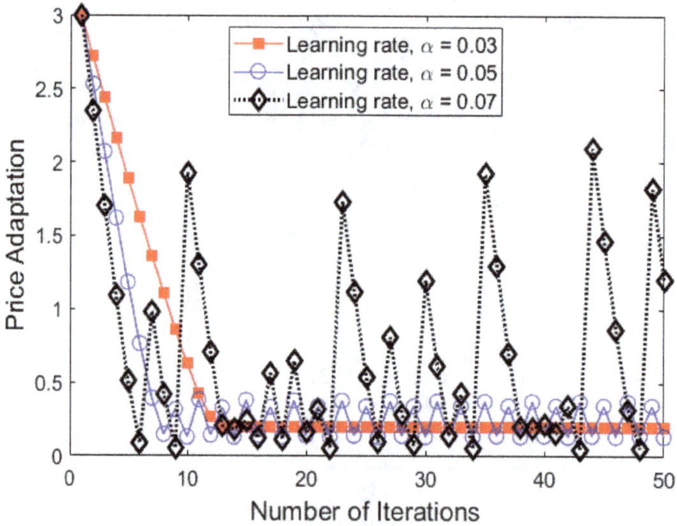

Figure 4. Price adaptation for learning rates 0.03, 0.05 and 0.07

It is important to choose an appropriate learning rate. Because if a very small value (e.g., 0.001) is chosen, it will take many iterations to reach to the market equilibrium. On the other hand, larger values of learning rate will prevent the algorithm from reaching the market equilibrium.

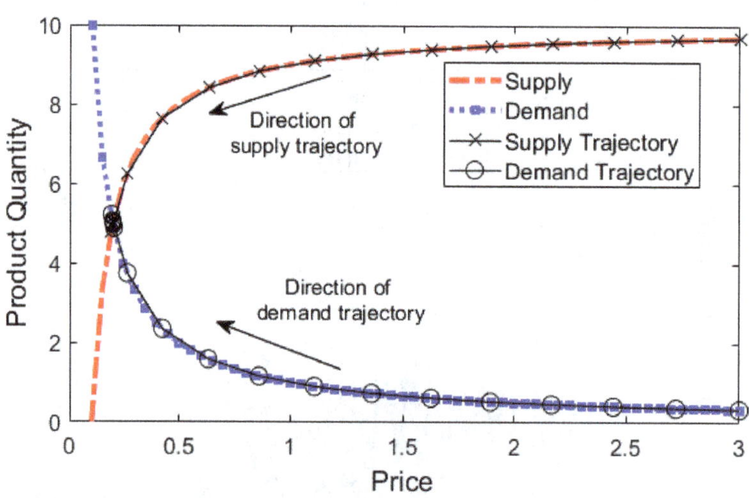

Figure 5. Trajectory diagram for different learning rates

Figure 5 shows the trajectory diagram of supply and demand for learning rate 0.03. It shows the steps for reaching the market equilibrium when the supply is equal to the demand. From the results, we can see that it takes only few iterations to determine the market equilibrium price. The time to reach the market equilibrium was found to be less than 1 millisecond.

CONCLUSION

In this paper, a new distributed price determination algorithm for agricultural products is proposed where the buyers can buy agricultural products from the sellers. This algorithm calculates the market equilibrium price based on the supply and demand of the sellers and buyers, respectively. Extensive analyses have been carried out to analyze the performance of the proposed model. The results show that the proposed method easily converges to the equilibrium solution within a very short time.

CONFLICT OF INTEREST

The author states no conflict of interest.

REFERENCES

1. Admati AR, Pfleiderer P and Zechner J, 1994. Large shareholder activism, risk sharing, and financial market equilibrium. Journal of Political Economy, 102:1097–1130.
2. Attaway S, 2013. Matlab: a practical introduction to programming and problem solving. Butterworth-Heinemann.
3. Ayuso J and Restoy F, 2006. House prices and rents: An equilibrium asset pricing approach. Journal of Empirical Finance, 13: 371–388.
4. Berry S, Levinsohn J and Pakes A, 1995. Automobile prices in market equilibrium. Econometrica: Journal of the Econometric Society, 841–890.
5. Besanko D and Kanatas G, 1993. Credit market equilibrium with bank monitoring and moral hazard. The Review of Financial Studies, 6: 213–232.
6. Black F, 1972. Capital market equilibrium with restricted borrowing. The Journal of Business, 45: 444–455.
7. Mankiw N, 2006. Principles of microeconomics (Vol. 10). Cengage Learning.
8. Niyato D and Hossain E, 2008a. Market-equilibrium, competitive, and cooperative pricing for spectrum sharing in cognitive radio networks: analysis and comparison. IEEE Transactions on Wireless Communications, 7: 4283–4723.
9. Niyato D and Hossain E, 2008b. Spectrum trading in cognitive radio networks: A market-equilibrium-based approach. IEEE Wireless Communications, 15: 71–80.
10. Niyato D and Hossain E, 2010. A microeconomic model for hierarchical bandwidth sharing in dynamic spectrum access networks. IEEE Transactions on Computers, 59: 865–877.
11. Pfitzner C, 1993. Mathematical Fundamentals for Microeconomics. Kolb Publication.
12. Pindyck RS and Rubinfeld DL, 2009. Microeconomics (7th ed.). Prentice Hall.
13. Schildt H, 2014. Java: The Complete Reference. McGraw-Hill Education Group.
14. Sharpe WF, 1964. Capital asset prices: A theory of market equilibrium under conditions of risk. The Journal of Finance, 19: 425–442.
15. Smith H. 2004. Supermarket choice and supermarket competition in market equilibrium. The Review of Economic Studies, 71: 235–263.
16. Varian HR, 1992. Microeconomic analysis (Vol. 2). Norton New York.

20

EFFECT OF NITROGEN AND PHOSPHORUS ON THE GROWTH AND YIELD PERFORMANCE OF SOYBEAN

Mst. Anjumanara Begum[1], Md. Aminul Islam[2*], Quazi Maruf Ahmed[3], Md. Anwarul Islam[1] and Md. Moshiur Rahman[1]

[1]Department of Agronomy, Bangladesh Agricultural University, Mymensingh-2202, Bangladesh; [2]On Farm Research Division, Bangladesh Agricultural Research Institute, Pabna-6600, Bangladesh; [3]Regional Horticultural Research Station, Bangladesh Agricultural Research Institute Narshingdi-1600, Bangladesh

***Corresponding author:** Md. Aminul Islam; E-mail: amin_bau@yahoo.com

ARTICLE INFO

Key words

Nitrogen
Phosphorus
Production
Soybean

ABSTRACT

The experiment was conducted at the Agronomy Field Laboratory of Bangladesh Agricultural University, Mymensingh to study the effects of nitrogen and phosphorus on the performance of soybean. Three levels of nitrogen (0, 25 and 40 kg N ha^{-1}) and four levels of phosphorus (0, 18, 36 and 54 kg P ha^{-1}) were considered as treatment for the experiment. Soybean responded remarkably to the added nitrogenous and phosphatic fertilizers as the crop characters were significantly influenced by different levels of nitrogen and phosphorus. Significant effect on number of branches and seeds $plant^{-1}$, plant height, number of filled pods $plant^{-1}$, weight of seeds $plant^{-1}$, dry weight of plant, stover weight $plant^{-1}$, 1000-seed weight, seed and stover yield were obtained from the combined application of 25 kg N with 54 kg P ha^{-1}.

INTRODUCTION

Soybean *(Glycine max L.* Merril) ranks first as an oilseed crop of the world. It has a tremendous value in agriculture as a good source of high quality plant protein and vegetable oils in one hand and nitrogen fixing ability on the other. It belongs to the family Leguminosae, sub family Papilionaceae. Soybean is quite wide spread in different regions of the world and grows well from the tropics to the temperate zones with greater production in the United States, Brazil, China, Mexico, Indonesia and Argentina. The world production of soybean as estimated in 2008 was 231.27 million ton from an area of 96.47 million hectares (FAO, 2009). For its nutritive value soybean has been called miracle golden bean, the golden nugget, the nugget of nutrition etc. soybean being a good source of protein, unsaturated fatty acids, minerals like Ca and P and vitamin A, B, C and D can meet up different nutritional needs of human being (USDA, 2009).

Prospects of soybean farming in Bangladesh is bright as it can successfully be grown under a wide range of climatic and edaphic conditions and cultivated throughout the year in Bangladesh (Rahman, 1982). Soybean helps to improve the soil by fixing the atmospheric nitrogen through *Rhizobium* bacteria. Steward (1966) observed that the soybean plants could fix 94 kg nitrogen ha^{-1} in soil in a season. In Bangladesh the area under soybean cultivation is about 5000 ha with a production of 4000 ton & the yield ranges from 1.50 to2.30 tha^{-1} (BARI, 2005). The lower yield of soybean at farmer's level is mainly attributed to the lack of improved agronomic management practices of which judicious fertilizer application is an important determinant for better yield of soybean.

Among the nutrients, nitrogen is a major essential plant nutrient element. It has the quickest and most pronounced effect on plant growth and yield of crops. It tends primarily to encourage above ground vegetative growth and to impart deep green colour to the leaves. In all plants, nitrogen governs a considerable degree of utilization of potassium, phosphorus and other nutrients. Plants receiving insufficient nitrogen are stunted in growth with restricted root systems. The leaves turn yellow or yellowish green and tend to drop off.

Phosphorus can play an important role in seed yield as it is one of the limiting plant nutrients for production of soybean (Rao *et al.,* 1995). Its uptake and utilization by soybean is essential for ensuring proper growth & improving yield and quality of the crop. It influences the growth of roots, helps uptake of more nutrients and nodule formation, balances the nitrogen deficiency in soil and assists in seed maturation. Thus, it is needed to find out proper amount of nitrogen and phosphorus required for achieving better yield of soybean. In view of the facts stated above, a field experiment was conducted to evaluate the effect of different levels of nitrogen and phosphorus and their interaction on the yield of soybean.

MATERIALS AND METHODS

The experiment was conducted at the Agronomy Field Laboratory, Bangladesh Agricultural University, Mymensingh during 5 January to 20 April, 2009 to study the effects of rate of nitrogen and phosphorus on the yield of soybean. The experimental field was a high land having sandy loam soil with pH 6.9. The initial soil (0-15 cm depth) test result showed that the soil contained 0.058% total N, 0.463% organic matter, 23 ppm available P, 5.0 ppm available S and 0.13 ppm exchangeable K. Three levels of nitrogen (0, 25 and 40 kg N ha^{-1}) and four levels of phosphorus (0, 18, 36 and 54 kg P ha^{-1}) were considered as treatment for the experiment. The experiment was laid out in a randomized complete block design with three replications. Urea & TSP were used as the source of N & P, respectively.

Half of the N and whole of P were applied as basal on 5 January 2009 as per treatment in the individual plots. Besides, MoP and gypsum were applied @ 120 and 115 kg ha^{-1}, respectively on the same date. Inoculum @ 50 g kg^{-1} of seed was mixed with seed prior to sowing. The remaining half of the N was top dressed at 25 days after sowing. Weeding followed by thinning was done simultaneously twice at 21 and 45 days after sowing. Irrigation was done simultaneously twice on 25 and 55 days after sowing. Sumithion 50 EC @ 3 $m1L^{-1}$ and Dimethion 40 EC @ 2 mlL^{-1} were sprayed to control leaf roller and hairy caterpillar. The crop was harvested on 20 April 2009 at full maturity. The collected data were analyzed statistically following the ANOVA technique and the mean differences were adjudged as per Duncan's Multiple Range Test (Gomez and Gomez, 1984).

RESULTS AND DISCUSSION

The results of the present experiment have been presented and discussed in the Table 1, 2 and 3.

Effect of nitrogen

Nitrogen had significant effect on yield and yield contributing characters of soybean (Table 1). Plant height was significantly influenced by nitrogen. Crop grown with 40 kg N ha^{-1} produced the tallest plant (34.18 cm) and with 0 kg N ha^{-1} treatment produced the shortest plants (30.01cm). The highest dry matter weight $plant^{-1}$ (17.89g), number of seeds pod^{-1} (1.94), number of seeds $plant^{-1}$ (94.93), seeds weight $plant^{-1}$ (3.41g), 1000 seed weight (111.26g), number of filled pods $plant^{-1}$ (47.59) but lowest empty pods $plant^{-1}$ (2.42) were observed with 25 kg N ha^{-1} followed by 40 kg N ha^{-1}application whereas, the crops showed poor performance with no nitrogen (0 kg N ha^{-1}) application. Among the treatments, 25 kg N ha^{-1} produced highest weight of 1000 seed (120.24g) which was statistically significant than other treatments. The lowest weight of 1000 seed (111.26g) was obtained with control treatment and subsequently lowers than others. The present result supports the report of Raju and Verma (1984) as they observed that increased N fertilizer has an advantageous role on 1000 seed yield increment. The highest seed yield (1.95 t ha^{-1}) was obtained in 25 kg N ha^{-1} and the lowest (1.41 tha^{-1}) was recorded in control (0 kg N ha^{-1}) treatment. The highest seed yield in 25 kg N ha^{-1} might have resulted due to cumulative favourable effects of number of seeds $plant^{-1}$, weight of seeds $plant^{-1}$ and 1000 seed weight. The result obtained is in agreement with the findings of Singh *et al.* (1992) as he reported yield of soybean increased with the increased rate of N fertilizer rate. The stover yield followed the similar trend as observed for seed yield. Leelavathi *et al.* (1991) obtained the similar findings in case of stover yield.

Effect of phosphorus

Phosphorus had tremendous effect on soybean (Table 2). The yield and yield contributing characters showed better response with the increased level of phosphorus. The tallest plant (34.26 cm), maximum dry matter weight $plant^{-1}$ (18.89 g) and maximum number of branches $plant^{-1}$ (3.37) was recorded with 54 kg P ha^{-1} which was significantly highest than those of other treatments. The number of filled pods $plant^{-1}$ (54.49) was the highest in 54 kg P ha^{-1} which was significantly highest than those of other treatments. The lowest number of filled pods $plant^{-1}$ was observed from 0 kg P ha^{-1}. The present results supports the reports of Singh and Bajpai (1990) who observed that increasing phosphorus rate increased the number of pods $plant^{-1}$. Among the treatments 0 kg P ha^{-1} produced the highest number of empty pods (3.98) which was statistically significant with other treatments and the lowest (2.16) was found in 54 kg P ha^{-1}. Islam et al. (2004) reported that the percentage of empty pods decreased with the increase of phosphorus application. Number of seeds pod^{-1} was escalated with the increased dose of phosphorus. Tomar *et al.* (2004) observed that number of seeds pod^{-1} increased with the increase of phosphorus application. Similar trend was found in case of seeds $plant^{-1}$. The obtained result is in agreement with the findings of Islam *et al.* (2004). Maximum weight of seed $plant^{-1}$ (3.49g) was observed in 54 kg P ha^{-1}. The highest weight of stover $plant^{-1}$ (9.07 g) and 1000-seed weight (122.2g) was observed in 54 kg P ha^{-1} which were statistically significant than those of other treatments whereas lowest from control (0 kg P ha^{-1}) treatment. Significantly higher seed yield (2.09 t ha^{-1}) was obtained in 54 kg P ha^{-1} and the lower (1.30 t ha^{-1}) was recorded in control (0 kg P ha^{-1}). The highest seed yield in 54 kg P ha^{-1} might have resulted due to cumulative effects of the number of seeds pod^{-1}, number of seed $plant^{-1}$ and weight of seeds $plant^{-1}$. The present results support the reports of Syafruddin *et al.* (1990) who observed that seed yield was highest 90 kg P ha^{-1}. The stover yield followed the similar trend as observed in seed yield. Tomar et al. (2004) reported higher trend of stover yield with higher dose of P fertilizer.

Table 1. Effect of nitrogen on the yield and related crop characters of soybean var. Shohag

Level of nitrogen (kg N ha^{-1})	Plant Height (cm)	Dry matter $plant^{-1}$ (g)	No. of branches $plant^{-1}$	No. of nodes $plant^{-1}$	No. of filled pods $plant^{-1}$	No. of empty pods $plant^{-1}$	No. of seeds pod^{-1}	No. of seeds $plant^{-1}$	Seed wt. $Plant^{-1}$ (g)	Stover wt. $Plant^{-1}$ (g)	1000-seed wt. (g)	Seed yield (t ha^{-1})	Stover yield (t ha^{-1})
0	30.01c	12.89c	2.17b	9.88	38.00c	3.31a	1.64c	64.98c	3.07b	6.14c	111.26c	1.41c	1.69c
25	31.78b	17.89a	2.73a	10.29	47.59a	2.42c	1.94a	94.93a	3.41a	8.48a	120.24a	1.95a	2.35a
40	34.18a	17.22b	2.65a	9.88	40.4b	2.75b	1.76b	73.20b	2.91b	8.20b	117.99b	1.89b	2.25b
Sx	0.35	0.11	0.05	0.34	0.65	0.06	0.03	1.75	0.11	0.07	0.55	0.02	0.02
Level of significance	**	**	**	NS	**	**	**	**	**	**	**	**	**

In a column figures with same letter or without letter do not differ significantly whereas figures with dissimilar letter differ significantly (as per DMRT)
Sx = Sample standard deviation; * = Indicates significant at 5% level of probability; ** = Indicates significant at 1% level of probability; NS = Indicates not significant

Table 2. Effect of phosphorus on the yield and related crop characters of soybean var. Shohag

Level of nitrogen (kg P ha^{-1})	Plant Height (cm)	Dry matter $plant^{-1}$ (g)	No. of branches $plant^{-1}$	No. of nodes $plant^{-1}$	No. of filled pods $plant^{-1}$	No. of empty pods $plant^{-1}$	No. of seeds pod^{-1}	No. of seeds $plant^{-1}$	Seed wt. $Plant^{-1}$ (g)	Stover wt. $Plant^{-1}$ (g)	1000-seed wt. (g)	Seed yield (t ha^{-1})	Stover yield (t ha^{-1})
0	30.95b	12.12d	1.85d	9.58	29.77d	3.98a	1.27c	38.07d	3.03b	5.67d	109.15d	1.30d	1.61d
18	31.3b	16.17c	2.09c	10.00	38.98c	2.67b	1.87b	73.31c	3.13ab	7.71c	115.06c	1.77c	2.11c
36	31.46b	16.83b	2.75b	10.19	44.74b	2.50b	1.91b	85.95b	2.86b	7.99b	119.58b	1.84b	2.21b
54	34.26a	18.89a	3.37a	10.31	54.49a	2.16c	2.07a	113.48a	3.49a	9.07a	122.2a	2.09a	2.46a
Sx	0.4	0.13	0.06	0.39	0.75	0.07	0.03	2.02	0.12	0.09	0.63	0.02	0.02
Level of significance	**	**	**	NS	**	**	**	**	**	**	**	**	**

In a column figures with same letter or without letter do not differ significantly whereas figures with dissimilar letter differ significantly (as per DMRT)
Sx = Sample standard deviation; * = Indicates Significant at 5% level of probability; ** = Indicates Significant at 1% level of probability; NS = Indicates Not significant.

Table 3. Interaction effect of nitrogen and phosphorus on the yield and related crop characters of soybean var. Shohag

Level of nitrogen and phophorus N x P (kg ha^{-1})	Plant Height (cm)	Dry matter plant^{-1} (g)	No. of branches plant^{-1}	No. of nodes plant^{-1}	No. of filled pods plant^{-1}	No. of empty pods plant^{-1}	No. of seeds pod^{-1}	No. of seeds plant^{-1}	Seed wt. Plant^{-1} (g)	Stover wt. Plant^{-1} (g)	1000-seed wt. (g)	Seed yield (t ha^{-1})	Stover yield (t ha^{-1})
0 x 0	31.73de	9.82j	1.63fg	9.27	23.5h	4.08a	1.25	29.36i	3.56abc	4.69h	105.13e	1.08i	1.28h
0 x 18	29.02fg	11.79h	1.43g	10.00	36.48ef	3.24b	1.63	59.62g	2.49ef	5.77f	111.23d	1.33g	1.50g
0 x 36	27.77g	13.28g	2.73bc	9.87	39.86de	3.12bc	1.78	70.94ef	2.64def	6.18f	114.47d	1.42f	1.77f
0 x 54	31.53de	16.69e	2.87b	10.40	52.14b	2.78cd	1.92	100.01c	3.60abc	7.94d	114.20d	1.83d	2.19d
25 x 0	31.22def	15.48f	2.03e	9.20	35.41f	3.85a	1.41	50.11h	2.30f	5.12g	107.60e	1.66e	2.07e
25 x 18	33.15cd	16.69e	2.48cd	10.07	44.33c	2.20ef	1.98	87.98d	3.15b-e	9.49b	114.27d	1.81d	2.21d
25 x 36	35.48ab	17.55d	2.70bc	10.83	52.34b	2.00f	2.05	107.15b	3.20a-e	9.30b	122.10bc	1.95c	2.26d
25 x 54	36.88a	20.81a	3.71a	11.07	58.28a	1.63g	2.31	134.48a	3.89a	10.01a	127.98a	2.30a	2.63a
40 x 0	30.95def	11.07i	1.88ef	10.27	30.41g	4.01a	1.14	34.75hi	3.23a-d	7.20e	114.73d	1.18h	1.49g
40 x 18	30.68ef	20.02b	2.34d	9.93	36.14ef	2.56de	2.00	72.34e	3.77ab	7.86d	119.67c	2.13b	2.48c
40 x 36	31.13def	19.66bc	2.83b	9.87	42.01cd	2.37ef	1.9b	79.76de	2.75def	8.49c	122.16bc	2.14b	2.59b
40 x 54	34.36bc	19.18c	3.55a	9.47	53.05b	2.07f	2.00	105.94bc	2.98c-f	9.26b	124.41b	2.18a	2.70a
Sx	0.69	0.22	0.10	0.68	1.3	0.12	0.06	3.49	0.22	0.15	1.09	0.03	0.03
Level of significance	**	**	**	NS	**	**	NS	*	**	*	**	**	**

In a column figures with same letter or without letter do not differ significantly whereas figures with dissimilar letter differ significantly (as per DMRT)
Sx = Sample standard deviation; * = Indicates significant at 5% level of probability; ** = Indicates significant at 1% level of probability; NS = Indicates not significant.

Effect of interaction of nitrogen and phosphorus

The interaction effect of nitrogen and phosphorus on plant height has been shown in table 3. It is evident from the results that there was no regular trend in plant height due to interaction of N and P. The highest plant height (36.88 cm) was obtained from the highest level of N and P. Whereas, the lowest plant height (27.77 cm) was obtained from the combination of 0 kg N with 36 kg P ha^{-1}. The highest dry weight of plant (20.81 g) was obtained from the highest level of nitrogen and phosphorus (25 kg N with 54 kg P ha^{-1}). Whereas the lowest dry weight of plant (9.82 g) was obtained from the combination of 0 kg N with 0 kg P ha^{-1}. It is evident from the results that there was no regular trend in number of branches $plant^{-1}$ and number of filled pods $plant^{-1}$due to interaction of N and P. There is a tendency of producing less number of empty pods $plant^{-1}$ due to the effect of interaction of highest levels of phosphorus irrespective of nitrogen. The highest empty pods $plant^{-1}$ (4.08) was recorded from the 0 kg N with 0 kg P ha^{-1} which was statistically significant with the 25 kg N with 0 kg P ha^{-1} and 40 kg N with 0 kg P ha^{-1}. The lowest number of empty pods $plant^{-1}$ (1.63) was obtained from 25 kg N with 54 kg P ha^{-1}. The number of seeds pod^{-1} due to interaction effect of nitrogen and phosphorus was statistically insignificant. The maximum (134.48) seeds $plant^{-1}$ was observed in 25 kg N with 54 kg P ha^{-1} and the minimum (29.36) was observed from 25 kg N with 54 kg P ha^{1}. Almost similar trend in the effect of interaction of N and P was observed on weight of seeds $plant^{-1}$ as it was exhibited on plant height, number of filled pods $plant^{-1}$, number of seeds $plant^{-1}$. Here also the highest weight of seeds $plant^{-1}$(3.89g), weight of stover $plant^{-1}$ (10.01 g) was obtained from 25 kg N with 54 kg P ha^{-1} and lowest (4.69 g) from the 0 kg N with 0 kg P ha^{-1}. The effect of interaction of nitrogen and phosphorus on 1000-seed weight was statistically significant. Almost the similar trend in the effect of interaction of N and P was observed on seed yield as it was exhibited on dry weight of plant, number of filled pods $plant^{-1}$, number of seeds $plant^{-1}$, and weight of seeds $plant^{-1}$ and the weight of stover $plant^{-1}$. The treatment combination of 25 kg N with 54 kg P ha^{-1} produced the highest seed yield (2.30 t ha^{-1}), which was similar with 40 kg N with 54 kg P ha^{-1} and significantly highest than those of other treatments. On the other hand 0 kg N with 0 kg P ha^{-1} produced the lowest (1.08 t ha^{-1}) seed yield. Stover yield varied significantly due to interaction of nitrogen and phosphorus. The results indicated that higher stover yield in soybean could be obtained by 40 kg N with 54 kg P ha^{-1}.

CONCLUSION

The results, therefore, indicated that nitrogen and phosphorus had significant effect on yield performance of soybean including other yield contributing parameters as well as suggested that the combined application of nitrogen and phosphorus @ 25 kg N with 54 kg P ha^{-1} might produce the best seed yield in soybean var. Shohag.

REFERENCES

1. BARI (Bangladesh Agricultural Research Institute) 2005. Handbook on Agro-technology. 2nd Edition, BARI, Gazipur. p. 155.
2. FAO (Food and Agriculture Organization). 2008. FAO Yearbook Production. Food and Agricultural Organization of the United Nations Rome. 57: 115.
3. Gomez, KA and AA Gomez 1984. Statistical Procedure for Agricultural Research. 2nd edition, International Rice Research Institute. John Wiley and Sons, Inc, Singapore. pp. 139-240.
4. Islam MK, MAA Mondal, MA Mannaf, MAA Mondal, MAH Talukder and MM Karim, 2004. Effects of variety, inoculum and phosphorus on the performance of soybean. Agricultural Research Station, Debigonj, Panchagar, Bangladesh. Pakistan Journal of Biological Science, 7: 2072-2077.
5. Leelavathi GSNS, GV Subbaiah and RN Pillai, 1991. Effect of different levels of nitrogen on the yield of mungbean (Vigna rakiata L. Wilczek). Andhra Agriculture Journal, 38: 93-94.
6. Mandal MR and MA Wahhab, 2001. Production Technology of Oil crops. Oilseed Research Centre, Bangladesh Agricultural Research Institute, Joydevpur, Gazipur. pp. 1-10.
7. Rahman L, 1982. Cultivation of Soybean and Its Uses. City press. Dhaka. pp. 5-7.

8. Raju MS and SC Varma, 1984. Response of mungbean (*Vigna radiata*) to Rhizobial inoculation in relation to fertilizer nitrogen. Legume Research 7: 73-76.
9. Rao AS, DD Reddy and PN Takkar, 1995. Phosphorus management-A key to boost productivity of soybean-wheat cropping system on swell-shrink soils. Fertilizer. News. 40: 87-95.
10. Singh HN, FW Prasad and JK Varshney, 1992. Effect of nitrogen and row spacing on nodulation, growth and yield of soybean (*Glycine max* L. Merr) var. Gaurav. New Agriculturist. 3: 31-34.
11. Singh VK and RP Bajpai, 1990. Effect of phosphorus and potash on the growth and yield of rainfed soybean. Indain Journal of Agronomy, 35: 310-311.
12. Steward WDP, 1966. Nitrogen fixation in plants Alhloe Press, University of London. P. 130.
13. Syafruddin RM, S Saenong and D Jamaluddin, 1990. Response of soybeans (*Glycine max* L. Merrill) to P and Zn application in calcareous alluvial soil. Soybean Abstract, 1992. 15: 220.
14. Tomar SS, R Singh and SP Singh, 2004. Response of phosphorus, sulphur and rhizobium inoculation on growth, yield and quality of soybean (*Glycine max*. L.). Progressive Agriculture, 4: 72-73.
15. USDA, 2009. Nutrient Database for Standard Reference Release 27 (Soybean). ndb.nal.usda.gov/ndb/

21

USABILITY OF BIOSLURRY TO IMPROVE SYSTEM PRODUCTIVITY AND ECONOMIC RETURN UNDER POTATO-RICE CROPPING SYSTEM

M. Asadul Haque[1*], M. Jahiruddin[2], M. Mazibur Rahman[2] and M. Abu Saleque[3]

[1]Department of Soil Science, Patuakhali Science and Technology University, Patuakhali [2]Department of Soil Science, Bangladesh Agricultural University, Mymensingh, [3]Bangladesh Rice Research Institute, Gazipur, Bangladesh

***Corresponding author:** M. Asadul Haque, E-mail: masadulh@yahoo.com

ARTICLE INFO

Key words

Bioslurry,
Integrated Plant Nutrition System
Manures
Potato
Rice

ABSTRACT

Establishment of huge number of biogas plant in the recent years in Bangladesh creates a burden to disposal of bioslurry. An attempt was undertaken to explore the usability of bioslurry in agricultural crop production under potato-rice system as well as to reduce bioslurry induced pollution. The experiment involved a sole chemical fertilizer treatment, four treatments based on integrated plant nutrition system with 5 t ha^{-1} cowdung and cowdung bioslurry and 3 t ha^{-1} poultry manure and poultry manure bioslurry, and a control. The potato crops received manures or slurries, and its residual effect was evaluated on the succeeding T.*Aman* crop. Poultry manure bioslurry, poultry manure, cowdung bioslurry and cowdung gave 22.5, 20.0, 9.9 and 2.9 % increase in total system productivity, respectively over sole chemical fertilizer treatment. Bioslurries had higher contribution compared to their respective original manure. Bioslurry was found very useful as manure for crop production.

INTRUDUCTION

The Government of Bangladesh has been implementing the National Domestic Biogas and Manure Program (NDBMP) since 2006. Under this program the Infrastructure Development Company Limited (IDCOL), with financial assistance from SNV, Netherlands and KfW Germany, has installed more than 31,000 biogas plants by the year 2013. Bioslurry produced from these biogas digesters is mostly being wastage, polluting the environment many cases and becoming a major burden in Bangladesh (Islam, 2011). This is happening due to less awareness and knowledge gap of the farmers about the nutritional value of bioslurry for crop production. If properly managed, bioslurry could play a major role in supplementing the use of expensive chemical fertilizers in Bangladesh (Yu et al., 2010; Abubaker, 2012).

The farmer needs to use chemical fertilizer to increase the crop production. However, if only chemical fertilizers are continuously applied to the soil without adding organic manure, productivity of land will decline. On the other hand, if only organic manure is added to the soil, desired increase in crop yield cannot be achieved. Some literature is available in Bangladesh on the effects of integrated use of aerobically decomposed manure and chemical fertilizers on various crops. The anaerobically digested manure like use of bioslurry under integrated plant nutrition system is very minimum addressed in the country as well as in the world. The experiment is therefore, undertaken to investigate the usability of bioslurry as manure in the potato-rice cropping system.

MATERIALS AND METHODS

The experiment was carried out during 2011 and 2012 at Soil Science Field Laboratory of Bangladesh Agricultural University (BAU) farm, Mymensingh. The BAU farm soil belongs to 'Sonatala' soil series an Inceptisol under the AEZ 9 (Old Brahmaputra Floodplain).The silty clay loam soil had the following properties at 0–15 cm depth: sand 96 g kg^{-1}, silt 700 g kg^{-1}, clay 204 g kg^{-1}, 6.4 pH, 2.13% organic matter, 1.1 mg g^{-1} total N, 5.00 mg kg^{-1} available P (Olsen), 0.11 cmol kg^{-1} exchangeable K, 12.0 mg kg^{-1} available S, 0.65 mg kg^{-1} available Zn and 0.24 mg kg^{-1} available B contents.

The experiment was laid out in a completely randomized block design with three replications, each plot size being 4 × 5 m and was separated by bunds. There were six treatments, namely recommended rate of nutrients from chemical fertilizers (T_1), and the next four treatments on integrated plant nutrition system (IPNS) basis i.e. recommended rate of nutrients from chemical fertilizer adjusted from manures - cowdung (T_2), cowdung bioslurry (T_3), poultry manure (T_4) and poultry manure bioslurry (T_5). The control treatment (T_6) received no fertilizer or manure. When N, P, K and S were supplied from organic and inorganic sources based on IPNS basis; the amount of all nutrients was equal for all the treatments except T_6. The rate of manure application was 5 t ha^{-1} for cowdung and cowdung bioslurry and 3 t ha^{-1} for poultry manure and poultry manure bioslurry. The treatments were tested on potato-rice cropping system in two consecutive years. In each crop cycle, manure was applied to the first crop (potato) and their residual effect was evaluated on the succeeding T. Aman rice. The nutrient content of the manures are given in Table 1. The rate of fertilizer for potato was 135, 25, 95 and 12 kg ha^{-1} of N, P, K & S, respectively for T_1- T_6.The T. Aman rice received equal amounts of N, P, K & S as chemical fertilizers at a rate of 75, 8, 30, 8 kg ha^{-1}, respectively. A blanket dose of Zn and B @ 2 and 1.5 kg ha^{-1}, respectively were applied in all plots of potato crops. Basal application of P, K, S, Zn and B, and treatment wise manures were made during final land preparation. Nitrogen was top dressed at 3 equal splits on final land preparation, 20 and 40 days after planting for potato and on 10, 25 and 45 days after transplanting for T. Aman rice.

The potato tubers were planted maintaining spacing of 20×50 cm. For T. Aman rice 25 days old 3-4 seedlings comprising a hill were transplanted in rows maintaining 20×20 cm spacing. The potato tubers were planted on 24 and 23 November of 2010 and 2011 and T. Aman rice seedlings were transplanted on 27 and 29 July of 2011 and 2012, respectively. The crop varieties were Diamant for potato and BINA dhan7 for T. Aman rice. Two weeding cum earthing-up followed by irrigation was made in potato whereas T. Aman rice was grown on rain-fed condition. All crop protection measures were taken to prevent insect and disease attacks. Yields of tuber/grain and stover/straw of each crop were recorded after harvest of a 6 m^2 area from each plot at physiological maturity. Observation was made in terms of yield and yield components. Total system productivity was calculated through addition of potato yield and potato equivalent rice yield (PERY).

The PERY was calculated according to the following equation (Ahlawat and Sharma, 1993):

$$\text{PERY}= \frac{Y_{rice} \times P_{rice}}{P_{potato}} \text{ ------------------------- (1)}$$

Where, Y_{rice} is the yield of rice (t ha^{-1}), P_{rice} is the price of rice (Tk 15000 t^{-1}), $Y_{potato}$is the yield of potato (t ha^{-1}) and P_{potato} is the price of potato (Tk 10000 t^{-1}).

Economic analysis was performed to identify the economically viable treatment(s). Marginal benefit-cost ratio (MBCR) is the indicative of the superior treatments. It is the ratio of marginal or added benefits and costs. Only variable costs i.e. manure and chemical fertilizer was taken into account as added cost. The benefit was calculated based on yield (main product and by-product). Data recorded on crop characters were subjected to statistical analysis through computer based statistical program Mstat-C following the basic principles, as outlined by Gomez and Gomez (1984).

RESULTS AND DISCUSSION

Contribution to growth and yield components of potato

Potato stem length, number of tubers plant^{-1}and tuber weight plant^{-1} were significantly influenced by the application of organic manure and inorganic fertilizers (Table 2). Stem length across the treatments ranged from 15.5 - 46.2 cm in 2011 and 21.8 - 52.5 cm in 2012.The tallest plants were found in PM bioslurry + IPNS basis chemical fertilizer (T_5) treatment in 2011. In 2012, the tallest plants were recorded with the T_4 treatment (PM + IPNS basis chemical fertilizer), however, it was statistically similar with T_6 treatment. Table 2 indicates that in both years the highest number of tubers plant^{-1} was recorded by T_5 treatment (6.08 in 2011 and 5.97 in 2012) and it was statistically similar with T_4 treatment (5.93 in 2011 and 5.77 in 2012). Regarding cattle manure sources, in both years higher number of tubers was produced by the T_3 treatment (CD-bioslurry + IPNS basis chemical fertilizer) compared to the original manure (T_2). The number of tubers plant^{-1} produced by the treatment T_1 (chemical fertilizer) was lower in relation to the IPNS treatments. In the first year trial, the highest and the next highest tuber weight plant^{-1} (g) were recorded by the treatments T_5 and T_4, respectively. Unlike first year, the highest tuber weight (302 g plant^{-1}) was produced by the T_4 treatment and it was statistically similar to T_5 treatment (295.8 g plant^{-1}). Treatment T_3 (CD bioslurry + IPNS basis chemical fertilizer) had higher tuber weight plant^{-1} in both years compared to the treatment T_2 (CD + IPNS basis chemical fertilizer).

Results of potato grading (A, B, C and D grade) based on tuber weight (%) are presented in Table 3. As observed in 2011, 27.8 % potato in treatment T_1 belongs to 'D' grade (extra-large size) and in the following year (2012), 14.0 % potato for the same treatment (T_1) fall under 'D' grade. The manure receiving treatments (T_2 to T_5) had 6.4 to 11.3 % 'D' grade potato over the years which were much lower compared to absolutely chemical fertilizer treatment (T_1). The 'C' grade (under size) potato was not much influenced by the treatments. Remarkable variation was observed in 'B' grade potato and was always higher in treatments that received manures (T_2-T_5). When 'A' (seed purpose) and 'B' (consumption purpose) grade potatoes were pooled, it was found that in 2011 about 60-65% potatoes produced by T_1 treatment belonged to these two grades and more than 80% potatoes of the T_2 – T_5 treatments fall in these groups. In 2012, 80% potatoes in T_1, and more than 90% in T_2 – T_5 fell into these groups. The results clearly indicated that integrated nutrient management had a distinct impact to produce the good size/quality of potato.

Contribution to tuber and stover yield of potato

Tuber yield of potato was influenced significantly by the different treatments (Table 4). In 2011, the range of tuber yield over the treatments was recorded as 3.00 to 21.80 t ha^{-1}, which was 5.40 to 25.28 t ha^{-1} in 2012. As recorded in the first year, the highest tuber yield was found in T_5 treatment (PM-bioslurry + IPNS basis chemical fertilizer) and it was statistically similar with the T_4 treatment (PM + IPNS basis chemical fertilizer). In the second year, the highest yield was observed in T_4 treatment and statistically similar yield was noted in T_5 treatment. When the two years' yields were pooled, the range came to 4.20 to 23.44 t ha^{-1}. Both the years, treatment T_3 (CD bioslurry + IPNS basis chemical fertilizer) showed statistically higher tuber yield than T_2 (CD + IPNS basis chemical fertilizer).Table 4 further indicates that sole chemical fertilizer treatment (T_1) had significantly lower yield compared to the IPNS treatments. The two years' pool data indicate that although not significant but higher stover yield was produced by poultry manure bioslurry than poultry manure. After poultry source, cowdung bioslurry (T_3) ranked the third position in relation to stover yield of potato. Poultry manure source produced the higher yield, as because some growth hormones and concentrates feed to poultry birds increased the growth of plants. Ullah et al. (2008) noted that the treatment where poultry manure bioslurry was used showed higher yield of different crops (tomato, cabbage, cauliflower, potato, maize and Boro rice). Singh (1995) reported that biogas slurry was found better than organic manure (FYM) in obtaining higher yield in pea, okra, corn and soybeans.

Table 1. Chemical composition of different manures used in potato 2011 and 2012

Manure	C(%)	N(%)	P(%)	K(%)	S(%)	C:N	C:P	C:K	C:S
				2011					
Cowdung (CD)	25.1	1.06	0.40	0.48	0.20	23.8	62.6	52.5	125.8
CD bioslurry	23.7	1.14	0.48	0.51	0.24	20.7	48.9	46.5	98.2
Poultry manure (PM)	19.7	1.42	0.81	0.67	0.30	13.9	24.3	29.4	66.1
PM bioslurry	13.9	1.54	1.44	0.58	0.39	9.0	9.65	24.0	35.6
				2012					
Cowdung (CD)	40.7	1.08	0.58	0.54	0.33	37.7	60.0	75.4	123.3
CD bioslurry	32.7	1.55	0.84	0.67	0.35	21.2	38.9	48.5	93.4
Poultry manure (PM)	23.9	1.64	0.95	0.53	0.41	14.6	25.2	45.1	58.3
PM bioslurry	16.8	1.68	1.34	0.64	0.49	10.0	12.5	26.3	34.3

Table 2. Effects of different manure and fertilizer treatments on the growth and yield components of potato (Diamant) in the potato-T. Aman rice cropping system

Treatments	Stem length(cm)		Tuber plant^{-1} (no.)		Tuber wt. plant^{-1} (g)	
	2011	2012	2011	2012	2011	2012
T_1: Chemical Fertilizer (CF)	37.1 c	41.1 b	4.10 c	5.13 a	210.6 cd	236.6 b
T_2: CD+IPNS basis CF	35.5 c	40.4 b	4.10 c	5.30 a	195.3 d	243.5 b
T_3: CD-bioslurry+IPNS basis CF	37.7 c	45.6 b	4.93 b	5.53 a	225.9 c	253.1 b
T_4: PM+IPNS basis CF	43.2 b	52.5 a	5.93 a	5.77 a	324.9 b	302.0 a
T_5: PM-bioslurry+IPNS basis CF	46.2 a	52.2 a	6.08 a	5.97 a	353.8 a	295.8 a
T_6: Control	15.5 d	21.8 c	1.93 d	3.23 b	40.03 e	59.60 c
SE (±)	0.8738	1.9683	0.091	0.2936	4.9330	10.938

Table 3. Effects of different manure and fertilizer treatments on the different grades of potato (%) in the potato-rice cropping system

Treatment	2011				2012			
	'A' grade	'B' grade	'C' grade	'D' grade	'A ' grade	'B' grade	'C' grade	'D' grade
T_1: Chemical Fertilizer (CF)	22.0	40.3	9.9	27.8	38.6	42.5	4.9	14.0
T_2: CD+IPNS basis CF	34.4	50.4	7.1	8.1	46.8	46.3	6.9	-
T_3: CD-bioslurry+IPNS basis CF	33.5	50.3	7.6	8.6	45.0	52.1	2.9	-
T_4: PM+IPNS basis CF	19.0	62.9	7.7	10.4	30.9	56.3	4.1	8.7
T_5: PM-bioslurry+IPNS basis CF	20.7	65.1	5.7	8.5	39.8	48.0	5.8	6.4
T_6: Control	81.3	-	18.7	-	63.3	-	36.7	-

'A' grade= 28-40 mm diameter, 'B' grade=41-55 mm, 'C' grade= <28 mm, 'D' grade= >55 mm

Table 4. Effects of different manure and fertilizer treatments on the grain and straw yields of potato (Diamant) in the potato-T. Aman rice cropping system

Treatments	Tuber yield (t ha^{-1})			Stover yield (t ha^{-1})		
	2011	2012	Mean	2011	2012	Mean
T_1: Chemical Fertilizer (CF)	16.34 c	20.85 d	18.60 c	0.803cd	0.944 c	0.874 c
T_2: CD+IPNS basis CF	15.91 c	22.48 c	19.19 c	0.746 d	1.010bc	0.878 c
T_3: CD-bioslurry+IPNS basis CF	17.79 b	23.59 b	20.69 b	0.843 bc	1.068 b	0.955 b
T_4: PM+IPNS basis CF	21.04 a	25.28 a	23.16 a	0.920ab	1.327 a	1.124 a
T_5: PM-bioslurry+IPNS basis CF	21.80 a	25.08 a	23.44 a	0.979 a	1.275 a	1.127 a
T_6: Control	3.00 d	5.40 e	4.20 d	0.080 e	0.311 d	0.195 d
SE (±)	0.4344	0.2817	0.3151	0.0242	0.0308	0.0216

Table 5. Residual effects of different manure and fertilizer treatments on yield contributing characters of T. Aman rice (BINA dhan7) in the potato -T. Aman rice cropping system

Treatments	Tillers hill^{-1} (no.)		Grains panicle^{-1} (no.)		1000-grain weight (g)	
	2011	2012	2011	2012	2011	2012
T_1: Chemical Fertilizer (CF)	13.13 a	12.03 b	81.17 a	89.10 b	22.13	22.82
T_2: CD+IPNS basis CF	15.07 a	11.70 b	87.90 a	89.60 b	22.57	21.76
T_3: CD-bioslurry+IPNS basis CF	15.43 a	13.03 a	88.67 a	92.87 ab	22.53	22.73
T_4: PM+IPNS basis CF	15.07 a	12.10 b	90.03 a	96.07 a	23.00	22.71
T_5: PM-bioslurry+IPNS basis CF	15.73 a	13.33 a	91.27 a	96.57 a	22.20	22.81
T_6: Control	9.133 b	7.933 c	67.13 b	74.23 c	21.87	22.04
SE (±)	0.8469	0.2726	3.1819	1.2902	0.2392	0.2907

Table 6. Residual effects of different manure and fertilizer treatments on the grain and straw yields of T. Aman rice (BINA dhan7) in the potato-T. Aman rice cropping system

Treatments	Grain yield (t ha^{-1})			Straw yield (t ha^{-1})		
	2011	2012	Mean	2011	2012	Mean
T_1: Chemical Fertilizer (CF)	4.110 c	4.194 b	4.152 c	4.619 b	4.473 c	4.546 c
T_2: CD+IPNS basis CF	4.223 bc	4.358 ab	4.291 bc	4.807 b	4.621 bc	4.714 bc
T_3: CD-bioslurry+IPNS basis CF	4.397 ab	4.440 ab	4.418 b	4.814 b	4.807 b	4.811 b
T_4: PM+IPNS basis CF	4.367 ab	4.404 ab	4.386 b	4.818 b	4.596 bc	4.707 bc
T_5: PM-bioslurry+IPNS basis CF	4.529 a	4.635 a	4.582 a	5.136 a	5.086 a	5.111 a
T_6: Control	2.728 d	2.633 c	2.680 d	3.036 c	3.155 d	3.096 d
SE (±)	0.0623	0.0838	0.0521	0.0965	0.0776	0.0737

Residual effects on yield components of T. *Aman* rice

In both years, the number of tillers hill^{-1}and grains panicle^{-1}was significantly varied due to fertilizers and manures added to the first crop (Tables 5). In 2011 and 2012, the number of tillers hill^{-1} varied from 9.1 to 15.7, and 7.9 to 13.3, respectively. In both years higher number of tillers hill^{-1} was recorded in T_5 treatment. The number of grains panicle^{-1} varied from 67.1 to 91.3 in 2011 and from 74.2 to 96.6 in 2012, the highest result being recorded further in treatment T_5.The 1000-grain weight ranged from 21.9-23.0 g in 2011 and from 21.8-22.8 g in 2012 with no statistical difference among the treatments (Table 5).

Table 7. Effects of different manure and fertilizer treatments on the total system productivity and economic return in the potato-T. *Aman* rice cropping system

Treatments	Total system productivity			Economic return			
	2011	2012	Mean	Gross return (Tk ha^{-1})	Added cost (Tk ha^{-1})	Added return (Tk ha^{-1})	MBCR
T_1: Chemical Fertilizer (CF)	22.51 c	27.14 d	24.82 c	255949	20394	168904	8.28
T_2: CD + IPNS basis CF	22.24 c	29.01 c	25.63 c	264193	20995	177148	8.44
T_3: CD-bioslurry+IPNS basis CF	24.39 b	30.25 b	27.32 b	281370	20504	194325	9.48
T_4: PM + IPNS basis CF	27.60 a	31.89 a	29.74 a	305639	21599	218594	10.12
T_5: PM-bioslurry+IPNS basis CF	28.59 a	32.04 a	30.31 a	311892	21433	224847	10.49
T_6: Control	7.093 d	9.347 e	8.223 d	87045	0	0	-
SE (±)	0.3965	0.2888	0.2808	-	-	-	

Price of inputs and outputs: Urea- 20 Tk kg^{-1}, TSP- 25 Tk kg^{-1},MoP- 25 Tk kg^{-1}, Gypsum- 8 Tk kg^{-1},Cowdung and cowdung-slurry- 1 Tk kg^{-1}, Poultry manure and poultry manure slurry- 2 Tk kg^{-1}, Potato tuber- 10 Tk kg^{-1}, Potato haulm- 1Tk kg^{-1}, Rice grain- 15 Tk kg^{-1}, and rice straw- 1.5 Tk kg^{-1}

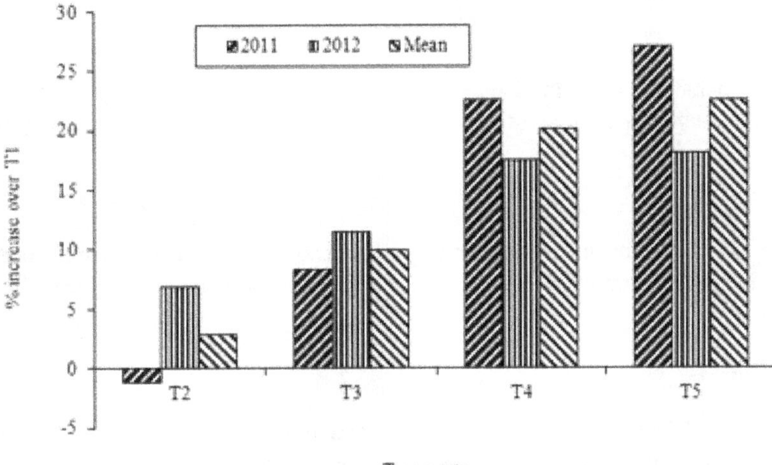

Figure 1. Percent increase of total system productivity by different IPNS treatments over chemical fertilizer treatment (T_1)

Residual effects on grain and straw yield of *T. Aman* rice

Application of manure and fertilizers to the previous potato crop significantly increased the grain and straw yield of *T. Aman* rice (second crop) in both years (Table 6).The grain production by different treatments ranged from 2.73 – 4.53 t ha^{-1} in 2011, and 2.63 – 4.64 t ha^{-1} in 2012. In both years, the highest grain yield was recorded by the T_5 treatment (PM-bioslurry + IPNS basis chemical fertilizer) and the second highest yield by the T_3 treatment (CD-bioslurry + IPNS basis chemical fertilizer). The straw yield ranged from 3.04 t ha^{-1} in T_6 treatment to 5.14 t ha^{-1} in T_5 treatment in 2011 and in 2012 the yield varied from 3.16 t ha^{-1} in T_6 treatment to 5.09 t ha^{-1} in T_5treatment.The two years' mean yield results indicated that the highest straw yield (5.11 t ha^{-1}) and the second highest (4.81 t ha^{-1}) yield were recorded in T_5 and T_3 treatments, respectively. Dwivedi and Thakur (2000) reported that among the organic manures applied to rice crop, biogas slurry and rice straw incorporation resulted in significant residual effect.

Contribution to total system productivity

Total system productivity was significantly influenced by different treatments ranging from 7.09 to 28.59 t ha^{-1} in 2011 and 9.35 to 32.04 t ha^{-1} in 2012 (Table 7). The poled TSP varied from 8.22 to 30.31 t ha^{-1} (Table 7). Both the years highest TSP was found in the treatment T_5 (PM-bioslurry+IPNS with chemical fertilizer) which was in 2012 statistically similar with T_4. Regarding cowdung source cowdung bioslurry had the higher TSP. Table 7 thus indicated that all the IPNS treatment had higher TSP compared to sole chemical fertilizer treatment (T_1) except T_2 in 2011. Among the IPNS treatments, PM-bioslurry, poultry manure, cowdung bioslurry and cowdung had the 22.5%, 20.0%, 9.9% and 2.9%increase in TSP over sole chemical fertilizer treatment (T_1) (Figure 1).Jeptooet al. (2013) reported that application of 7.8 t ha^{-1} of bioslurry increased yields of carrot by 8.8% in season 1 and 23.5% in season-2 compared to the control.

Contribution to economic profitability

It appears from Table 7 that the T_5 treatment had the highest marginal benefit-cost ratio (MBCR) (10.49), which was followed by T_5 treatment (10.12). The highest gross return was also found in T_5 treatment. Other IPNS treatments exhibited higher economic performance compared to sole chemical fertilizer treatment. Higher economic profitability in IPNS treatments was associated with lower market price of manures due to local availability, whereas purchasing chemical fertilizers require much higher money in one way; on the other way higher yield had a positive reflection in economic performance of IPNS treatments. Nevertheless if the benefit of manure use to the improvement of soil properties is added, all the manure based treatments (T_2-T_5) would produce higher benefits over all other treatments. Indeed, for achieving sustainable crop yield without incurring loss to soil fertility, the IPNS approach i.e. combined application of manure and fertilizers deserves attention.

CONCLUSION

Poultry manure bioslurry or cowdung bioslurry gave greater amount of tuber/grain and stover/straw yield compared to their respective original state. Poultry manure source produce significantly greater yield than cowdung.

ACKNOWLEDGEMENT

This work was supported by the World Bank funded Higher Education Quality Enhancement Project (HEQEP) executed in the Department of Soil Science, Bangladesh Agricultural University, Mymensingh-2202.

REFERENCES

1. Abubaker J, 2012. Effects of fertilization with biogas residues on crop yield, soil microbiology and greenhouse gas emissions. ActaUniversitatis agriculturae Sueciae, Department of Microbiology, Swedish University of Agricultural Sciences, 46: 1-79.
2. Ahlawat IPS and RP Sharma, 1993. Agronomic terminology. 3rd ed. New Delhi: Indian Society of Agronomy.
3. Dwivedi DK and SS Thakur, 2000. Effect of organic and inorganic fertility levels on productivity of rice (*Oryza sativa*) crop. Indian Journal of Agronomy, 45: 568-574.
4. Gomez KA and AA Gomez, 1984. Statistical Procedures for Agricultural Research, John Wiley and Sons, New York.
5. Islam MF, 2011. Bioslurry: An untapped black gold. http://www.biocompostbd.blogspot.com
6. Jeptoo A, JN Aguyoh and M Saidi, 2013. Improving Carrot Yield and Quality through the Use of Bio-slurry Manure. Sustainable Agricultural Research, 2: 164-172.
7. Singh JB, 1995. Effect of biogas bioslurry manure and mineral fertilizer on pea, okra, soybean and maize. Agriculture Wastes, 9: 73-79.
8. Ullah MM, R Sen, MK Hasan, MB Isalm and MS Khan, 2008. Project report on bioslurry management and its effect on soil fertility and crop production, Bangladesh Agricultural Research Institute. Gazipur.
9. Yu FB, XP Luo, CF Song, MX Zhang and SD Shan, 2010. Concentrated biogas bioslurry enhanced soil fertility and tomato quality. Plant Soil Science, 60: 262-268.

22

INFLUENCE OF DIFFERENT STANDS OF SAL (*Shorea robusta* C. F. Gaertn.) FOREST OF BANGLADESH ON SOIL HEALTH

Mohammad Kamrul Hasan* and Md. Bayeazid Mamun

Department of Agroforestry, Faculty of Agriculture, Bangladesh Agricultural University, Mymensingh-2202, Bangladesh

*Corresponding author: M. Kamrul Hasan, E-mail: mkhasanaf@gmail.com

ARTICLE INFO

Key words
Soil health
Sal forests
Chemical properties
Fungi

ABSTRACT

The study was conducted in Dukhula sadar and Gasabari forest range under Madhupur Sal Forest of Bangladesh to determine the soil nutrient composition and isolation of fungi with varying stands. Three stands viz. pure sal, plantation and mixed were considered as treatment of the study. A quadrate sample plot of 10×10 m^2 size was measured to collect soil samples for both chemical analysis and fungi isolation. Soil pH, electrical conductivity, organic matter content, total N, available P, exchangeable K, available S, fungal abundance and colony character (cm) were determined to achieve the objective of the study. The results revealed that soil pH and electrical conductivity were highest (6.61 and 21.10µS/cm) in mixed stand and lowest (6.38 and 10.75µS/cm) in pure stand. Organic matter content and total N were highest (2.24 and 0.145%) in plantation stand and lowest (1.65 and 0.112%) in mixed and pure stand, respectively. Available P, exchangeable K and available S were highest (3.65, 98.66 and 17.53ppm) in pure stand and lowest (1.97, 79.49 and 10.25ppm) in plantation stand. In addition, four fungal genera *Sclerotium, Rhizoctonia, Pythium* and *Verticillium* were identified in the study area soils. The highest fungal population (entire genus except *Verticillium)* (colony number/g soil) was found in mixed stand while it was found lowest in pure (*Sclerotium*) and plantation stand (*Rhizoctonia* and *Pythium*). There was no significant variation in colony diameter of the fungi among the treatments. Therefore, it can be concluded that better soil health was maintained in natural forest rather than plantation forest.

INTRODUCTION

Every country has needs at least 25% of forest coverage of its total area for balance environmental condition. But Bangladesh has only 17.08% of forest coverage (Bangladesh Forest Department, 2013). For sustainability of existing forest coverage, it is essential to ensure the soil fertility of the forest. Soil nutrient content and the maintenance of site fertility is one of the most important factors to consider when designing forest management systems. Although forest productivity is influenced by other factors such as temperature, water availability, and incoming radiation, nutrients are essential to forest growth and directly influenced by forestry operations. The most important nutrients in forests are the macronutrients such as N, P, S, K, Ca, and Mg, each of which is needed directly for plant growth. Micronutrients such as Mn, Fe, Cl, Cu, Zn, B, Mo, and Co are also required by plants, but are usually abundant in soils and rarely limit plant growth (Binkley, 1986). Within a forest stand, nutrients exist in many forms and distinct pools and are cycled between soils and plants. Plants uptake nutrients from the soil solution and incorporate them into biomass, which is then returned to the soil through litterfall, root turnover, and tree mortality. This biomass or organic matter is then decomposed by soil organisms such as bacteria and fungi that excrete enzymes to breakdown organic molecules into smaller units, liberating nutrients and making them available to plants again (Chapin et al., 2002). This cycle regulates fluxes between individual nutrient pools, which vary in size and turnover rates.

Forest soils influence the composition of the forest stand and ground cover, rate of tree growth, vigor of natural reproduction and other silviculturally important factors (Bhatnagar, 1965). For instance, growth of *Shorea robusta* (Sal) and other tree species, such as *Terminalia alata* and *Syzygium cumini*, in tropical forests is highly influenced by nitrogen, phosphorus, potassium, and soil pH (Bhatnagar, 1965). Physiochemical characters of forest soils vary in space and time due to variations in topography, climate, physical weathering processes, vegetation cover, microbial activities, and several other biotic and abiotic factors. Vegetation plays an important role in soil formation. For example, plant tissues both from aboveground litter and belowground root detritus are the main source of soil organic matter, which influences physiochemical characteristics of soil such as pH, texture and nutrient availability (Johnston, 1986).

The plain land 'Sal' forest is situated in central and northern parts of the country covering an area of 1, 20,000 ha about 0.81% of total land mass of the country and 7.8% of the country forest land. Most of the Sal forest areas are covered by Madhupur Sal forest. Sal (*Shorea robusta*) is the dominant species of this forest. The importance of Sal forests lies in the fact that these are the only natural forest resources of the central and northern parts of Bangladesh where the vast majority of the population dwells. These forests have a high economical and ecological significance in the central part of Bangladesh. Historically, the agrarian rural people around the forests have been heavily dependent on Sal forest resources for their livelihood. People living in close proximity to the Sal forest, particularly various ethnic groups such as the Garos, Koch and Hajongs totally depend on its resources to meet their subsistence needs (Rahman et al., 2010). The study area of this research has high population density, 1485 persons per km^2 (BBS, 2011). As a result, demand for land for both settlement and agricultural uses within forested areas have accelerated the rate of deforestation with loss of ecosystem productivity and biological diversity, leading to overall environmental degradation in the area. Various forestry activities, human disturbances, industrialization and climate change have significant impacts on Sal forest ecosystems of Bangladesh. Therefore, the Forest Department of Bangladesh has been taken some initiatives to save the Sal forests of Bangladesh like reforestation, afforestation, etc. through participatory forestry programme under the social forestry with *Acacia* spp. and *Eucalyptus* spp. (Bangladesh Forest Department, 2013). This social forestry or plantation forestry activities can bring change to the rural people lifestyle, but it is not sure that it can sustain the soil fertility and microbial activities of natural forest. Therefore, it is necessary to examine the soil health besides the plantation forestry for sustainable forest coverage. This is indeed needed for understanding the biogeochemical processes that shape the forests in order to ensure their protection specially the soil properties of the forests. An understanding of these conditions is also essential for developing forest management policies in Bangladesh. Several studies have been investigated on several aspects of soils in various forest types throughout the world and in Indian sub-continent by many more researchers (Rawat et al., 2009). Some studies on soil and leaf litter nutrients and their effects on crops have been carried out for individual forest by Iltuthmish et al. (2006), Chowdhury et al. (2007), Zaman et al. (2008), Haider et al. (2009), Sarker et al. (2010), Hossain et al. (2011) in Bangladesh. However, the information on soil analysis of Madhupur Sal forest of Bangladesh is still in small pockets.

Therefore, it thought necessary to analyze biological and chemical characteristics of soils of the Madhupur Sal forest of Bangladesh. Keeping in view of the above aspects the study was undertaken to characterize the nutritional composition and isolation of major fungi present in soils of Madhupur Sal forest of Bangladesh with varying stands.

MATERIALS AND METHODS

Study area and sampling design

The study was conducted in Dukhula sadar and Gasabari forest range under Madhupur Sal Forest of Tangail district. The soils of the areas are highly oxidized reddish brown clay with moderate to strong acidic reaction. Sal (*Shorea robusta*) is the dominant species of this forest and usually it forms 75% to 25% of the upper canopy in the natural habitat (Alam, 1995). Besides, other species like mixture of Sal (*Shorea robusta*), Koroi (*Albizzia* spp.), Azuli (*Dillenia pent*agyna), Sonalu (*Cassia fistula*), Bohera (*Terminalia bellirica*), Haritaki (*Terminalia chebula*), Kanchan (*Bauhinia acuminata*), Jarul (*Lagerstroemia speciosa*), Jam (*Syzygium* spp.), etc. are grown in this forest. Presently participatory forestry programs are being implemented here under the social forestry initiatives with *Acacia* spp. and *Eucalyptus* spp. According to the above information and species distribution, three stands were distinguished from the selected forest ranges which were considered as treatments of the study. These are:

T_1= Pure stand (including *Shorea robusta*)

T_2= Plantation stand ((including *Acacia* spp., *Eucalyptus* spp.)

T_3= Mixed stand (including *Shorea robusta, Albizzia* spp., *Dillenia pentagyna, Cassia fistula, Terminalia bellirica, Terminalia chebula*, *Bauhinia acuminata*, *Lagerstroemia speciosa*, *Syzygium* spp., etc.).

Within the above mentioned stands, a 10×10 m^2 quadrate plot with four replications following random sampling was made for each stands of the selected forest ranges in Madhupur Sal forest to collect soil sample.

Soil samples collection

From each quadrate plots, five soil cores was taken with an auger at 0-15 cm depth randomly. A total of 20 soil cores were collected from four replicated plots for each stand. Then the collected soil cores from the same plot were well mixed to make one composite sample. Each sample was placed in a sterile plastic bag, sealed and transported to the laboratory. All soils were air dried and grounded and passed through a 2.0 mm sieve and stored at 4^0C until the chemical analysis was conducted. For fungal isolation same procedure was followed to collect soil cores from each quadrate plots. Then the collected soil cores were placed plot wise in a sterile polybag, wrapping with brown paper, sealed and carefully transported to the laboratory and stored at 4^0C until the isolation was started.

Soil chemical analysis

From the collected soil samples, soil pH was determined by Glass electrode pH meter (WTW pH 522) at a soil-water ratio of 1:2.5 as described by Ghosh et al. (1983) and electrical conductivity by a conductivity meter method as described by Jackson (1958). Percentage soil organic carbon was determined using Walkley–Black method modified by Anderson and Ingram (1989). Organic matter was calculated from the content of organic carbon by Van Bemmelen factor, 1.73. Assuming that organic matter contains approximately 58% C. Total N was determined by semi-micro Kjeldahl method. Available P was measured by Bray and Kurtz method outlined by Tandon (1995) where phosphorus extracted from soil using 0.5M $NaHCO_3$ at a nearly constant pH 8.5. Exchangeable K was extracted by ammonium acetate (1NNH_4OAc) and measured by digital flame photometer and S (extractable sulphate) was extracted by 0.15% $CaCl_2$ and measured following turbidimetric procedure improved by Hunter (1984) where the turbidity measured by spectrophotometer at 420 nm wavelengths.

Soil biological analysis

Soil fungi were isolated by using dilution plate technique (Warcup, 1955). For making dilution of soil samples, 1 g working sample was prepared from the composite soil sample and dilution was made up to 10^{-4}.

Then diluted soil sample (10^{-4}) was placed at the center of Potato dextrose agar (PDA) in the glass petri-dish and spreaded well with a glass rod. Five petri-dishes for each treatment were inoculated with 1 ml of diluted sample. This was repeated with every soil sample. Potato dextrose agar (PDA) media was prepared with potato (peeled and sliced), dextrose, agar and water. The inoculated PDA plates were incubated for 4-5 days at room temperature (25±1°C). The colonies grow out on PDA medium was recorded every day after 3-5 days of incubation. The number and diameter of colonies developed in each PDA plates were counted and average values were calculated for each sample. Number of colonies per ml of soil suspension was calculated by its colony forming units (CFU). The number of CFU/g sample was calculated by using following formula:

Population of fungi= Average number of total colonies/ml in five Petri-dishes × Dilution factor (10^{-4})

Most of the isolated fungi were identified up to genera level with the help of the book "Illustrated Genera of Imperfecti Fungi" (Barnett, 1965) and a manual of soil fungi (Gilman, 1957). Moreover, morphology and taxonomy of Fungi (Bessay, 1964), Fungi in Agricultural Soils (Domsch and Gams, 1972) were also consulted to identify fungi.

Data analysis

The collected data were tabulated and analyzed through a standard computer package statistical procedure MSTAT-C (Gomez and Gomez, 1984). Duncan's Multiple Range Test (DMRT) was used to rank the results of soil samples analysis.

RESULTS AND DISCUSSION

Soil chemical properties

Soil pH

The result shows that chemical and biological reactions of soils are regulated by soil pH. The average pH values of the study soils such as pure stand, plantation stand and mixed stand were recorded as 6.38, 6.55 and 6.62, respectively (Table1). Soil pH of mixed stand was the highest (6.62) which was statistically higher compare to the pH of pure stand and plantation stand. Similar results was also found by Khan et al. (1997) who reported that the pH values of the different soil series of the Madhupur clay soils of Madhupur tract ranged from 5.6 to 6.1.

Electrical Conductivity (EC)

The average electrical conductivity values of the study soils such as pure stand, plantation stand and mixed stand were recorded as 20.11, 10.75 and 21.10µS/cm, respectively (Table 1). The highest (21.10µS/cm) soil electrical conductivity was measured in mixed stand which was statistically different to the electrical conductivity of pure stand and plantation stand. The lowest EC was found in plantation stand (Table 1). Gomes (2005) also found that the electrical conductivity of Madhupur Sal forest soil ranged from 6 to 57µS/cm which was supportive to the present study.

Organic matter (OM)

The result shows that the highest (2.24%) organic matter content of Madhupur Sal forest was appeared in plantation stand compare to both pure (1.72%) and mixed stand (1.65%) (Table 1). Accumulation of plant leaves might have caused to the increment of organic matter in case of plantation stand. Leaf litter of plantation species might add huge amount of organic matter to the soil in plantation stand. It is in line with Gomes (2005) who found that organic matter content of the surface soil (0-15 cm depth) of Madhupur Sal forest varies from 0.70% to 2.11%.

Total nitrogen (N)

Soil total N content in different treatments followed the similar trend like soil organic matter content (Table 1). The highest total nitrogen content (0.145%) was measured in plantation stand soil which was statistically different from the total nitrogen in soil of pure stand (0.112%) and mixed stand (0.112%). Basically there was no significant difference between pure stand and mixed stand total N content. Gomes (2005) found that the total nitrogen content of Madhupur Sal forest soil varies from 0.026% to 0.105% at 0-30 cm soil depth which was strongly supported by the present findings.

Available phosphorus (P)

The average available phosphorus content of the study soils such as pure, plantation and mixed stand was recorded as 3.65, 1.97 and 2.39ppm, respectively (Table 1). The available phosphorus in soil of pure Sal stand was the highest (3.65ppm) which was statistically differ to available phosphorus content in soil of plantation and mixed stand. The higher levels of soil nutrients in the pure forest were due partly to reduction in the loss of top soil and partly to the increased supply of nutrients in the form of leaf litter and biomass from the larger number of Sal trees and their saplings rather than plantation stand and mixed stand.

Exchangeable potassium (K)

The result shows that the exchangeable potassium content in soil of pure Sal stand was the highest (98.66ppm) which was statistically dissimilar from exchangeable potassium content in soil of plantation and mixed stand soils of Madhupur Sal forest area. Paudel and Sah (2003) also found the same kind of result that higher values of humus, organic matter, nitrogen and potassium (7.34%, 2.42%, 0.117%, 267.73 kg/ha, respectively) were found in pure forest soils.

Available sulphur (S)

The available sulphur in soil of pure Sal stand was the highest (17.53 ppm) which was statistically differing to available sulphur content in soil of plantation (10.25 ppm) and mixed stand (13.85 ppm) (Table 1). Similar results also found by Akter (2009) who stated that the available S content of Madhupur forests soil was 13.41 ppm at upper 0-15 cm soil layer which was highly supported by the above findings.

Table 1. Chemical properties such as Soil pH, Electrical Conductivity (EC), Organic matter, Total N, Available P, Exchangeable K and Available S of study areas soils in Madhupur Sal forests of Bangladesh.

Treatments	Soil pH	EC (µS/cm)	OM (%)	N (%)	P (ppm)	K (ppm)	S (ppm)
T_1	6.38 c	20.11 b	1.72 b	0.112 b	3.65 a	98.66 a	17.53 a
T_2	6.55 b	10.75 c	2.24 a	0.145 a	1.97 c	79.49 c	10.25 c
T_3	6.62 a	21.10 a	1.65 c	0.115 b	2.39 b	84.57 b	13.85 b
LSD (0.05)	0.024	0.121	0.014	0.005	0.004	0.033	0.043
CV (%)	0.234	0.469	0.484	2.809	0.226	0.023	0.200
Level of significance	**	**	**	**	**	**	**

T_1= Pure stand, T_2= Plantation stand, T_3= Mixed stand. **= significant at 1% level of probability. In the column figure(s) having same letter(s) do not differ significantly. LSD= Least Significant Difference; CV= Co-efficient of variation

Soil biological properties

Biological properties of studied three soils such as pure stand, plantation stand and mixed stand were presented by the diversity and abundance of major fungal genera isolated from soil samples. Based on the colony color and microscopic observation of different structures, four genera of fungi were identified in the studied soils. The genera were *Sclerotium*, *Rhizoctonia*, *Pythium* and *Verticillium*.

Fungal abundance (Colony number/g soil)

For *Sclerotium* spp. colony number/g soil at 5 Days after incubation (DAI) was significantly higher (9.75) in mixed stand than pure (3.75) and plantation stand (5.50). At 6 DAI, colony number/g soil was also significantly higher in mixed stand (12.00) than pure (8.00) and plantation stand (8.00). Accordingly at 7 DAI colony number/g soil was significantly higher in mixed stand (12.50) than pure (8.25) and plantation stand (8.25) (Figure 1, Figure 2 and Figure 3).

In case of *Rhizoctonia* spp. at 5 DAI colony number/g soils was significantly highest (7.50) in mixed stand and lowest (4.00) in plantation stand which was statistically similar to pure stand (5.50). At 6 DAI, colony number/g soil was significantly higher in mixed stand (10.75) compare to pure (6.00) and plantation stand (6.00). At 7 DAI, colony number/g soil was found in similar pattern stated above at 6 DAI (Figure 1, Figure 2 and Figure 3).

In case of *Pythium* spp. highest (5.25) colony number/g soil was observed in mixed stand and lowest (2.25) in plantation stand at 5 days after interval. At 6 and 7 DAI colony number/g soil was significantly higher in mixed stand (5.25 and 6.00) than pure (3.00 and 3.50) and plantation stand (3.50 and 3.75) (Figure 1, Figure 2 and Figure 3).

For *Verticillium* spp. colony number/g soil was found similar in case of all treatments at 5 DAI, 6 DAI and 7 DAI. It means that there was no significant variation appeared among pure, plantation and mixed stand soils of Madhupur Sal forest areas (Figure 1, Figure 2 and Figure 3).

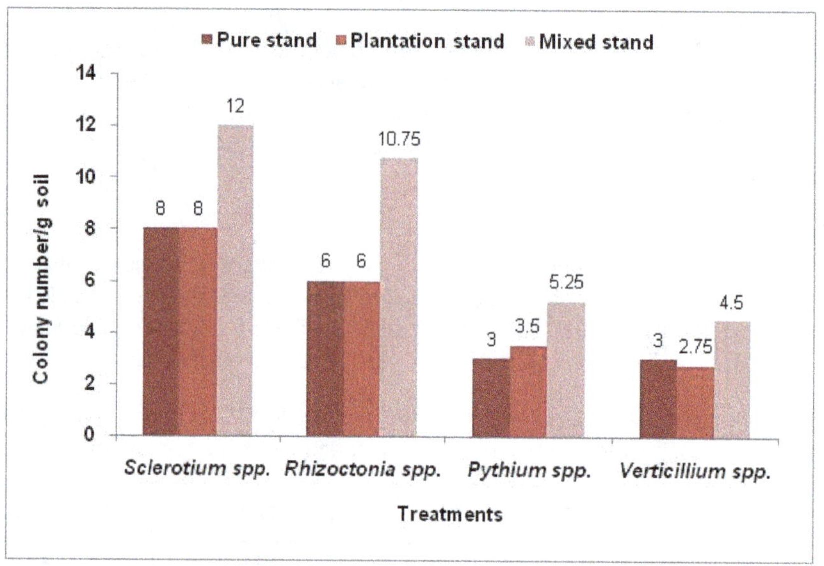

Figure 1. Bar graph showing colony number/g soil of four identified fungi (*Sclerotium* spp., *Rhizoctonia* spp., *Pythium* spp., *Verticillium* spp.) of study area soils at 5 DAI

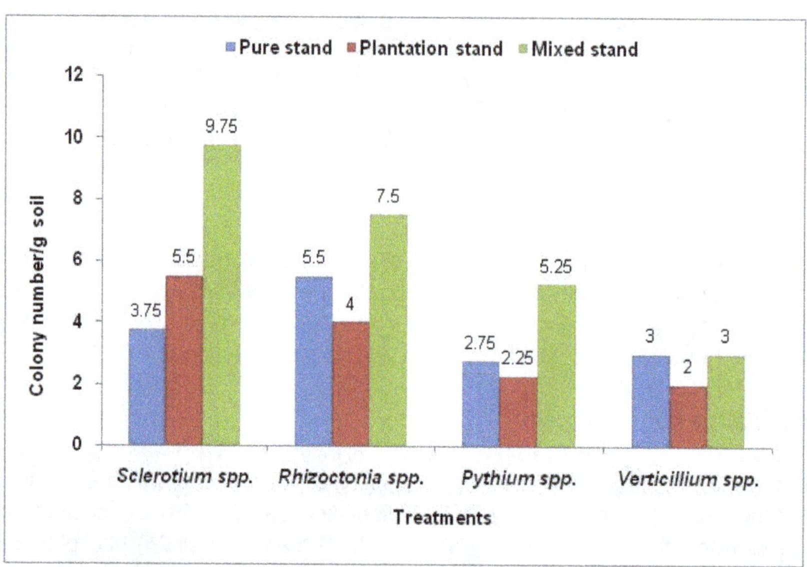

Figure 2. Bar graph showing colony number/g soil of four identified fungi (*Sclerotium* spp., *Rhizoctonia* spp., *Pythium* spp., *Verticillium* spp.) of study area soils at 6 DAI

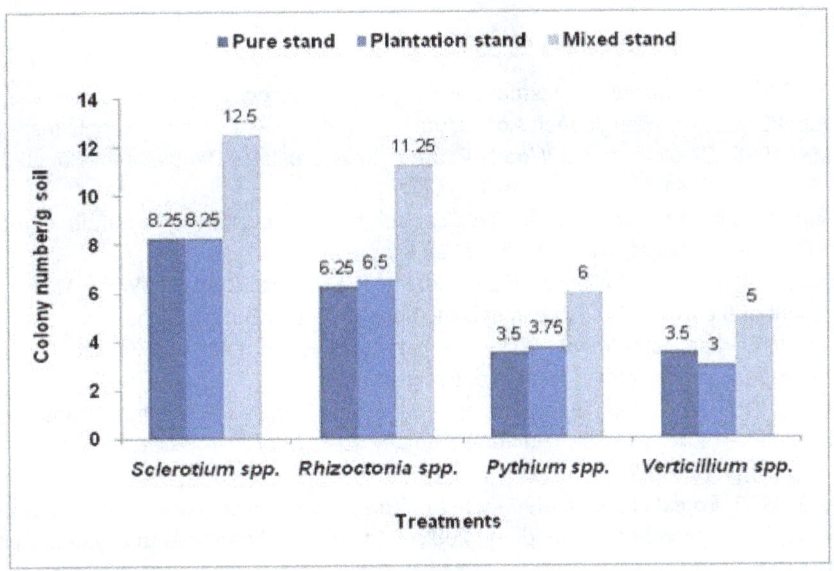

Figure 3. Bar graph showing colony number/g soil of four identified fungi (*Sclerotium* spp., *Rhizoctonia* spp., *Pythium* spp., *Verticillium* spp.) of study area soils at 7 DAI

Colony character

For the all four genera (*Sclerotium, Rhizoctonia, Pythium* and *Verticillium*) stated in the above section, colony diameter (cm) or colony growth was observed similar and there was no significant variation among the treatments at 5 DAI, 6 DAI and 7 DAI (Table 2).

Table 2. Colony diameter (cm) of four identified fungal genera at 5 DAI, 6 DAI and 7 DAI

Fungal genera	Colony diameter (cm)								
	5 DAI			6 DAI			7 DAI		
	T_1	T_2	T_3	T_1	T_2	T_3	T_1	T_2	T_3
Sclerotium	1.75	2.25	1.75	2.5	2.75	2.25	2.75	3.0	3.25
Rhizoctonia	1.5	2.0	2.5	2.0	2.0	2.75	2.0	2.25	3.0
Pythium	1.0	1.25	1.0	1.5	2.0	1.75	2.0	2.25	2.0
Sclerotium	1.0	1.0	1.25	1.75	1.75	1.75	2.0	2.0	2.0
Level of significance	NS	NS	NS	NS	NS	NS	NS	NS	NS

T_1=Pure stand, T_2=Plantation stand, T_3=Mixed stand; 'NS' mean non-significant; CV=Co-efficient of variation

CONCLUSION

From this experiment it can be concluded that the overall nutrient content was higher in pure and mixed stands compare to plantation stand of Madhupur Sal forests. Similarly total fungal population was higher in pure and mixed stands characterized as less undisturbed forest sites rather than highly disturbed plantation stand. The higher amount of soil nutrients and fungal population in the pure forest and mixed forest were due partly to reduction in the loss of top soil and partly to the increased supply of nutrients in the form of leaf litter and biomass from the larger number of trees and their saplings rather than plantation stand.

REFERENCES

1. Akter M, 2009. Soil quality assessment and its impact on ecosystem. M.S. Thesis, Department of Agricultural Chemistry, Bangladesh Agricultural University, Mymensingh, Bangladesh, pp: 69-86.
2. Alam MK, 1995. Diversity in the Woody Flora of Sal Forests of Bangladesh. Bangladesh Journal of Forest Science, 24: 41-52.
3. Anderson JM and JSI Ingram, 1996. Tropical soil biology and fertility a handbook of methods. 2nd edition, CAB International, Wallingford, United Kingdom, pp: 57-58.
4. Bangladesh Bureau of Statistics (BBS), 2011. Population Census Wing, Ministry of Planning, Government of the People's Republic of Bangladesh, Dhaka, Bangladesh.
5. Bangladesh Forest Department, 2013. Ministry of Environment and Forest, Government of the People's Republic of Bangladesh, Dhaka, Bangladesh.
6. Barnett HL, 1965. Illustrated Genera of Imperfect Fungi. Burgess publishing company, p: 225.
7. Bhatnagar HP, 1965. Soils from different quality Sal (*Shorea robusta*) forests of Uttar Pradesh. Tropical Ecology, 6: 56-62.
8. Binkley D, 1986. Forest Nutrition Management. Wiley Interscience. New York, NY, p: 290.
9. Chapin FS, PA Matson and HA Mooney, 2002. Principles of Terrestrial Ecosystem Ecology. Springer Verlag.
10. Chowdhury MAH, KM Mohiuddin, R Fancy and AHMRH Chowdhury, 2007. Micronutrient release pattern of some tree leaf litters in Rajendrapur forest soil. Bangladesh Journal of Crop Science, 18: 381-386.
11. Domsch KH and W Gams, 1972. Fungi in agricultural soils. Published by Longman group Ltd., London, p: 290.
12. Ghosh AB, JC Bajaj, R Hassan, and D Singh, 1983. Soil and Water Testing Methods Laboratory manual. Division of Soil Science and Agricultural Chemistry, IARI, New Delhi, India, pp: 221-226.
13. Gilman JC, 1957. A manual of soil fungi. 2nd edition, The Iowa State college press, p: 450.
14. Gomez KA and AA Gomez, 1984. Duncan's Multiple Range Test. Statistical Procedures for Agricultural Research, 2nd edition, A Wiley Inter-Science Publication, Johan and Sons, New York, pp: 202-215.
15. Gomes SA, 2005. Study on soil characteristics and species diversity of Madhupur Sal forest. M.S. Thesis, Department of Agroforestry, Bangladesh Agricultural University, Mymensingh, Bangladesh, pp: 21-55.
16. Haider SMS, MF Khatun, MAH Chowdhury, MS Islam and MZA Talukder, 2009. Effects of different tree leaf litters on growth and yield of okra in Madhupur Forest Soil. Journal of Innovative Development Strategy, 3: 55-60.
17. Hossain M, MRH Siddique, MS Rahman, MZ Hossain and MM Hasan, 2011. Nutrient dynamics associated with leaf litter decomposition of three agroforestry tree species (*Azadirachta indica*, *Dalbergia sissoo* and *Melia azedarach*) of Bangladesh. Journal of Forestry Research, 22: 577-582.
18. Hunter AH, 1984. Soil Fertility Analytical Services in Bangladesh. Consultancy Report, Bangladesh Agricultural Research Council, Dhaka, Bangladesh.
19. Iltuthmish AHM, MAH Chowdhury, KM Mohiuddin and AHMRH Chowdhury, 2006. Macronutrient release pattern from soils added with leaf litters of some forest trees. Progressive Agriculture, 17(1): 81-87.
20. Jackson ML, 1958. Soil Chemical Analysis. Prentice-Hall Incorporation, Englewood Cliffs, New Jersey, USA.
21. Johnston, AE, 1986. Soil organic matter; effects on soil and crops. Soil Use Manage, 2: 97-105.
22. Khan ZH, AR Mazumder, ASM Moduddin, MS Hussain and SM Saheed, 1997. Physical properties of some Benchmark soil from the floodplains of Bangladesh. Journal of Indian Society of Soil Science, 46: 442-446.
23. Paudel S and JP Sah, 2003. Physiochemical characteristics of soil in tropical Sal (*Shorea robusta*) forest in eastern Nepal. Hymalaya Journal of Science, 1: 107-110.

24. Rahman MM, MR Motiur, Z Guogang and KS Islam, 2010. A review of the present threats to tropical moist deciduous Sal (*Shorea robusta*) forest ecosystem of central Bangladesh. Tropical Conservation Science, 3: 90–102.
25. Rawat N, BP Nautiyal and MC Nautiy, 2009. Litter production pattern and nutrients discharge from decomposing litter in a Himalayan alpine ecosystem. New York Science Journal, 2: 54-67.
26. Sarkar UK, BK Saha, C Goswami and MAH Chowdhury, 2010. Leaf litter amendment in forest soil and their effect on the yield quality of red amaranth. Journal of Bangladesh Agricultural University, 8: 221-226.
27. Tandon HLS, 1995. Methods of analysis of soils, plants, waters and fertilizers. Fertilizer development and consultation organization, New Delhi, pp: 19-23.
28. Warcup JH, 1955. On the origin of colonies of fungi developing on the soil dilution plates. Transaction of the British Mycological Society, 38: 298-301.
29. Zaman AHMR, KM Mohiuddin, R Fancy, AHMRH Chowdhury and MAH Chowdhury, 2008. Decomposition of some tree leaf litters in Rajendrapur forest soils under laboratory condition. Bangladesh Journal of Progressive Science and Technology, 6: 133-136.

Permissions

All chapters in this book were first published in RALF, by AgroAid Foundation; hereby published with permission under the Creative Commons Attribution License or equivalent. Every chapter published in this book has been scrutinized by our experts. Their significance has been extensively debated. The topics covered herein carry significant findings which will fuel the growth of the discipline. They may even be implemented as practical applications or may be referred to as a beginning point for another development.

The contributors of this book come from diverse backgrounds, making this book a truly international effort. This book will bring forth new frontiers with its revolutionizing research information and detailed analysis of the nascent developments around the world.

We would like to thank all the contributing authors for lending their expertise to make the book truly unique. They have played a crucial role in the development of this book. Without their invaluable contributions this book wouldn't have been possible. They have made vital efforts to compile up to date information on the varied aspects of this subject to make this book a valuable addition to the collection of many professionals and students.

This book was conceptualized with the vision of imparting up-to-date information and advanced data in this field. To ensure the same, a matchless editorial board was set up. Every individual on the board went through rigorous rounds of assessment to prove their worth. After which they invested a large part of their time researching and compiling the most relevant data for our readers.

The editorial board has been involved in producing this book since its inception. They have spent rigorous hours researching and exploring the diverse topics which have resulted in the successful publishing of this book. They have passed on their knowledge of decades through this book. To expedite this challenging task, the publisher supported the team at every step. A small team of assistant editors was also appointed to further simplify the editing procedure and attain best results for the readers.

Apart from the editorial board, the designing team has also invested a significant amount of their time in understanding the subject and creating the most relevant covers. They scrutinized every image to scout for the most suitable representation of the subject and create an appropriate cover for the book.

The publishing team has been an ardent support to the editorial, designing and production team. Their endless efforts to recruit the best for this project, has resulted in the accomplishment of this book. They are a veteran in the field of academics and their pool of knowledge is as vast as their experience in printing. Their expertise and guidance has proved useful at every step. Their uncompromising quality standards have made this book an exceptional effort. Their encouragement from time to time has been an inspiration for everyone.

The publisher and the editorial board hope that this book will prove to be a valuable piece of knowledge for researchers, students, practitioners and scholars across the globe.

List of Contributors

Md. Mahfuzur Rahman, MA Khaleque Mian, Asgar Ahmed and Md. Motiar Rohman
Bangabandhu Sheikh Mujibur Rahman Agricultural University, Salna, Gazipur-1706, Bangladesh Breeding Division, Bangladesh Agricultural Research Institute, Joydevpur, Gazipur-1701, Bangladesh

Md. Abdul Alim
Assistant Production Manager, ACI (Seed) Ltd.

Bhabendra Kumar Biswas and Md. Hasanuzzaman
Professor, Department of Genetics and Plant Breeding, Hajee Mohammad Danesh Science and Technology University (HSTU), Dinajpur, Bangladesh

Pronay Bala
Department of Crop Physiology & Ecology, HSTU, Dinajpur

Santanu Roy
Program officer, Natural Resources Management Project, Caritas Fisheries Program, Mirpur, Dkaka, Bangladesh

Shanjida Rahman, Md. Mukul Mia, Tamanna Quddus, Lutful Hassan and Md. Ashraful Haque
Department of Genetics and Plant Breeding, Faculty of Agriculture, Bangladesh Agricultural University, Mymensingh- 2202, Bangladesh

Md. Nasir Uddin Khan and Mohammad Kamrul Hasan
Department of Agroforestry, Faculty of Agriculture, Bangladesh Agricultural University, Mymensingh-2202, Bangladesh

Firoj Alom
Department of Animal Husbandry and Veterinary Science, University of Rajshahi, Rajshahi Sadar-6205

Mahbub Mostofa, M. Nurul Alam, M.Golam Sorwar, Jashim Uddin and M. Mizanur Rahman
Department of Pharmacology, Bangladesh Agricultural University, Mymensingh-2202, Bangladesh

Mahfuza Afroj
Department of Agribusiness and Marketing

Mohammad Mizanul Haque Kazal
Department of Development and Poverty Studies, Faculty of Agribusiness Management, Sher-e-Bangla Agricultural University, Sher-e-Bangla Nagar, Dhaka-1207, Bangladesh

Md. Mahfuzar Rahman
Department of Agronomy, Faculty of Agriculture, Sher-e-Bangla Agricultural University, Sher-e-Bangla Nagar, Dhaka-1207, Bangladesh

Fauzan Zakaria and Nurdin
Faculty of Agriculture, State University of Gorontalo, Indonesia

Muntaha Rakib
Department of Economics, Shahjalal University of Science and Technology, Sylhet, Bangladesh

Shah Mohammad Hamza Anwar
Department of Economics, Metropolitan University, Sylhet, Bangladesh

KM Mohiuddin, Md. Mehediul Alam, Md. Shahinur Rahman, Md. Shafiqul Islam and Istiaq Ahmed
Department of Agricultural Chemistry, Faculty of Agriculture, Bangladesh Agricultural University, Mymensingh-2202, Bangladesh

MAB Khalil Rahad, M. Ashraful Islam, M Abdur Rahim and S Monira
Department of Horticulture, Faculty of Agriculture, Bangladesh Agricultural University, Mymensingh-2202, Bangladesh

Md. Maniruzzaman, Md. Akhter Hossain Chowdhury and KM Mohiuddin
Department of Agricultural Chemistry

Tanzin Chowdhury
Department of Agronomy, Faculty of Agriculture, Bangladesh Agricultural University, Mymensingh-2202, Bangladesh

Mohammad Mosharraf Hossain
Department of Soil Science, Sher-e-Bangla Agricultural University, Dhaka-1207, Bangladesh

Keshob Chandra Das
SSO, Molecular Biotechnology Division, National Institute of Biotechnology, Savar-1349, Dhaka, Bangladesh

Sabina Yesmin
Mushroom Development Officer, Mushroom Development Institute, Savar-1340, Dhaka, Bangladesh

Syfullah Shahriar
Department of Soil Science, Sher-e-Bangla Agricultural University, Dhaka-1207, Bangladesh

Elina Aziz, Md. Younus Mia and Nowara Tamanna Meghla
Department of Environmental Science and Resource Management, Mawlana Bhashani Science and Technology University, Santosh, Tangail-1902, Bangladesh

Abdullah-Al-Masud, Md. Shawkat Ali and Muslah Uddin Ahammad
Department of Poultry Science, Faculty of Animal Husbandry, Bangladesh Agricultural University, Mymensingh-2202, Bangladesh

Mai Thanh Vu
Hung Vuong University of Phu Tho, Vietnam
Thai Nguyen University of Agriculture and Forestry, Vietnam

Van Thanh Tran and My Thi Thuy Nguyen
Thai Nguyen University, Vietnam
Thai Nguyen University of Agriculture and Forestry, Vietnam

Van Cao and Tuan Ngoc Minh Nguyen
Hung Vuong University of Phu Tho, Vietnam

Md. Mizan Ul Islam, Parth Sarothi Saha, Tusher Chakrobarty, Nibir Kumar Saha, Md. Sirajul Islam and MA Salam
Agriculture and Food Security Program, BRAC Agricultural Research and Development Centre, Gazipur-1701, Bangladesh

Suzan Khan, M Hammadur Rahman and Mohammed Nasir Uddin
Department of Agricultural Extension Education, Faculty of Agriculture, Bangladesh Agricultural University, Mymensingh-2202, Bangladesh

Md. Maniruzzaman
Soil Resources Development Institute, Farmgate, Dhaka, Bangladesh

Tanzin Chowdhury
Department of Agricultural Chemistry, Sher-e-Bangla Agricultural University, Sher-e-Bangla Nagar, Dhaka-1207, Bangladesh

Md. Arifur Rahman and Md. Akhter Hossain Chowdhury
Department of Agricultural Chemistry, Faculty of Agriculture, Bangladesh Agricultural University, Mymensingh-2202, Bangladesh

Md Rakib Hassan
Department of Computer Science and Mathematics, Faculty of Agricultural Engineering and Technology, Bangladesh Agricultural University, Mymensingh 2202, Bangladesh

Mst. Anjumanara Begum, Md. Anwarul Islam and Md. Moshiur Rahman
Department of Agronomy, Bangladesh Agricultural University, Mymensingh-2202, Bangladesh

Md. Aminul Islam
On Farm Research Division, Bangladesh Agricultural Research Institute, Pabna-6600, Bangladesh

Quazi Maruf Ahmed
Regional Horticultural Research Station, Bangladesh Agricultural Research Institute Narshingdi-1600, Bangladesh

M. Asadul Haque
Department of Soil Science, Patuakhali Science and Technology University, Patuakhali

M. Jahiruddin and M. Mazibur Rahman
Department of Soil Science, Bangladesh Agricultural University, Mymensingh

M. Abu Saleque
Bangladesh Rice Research Institute, Gazipur, Bangladesh

Mohammad Kamrul Hasan and Md. Bayeazid Mamun
Department of Agroforestry, Faculty of Agriculture, Bangladesh Agricultural University, Mymensingh-2202, Bangladesh

Index

www.ingramcontent.com/pod-product-compliance
Lightning Source LLC
Chambersburg PA
CBHW082026190326
41458CB00010B/3283